Practical Control Engineering for Mechatronics and Automation

Fernando Martell

Research Center in Optics
Aguascalientes, Mexico

Irma Y. Sanchez

Professor at Tecnologico de Monterrey
Monterrey, Mexico

CRC Press

Taylor & Francis Group

Boca Raton London New York

CRC Press is an imprint of the
Taylor & Francis Group, an **informa** business

A SCIENCE PUBLISHERS BOOK

First edition published 2024
by CRC Press
2385 NW Executive Center Drive, Suite 320, Boca Raton FL 33431

and by CRC Press
4 Park Square, Milton Park, Abingdon, Oxon, OX14 4RN

CRC Press is an imprint of Taylor & Francis Group, LLC

Library of Congress Cataloging-in-Publication Data (applied for)

ISBN: 978-1-032-41390-7 (hbk)
ISBN: 978-1-032-41391-4 (pbk)
ISBN: 978-1-003-35788-9 (ebk)

DOI: 10.1201/9781003357889

Typeset in Times New Roman
by Radiant Productions

To Joana and Michelle

"Everything should be made as simple as possible, but not simpler".

—Albert Einstein

Preface

Control systems are essential components in mechatronic products as well as in the automation of production systems. The design of control systems is generally required for the development of multidisciplinary engineered systems. Mechatronic design methodologies rely heavily on system modeling and real time computer simulation, which are fundamental engineering tools for the design and verification of control systems.

The theory of control systems is often presented with high mathematical formalism suitable for specialized college education in control engineering. Design of control systems needs to be addressed comprehensibly for diverse engineering careers. This book covers control engineering with a practical approach. Mathematical concepts of signals and systems are introduced and applied to system or process modeling. Discretization of mathematical models is explained for their computational implementation. Difference equations are necessary to reproduce and simulate the dynamical response of physical systems but they are also required for implementing computer based control algorithms.

One relevant concept for automation and control is the Proportional-Integral–Derivative (PID) controller. PID is considered the workhorse of industrial process control but it is also a suitable control algorithm for most mechatronic applications. PID is a reliable and robust algorithm for the design of closed loop feedback controllers. Particularly, the model based PID techniques are widely used for the synthesis and tuning of PID controllers and, therefore, applied to industrial process control and to the design of mechatronic systems. PID is explained in detail, and PID tuning is deduced with the direct synthesis and the internal model principle. Model based PID concepts are emphasized to introduce more advanced control strategies such as model reference control, multiple loop control and digital controllers.

Another key concept is the Finite State Automata (FSA). The FSA or Finite States Machines (FSM) are used for the automation of industrial manufacturing systems that can be modeled as discrete-event systems. FSM is explained as a model based tool for the design of automatisms and sequential logic control. FSM methods are conveniently used for structured PLC programming. The familiarity with FSM enables the introduction of hybrid automata. FSM based design and programming are used as practical foundations for the implementation of hybrid control systems.

Industrial automation is in continuous evolution incorporating process control techniques and automation technologies. Some technological trends such as cyber-physical systems and digital twins apply real time computational simulation of the physical processes and systems to improve their performance. This book also presents a basic but functional approach to cyber-physical systems that is conceptualized as an enhanced supervisory control system, and explains how it can be implemented using model based control design and hardware in the loop simulation.

The purpose of the book is to support a progressive comprehension of applied control engineering though examples and case studies, as didactic resources. First, a general overview of the development of mechatronic and automation projects is offered as a reference frame, then continuous-time and discrete-time process modeling and control are boarded, as well as the logical and sequential control, and finally hybrid systems are presented to illustrate the potential and synergy of fundamental concepts and tools. Moving from basics to advanced applications is a pathway of perseverance and development that takes place personally and in learning groups, and also in a knowledge field, from a broader perspective. Trying to present contents in a simple way does not eliminate the challenge inherent to them and to such a growth process, but hopefully this book may be a valuable guide.

Contents

Chapter 1
Mechatronic Systems

1. Introduction

Mechatronics is "the synergistic integration of mechanical engineering, with electronics and intelligent computer control in the design and manufacturing of industrial products and processes" (Harashima et al., 1996). Mechatronics exploits systems engineering to guide product realization through different stages: design, modeling, simulation, analysis, prototype, optimization, and validation (Shetty and Kolk, 2011). Mechatronic systems with highly efficient performance are developed by adequately selecting and integrating sensors, actuators, computer hardware, and control algorithms. The integration within a mechatronic system is performed through the combination of hardware (components) and software (information processing). Hardware integration results from designing the mechatronic system as an overall system and combining the sensors, actuators, and microcomputers into the mechanical system. Software integration is primarily based on advanced control functions (Isermann, 2005).

This chapter reviews the design process, methodologies, and good practices used to develop mechatronic systems. It explains how mathematical modeling and real-time computational simulations are required for the concurrent interdisciplinary design of the mechatronic system and its hardware and software integration and verification stages. In Section 2, the mechatronic system components are described. Section 3 presents a brief introduction to Mechatronic Design Methodologies; mainly, in subsection 3.5, an alternative mechatronic design methodology based on the V-model and essential concepts from the VDI 2206 standard is discussed. This methodology reinterprets the mechatronic design process and rearranges the stages to impose a logical continuity and connection among them. Section 4, Model-Based Design, describes how the model is required at the mechatronic design process's design stage to understand better the system components, process, and control variables. Section 5, Hardware-in-the-loop simulation, explains how real-time computational simulation schemes are used for hardware and software integration, particularly for verifying the control system. Section 6 provides the conclusions.

2. Mechatronic System Components

Mechatronic systems and processes are composed of mechanical, electrical, electronic, and computer hardware devices classified as processes, sensors, actuators, and controllers based on their functionality. Figure 1.1 shows a schematic diagram of the mechatronic system components. The actuators used in the mechatronic systems are pneumatic and hydraulic valves and cylinders, electrical motors such as DC motors, AC motors, stepper motors, servomotors, and electro-mechanical and piezoelectric actuators. The diverse types of sensors employed in mechatronic systems are linear, rotational, acceleration, force, torque, and many others, such as pressure, flow, temperature, proximity, light, etc.

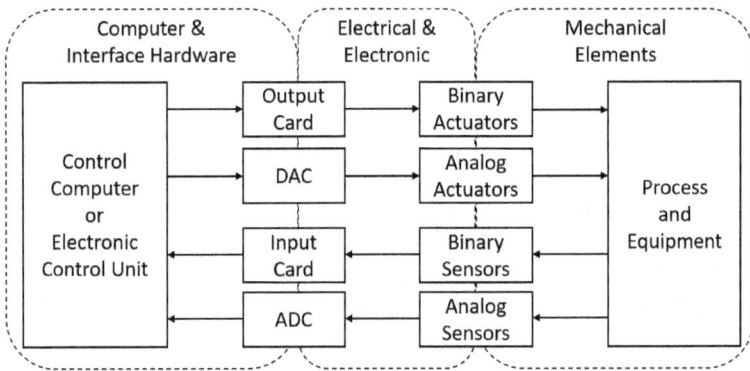

Fig. 1.1. Mechatronic system components.

2.1 Mechanical Elements

Mechanical elements refer to mechanical structures and mechanisms, electromechanical, thermal, fluidic processes, and hydraulic systems. Mechanical elements may have static and dynamic characteristics. A mechanical element is designed to interact with its environment. Therefore, mechanical elements require physical power to produce motion, force, heat, etc. Mechanical systems require sensors and actuators, and controllers to be automatically operated.

2.2 Sensors

The sensors are input devices that measure process variables and detect events or alarm conditions. In mechatronic systems, sensors detect movement or displacement, indicate a limit or threshold has been reached, sense commands by operators, etc. Typical input devices may include limit switches, photoelectric sensors, pushbuttons, proximity sensors, etc. There are binary (on- or off-state) input signals and analog signals to sense physical variables using sensors such as photo-resistor, level and displacement potentiometer, direction/tilt using, stress, and pressure sensors, touch using a micro-switch, temperature using thermistor, and humidity using conductivity sensors.

2.3 Actuators

Actuators are output devices used to produce motion or cause some control actions such as start/stop operation of equipment such motors and pumps, on/off control of valves, operator status indications and alarms, etc. The output signals, like input signals, can be discrete or continuous. Actuators apply commanded action on the physical process; typical actuator devices include AC and DC motors and servomotors, stepper motors, relays, solenoids, electromagnets, pumps, hydraulic and pneumatic valves, and cylinders. There are binary actuators, including relays, motor starters, pilot lights, etc.

2.4 Electrical and Electronic Components

The electrical and electronic elements interface electro-mechanical sensors and actuators to the control hardware devices. Electrical components are, for example, resistors (R), capacitors (C), inductors (L), transformers, etc. Electrical and electronic circuits include transistors, thyristors, optoisolators, operational amplifiers, power electronics, signal conditioning, analog-to-digital-converters (ADC), digital-to-analog converters (DAC), digital input/output (I/O), counters, timers, and data acquisition.

2.5 Control Hardware

Control hardware can be computers, microprocessors, microcontrollers, digital signal processors (DSP), and programmable logic controllers (PLC). Their function is to implement control algorithms, which use sensor measurements to compute control actions to be applied by the actuators. Control interface hardware (analog/digital interfacing) allows the communication of sensor signal to the controller and communication of control signal from the controller to the actuator.

2.6 Information Systems

Computers store a large number of data for further processing through different specialized software. The desired performance can be achieved with the control strategy implemented in control software. However, advanced control and connectivity functions are required to add value to a mechatronic system and to incorporate competitive advantages. A computer-aided design (CAD) environment is essential for mechatronics design. CAD systems provide analysis, optimization design, simulation, virtual instrumentation, rapid control prototyping, hardware-in-the-loop simulation, and PC-based data acquisition and control.

2.7 Mechatronic Areas of Expertise

Mechatronics can be considered a discipline but also a methodology used for the optimal design of electromechanical products for varied applications. Mechatronics is essentially a multidisciplinary field. A typical knowledge base for optimal design and integration of a mechatronic system comprises dynamic system modeling and analysis, decision and control theory, sensors and signal conditioning, actuators,

Table 1.1. Mechatronic areas and topics of expertise.

Areas/Disciplines	Disciplines/Topics
1) Mathematical Signals and Systems	• Linear Time-Invariant Systems • Response of Dynamical Systems • Frequency Response • Transfer Functions • State Space Systems
2) Physical Systems Modeling	• Mechanism • Precision Machines • Electromechanical Systems • Thermal Systems • Fluid Systems • System Identification
3) Instrumentation (Sensors and Actuators)	• Measurement Systems • Transducers • Signal Conditioning • Power Electronics • Data Acquisition • Motors, Pneumatic and Hydraulic actuators
4) Digital Systems	• Combinational and Sequential Logic • FPGAs • Microcontrollers • Programmable Logic Controller
5) Control Engineering	• Discrete Event Systems • Feedback Regulatory Control • Supervisory Control
6) Data Communications	• Communications Systems • Data Communication Protocols • Telemetry Systems • LAN, WAN, and Industrial Networks
7) Information Systems	• Data Logging • Human Machine Interfaces • Web Interfaces

power electronics, data acquisition (ADC, DAC, digital I/O, counters, timers, etc.), hardware interfacing, and embedded computing. The topics mainly involved in Mechatronic design within the areas or disciplines are shown in Table 1.1. In practice, engineering teams must support all the activities to develop mechatronic systems.

3. Mechatronic Design Methodologies

Mechatronic systems are comprehensive mechanical systems with fully integrated electronics and information technology (IT). Such systems require another design approach for efficient development, different from the design of separate and purely mechanical, electronic, electric, and IT products (Isermann, 2005). Mechatronics is a methodology that applies an interdisciplinary approach of concurrent engineering for the optimal design of electromechanical products. In mechatronics-based product realization, the mechanical, electrical, computer engineering, and information systems are integrated throughout the design process so that the final products can be better than the sum of their parts. Applying a good mechatronic methodology should result in a better product optimized in cost, performance, and fabrication time.

3.1 Multidisciplinary Systems

Multidisciplinary systems have been successfully designed and used for many years. Electromechanical devices are examples of multidisciplinary systems which often use a computer algorithm to modify the behavior of a mechanical system. Electronics are used to transduce information between the computer and mechanical devices.

Traditionally an electromechanical system employed a sequential design-by-discipline approach. The design is often accomplished in three steps: (1) mechanical design, (2) electrical/electronic design, and (3) control algorithm design and implementation. The disadvantage is that by fixing design aspects at various points in the sequence, new constraints are created and passed on to the following discipline. Mechatronics uses an interdisciplinary instead of a multidisciplinary approach.

3.2 Concurrent Engineering

Concurrent engineering is a design approach in which the design and manufacture of a product are merged. Knowledge and information need to be coordinated amongst different experts. The characteristics of concurrent engineering are: (1) a better product definition without late changes; (2) a product development process is well-defined; (3) better cost estimates; and (4) design for manufacturing and assembly considered in the early design stage. The model-based design makes concurrent engineering possible.

3.3 Mechatronic Design Process

Mechatronics implies the combination of actuators, sensors, control systems, and computers in the design process. The primary goal is to apply new controls to extract new performance levels from a mechanical device. The integration and interface among the subsystems are essential—mechatronic design results in processes or products with more synergy. Synergy is the combined working of two or more parts of a system so that the combined effect is greater than the sum of the efforts of the parts.

The design methodologies for developing mechatronic products provide means to organize better and execute the diverse design and development tasks such as specification, design, fabrication, prototyping, integration, and validation and verification tests. Several mechatronic design methods suggest using good design practices, such as model-based design, system prototyping, design optimization, and functional tests. The mechatronic design process consists of three phases: modeling and simulation, prototyping, and deployment, as depicted in Fig. 1.2, adapted from (Shetty and Kolk, 2011).

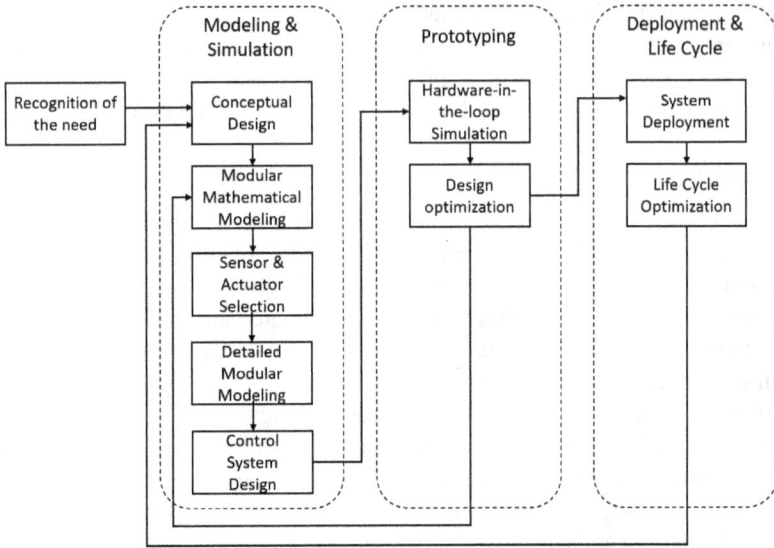

Fig. 1.2. Mechatronic design process (adapted from Shetty and Kolk 2011).

3.4 VDI 2206 Standard

An example of a mechatronics design methodology is the VDI 2206 standard for Mechatronic Systems Design. VDI stands for "Verein Deutsche Ingenieure" (Society of German Engineers). VDI 2206 merges two design guidelines: the standard VDI 2201 for designing mechanical systems and the V-model for developing IT systems. The V-model represents a software and hardware development process. It demonstrates the relationships between each phase of the development life cycle and its associated testing phase. The horizontal and vertical axes represent time or project completeness (left to right) and level of abstraction (bottom-up).

VDI 2206, see Fig. 1.3, is a widely accepted industrial guideline that defines crucial steps and measures to finalize efficient and cost-efficient mechatronic products. This guideline aims to provide methodological support for cross-domain development, especially in the early development phase, concentrating on system design (Bishop, 2012). The guideline consists of three essential elements: a general problem-solving cycle as a micro-cycle, the V-model as a macro-cycle, and predefined process modules for recurrent working steps. In the description of the micro-cycle, the guideline VDI 2206 refers to the problem-solving cycle used in systems engineering.

At the domain-specific level, the design process has to be concurrent since "the design of the mechanical system influences the electronic system, and the design of the electronic system also influences the mechanical system" (Bishop, 2012). VDI 2206 also specifies the need for validation and verification during the system integration at each system, subsystem, and component level. The objective of verification and validation is to gain confidence in the correctness of a system concerning its specification (verification) and when it is placed into its target environment (validation).

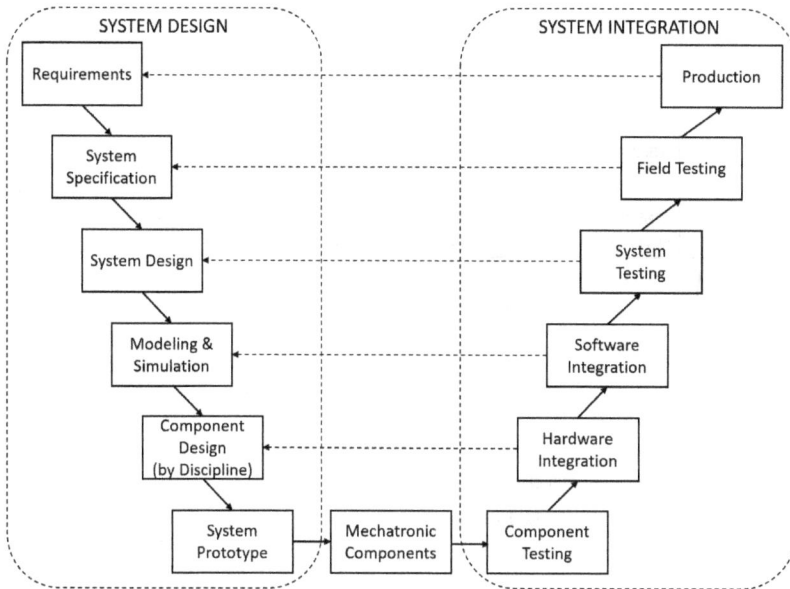

Fig. 1.3. V-model for mechatronic product development (adapted from Isermann 2005).

3.5 *Alternative Design and Integration Process*

In the mechatronic design process and methodologies, the recurrent recommendation is to use good design practices such as "model-based design" and "real-time simulation" for better product design and integration. The alternative methodology depicted in Fig. 1.4 is based on the V-model and takes essential concepts and considerations from the VDI 2206 standard. This methodology reinterprets the design process and rearranges the stages to impose a logical continuity and connection. It recognizes the need to define concurrent and sequential activities and implies precedence and subsequence relations. The three levels represent a hierarchy: the higher level is the system design and integration inputs and outputs, the second level is software centered, and the lower level is related to the hardware components. The

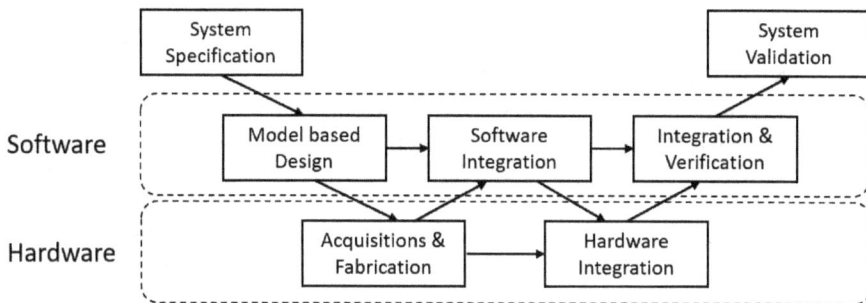

Fig. 1.4. Alternative mechatronic design process.

Table 1.2. Alternative methodology stages.

Stages	Activities/Products
System Specification	This entry stage is user and application centered; it defines the user interface and embodiment considerations. It also identifies critical functions and process variables. A functional specification or description is the product of this stage.
System Design	Model-based system design, functional specification, CAD-based 3D modeling, signals and system response modeling and analysis, bill of materials, 3D mechanical CAD design, and PCB CAD design.
System Fabrication	Acquisitions and fabrication: components purchase and procurement for different subsystems; fabrication of mechanical and electronic components; selection of sensors and actuators.
Hardware Integration	Component-level integration of the hardware devices, such as the mechanisms, sensors, and actuators. Prototype assembly (including sensors, actuators, and control system hardware).
Software Integration	Software or, preferably, hardware in the loop simulation (to verify modeling and control), connectivity, and user interface functional test.
System Integration and Verification	Integration of Hardware and Software Components; Verification of Functional Specifications; Control program and user interface application integration with the hardware components. Hardware Integration. Functional Verification Test.
System Validation and Deployment	Integration of hardware and software components. Functional and operational validation. Operational Validation Test to final deployment.

activities on the left are related to the design, and the activities on the right are related to the integration, such as in the VDI 2206 standard.

The alternative methodology is intended for fast system prototype development (Martell et al., 2022). An advantage of this design methodology is that each stage has clearly defined outputs for the next stage. The description of each proposed design methodology stage is shown in Table 1.2.

4. System Modeling and Simulation

The mechatronics design approach relies heavily on system modeling and simulation throughout the design and prototyping stages. Modeling is the process of representing the behavior of a real system by a collection of mathematical equations and logic. In Mechatronics, the process and the control system are simultaneously designed. The static and dynamic behavior of the process, the type and placement of the actuators, and the type and position of the sensors are designed appropriately, resulting in better overall system response. The mechatronics design process and the VDI 2206 methodology consider employing model-based design and hardware in the loop simulations as a part of the design process. The development of mechatronic systems starts with modeling and simulation: (1) system design by building static and dynamic models; (2) model transformation into discrete-time simulation models; (3) programming the computer-based control; (4) hardware in the loop simulation for verification of control functions; and (5) complete hardware-software integration.

4.1 Model-based System Design

Before developing a prototype of a mechatronic system, it is convenient to model the behavior of the system components for a better understanding of the critical functions and the system dynamics. Mathematical modeling is intended to represent the behavior of the physical system, but system modeling may improve the design of the physical systems. Models are cause-and-effect structures; they accept external information and process it with logic and equations to produce outputs. There are two ways to obtain models: (1) theoretical modeling based on first principles (physical laws), and (2) experimental modeling with measured input and output variables (system identification). Physical laws predict the behavior of engineered systems: electrical engineering uses Ohm's and Kirchoff's laws; mechanical engineering applies Newton's law; electromagnetics uses Faraday's and Lenz's laws; fluids dynamics apply Bernoulli's laws, etc.

The modeling of the physical system or process becomes relevant for designing and implementing model-based control techniques. A detailed description of components and their interconnections is needed (preferably in a computable form). Models can also be text-based programs or visual diagrams (block diagrams, flow charts, state transition diagrams, bond graphs). Most continuous time systems represent how continuous signals are transformed via differential equations. Simulation is the process of solving the model and is performed on a computer. Numerical simulation relies on numerical methods for solving models containing differential, discrete, linear, and nonlinear equations. Most discrete-time systems represent how discrete signals are transformed via difference equations. The order of the discrete-time system is the highest number in the differential equation by which the output is delayed.

4.2 Model-based Control Design

The control algorithm becomes a key component of the mechatronic products. Basic and advanced automation functions can be implemented in the control systems. In highly integrated mechatronic devices, optimal, safe, and robust performance can be obtained with a proper control system design. Model-based control design and verification can be incorporated into the mechatronic design process. Linear system theory is a powerful tool for mechatronic system design. A dynamical response can be designed for better linearity, low order, small or no dead times, good natural damping, or other essential items for control design, see Fig. 1.5.

Sequential logic control problems can be addressed with conventional logic control techniques or, with more formalism, using finite state machines. Regulatory and reference tracking control (servo control) can be implemented with feedback control techniques, either proportional integral-derivative control or more advanced control schemes such as model reference-adaptive control, model predictive control, or even soft computing techniques such as fuzzy logic or rule-based control.

Control Software Design   Control Software Verification

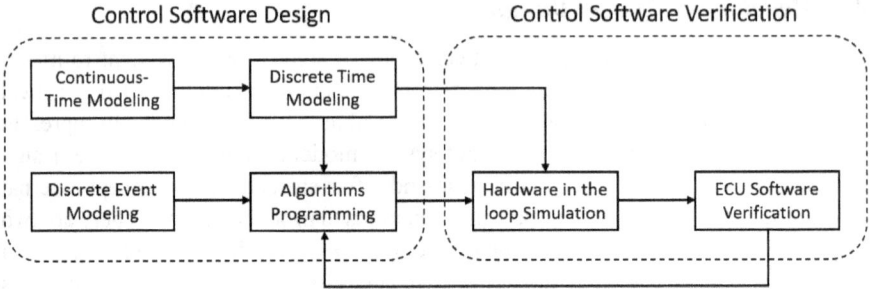

Fig. 1.5. Control software development and verification.

4.3 Hardware-in-the-Loop Simulations

The validation and verification of the control algorithms to be implemented in the Electronic Control Unit (ECU) can be carried out with real-time simulations, either software-in-loop simulations (SiL) or, preferably, with hardware-in-the-loop simulations (HiL). These are the primary schemes for real-time simulations to verify the control software: (1) control system prototyping, (2) software-in-the-loop simulations, and (3) hardware-in-the-loop simulations. Control prototyping is a simulation scheme when the process, the actuators, and the sensors may be real components. It helps design and test complex control systems and their algorithms under real-time constraints. A real-time controller simulation (emulation) with hardware (a development board) other than the final series production hardware (particularly ASICS) is an example of control prototyping.

HiL simulation is increasingly used for the testing of control systems, where some of the control loop components are real hardware, and some are simulated. Usually, a process is simulated because it is not available (is under construction) or because experiments with the real process are too costly or require too much time (Iserman, 1999). In the HiL, the control system hardware and software are the same as those used for series production (see Fig. 1.6). The controlled process (consisting of actuators, physical processes, and sensors) can comprise simulated or real components. The HiL simulation aims to control and verify the control functions and correct operation of the user interface and connectivity functions such as data visualization and logging.

Real Control System   Simulated Process & Instrumentation

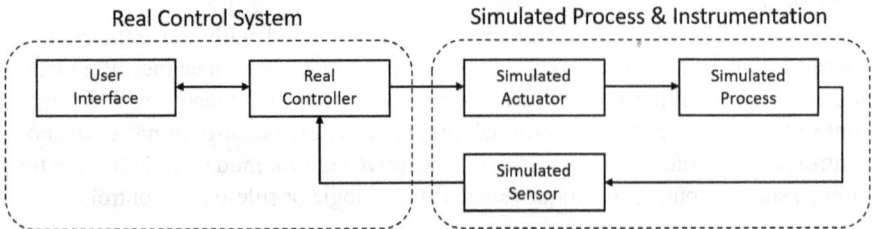

Fig. 1.6. Real time HiL scheme.

5. Case Study – DDR Mobile Robot

In mechatronics systems, the design of the control systems is critical to achieving the required functionality, and this task can be carried out following a proper software integration process. The development of the control systems of a differential drive (DDR) mobile robot is presented as a case study. The virtual robot comprises 3D graphics and a numerical simulation of the wheels and kinematics incorporated in an HiL simulation, whose inputs and outputs are communicated to a microcontroller. This example shows the relevance of the modeling of the locomotion and kinematics of the mobile robot, from its mathematical modeling to the implementation of hardware-in-the-loop simulation for the verification of the control system.

5.1 Control System Design Methodology

The development of the control system requires the design of a virtual DDR mobile robot for the Hardware in the loop simulation. The virtual robot is a computational representation of a physical robot instrumented with conventional DC motors as actuators and photo resistors as proximity sensors. Developing a virtual DDR mobile robot aims to verify the control system algorithms. Since the control of the virtual robot is the primary concern, only the functional characteristics of the real robot are considered.

The design methodology consists of the following steps: (1) modeling and simulation of the virtual robot components considering sensors and actuator dynamics; (2) design of the 3D graphics of the virtual robot; (3) programming the control functions in the microcontroller; (4) development of the bidirectional connection to the microcontroller using the Modbus RTU protocol; and (5) verification of the control systems functionality.

The virtual robot is composed of 3D graphics and a numerical simulation of the wheels. The robot kinematics is incorporated in an HiL simulation, with inputs and outputs communicated to a microcontroller. The selected communication protocol was Modbus RTU, an open-source protocol based on a master/slave architecture. The Modbus communication protocol allows the virtual robot control signals to be available to an Arduino® card, a Raspberry Pi®, or some other controller with a Modbus connection. Figure 1.7 shows the scheme of the system. More details on the design of the mobile robots can be found in Cardenas et al. (2021).

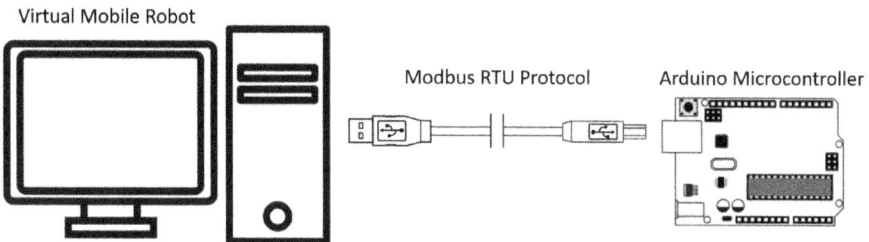

Fig. 1.7. HiL simulation with the virtual robot.

5.2 Mobile Robot Modeling and Simulation

The robot's angular, ω, and linear, v, speeds can be computed from the angular velocity of the left, ω_l, and right, ω_r, wheels, and the distance between them, L, according to the following kinematic equations:

$$V_l = r\omega_l; \, V_r = r\omega_r \tag{1.1}$$

$$\omega = \frac{V_r - V_l}{L}; \, v = \frac{V_r + V_l}{2} \tag{1.2}$$

The kinematics of the virtual robot is well known and can be represented as a mobile coordinate frame with respect to a fixed coordinate system, Fig. 1.8. The center of the robot has a position given by the coordinates (x_m, y_m) and an orientation given by θ.

A simple way to take the robot to a new desired position (x_n, y_n) and orientation θ_n is to make a longitudinal displacement, d, which can be computed by (1.3) and rotate an angle, θ_r, clockwise (CW) or counterclockwise (CCW) according to the subtraction of the current orientation angle from the desired orientation angle, as expressed by (1.4):

$$d = \sqrt{(y_n - y_m)^2 - (x_n - x_m)^2} \tag{1.3}$$

$$\theta_r = \theta_n - \theta \tag{1.4}$$

DC motors provide the motion of the wheels. Although an instantaneous response of the DC motors can be assumed, this modeling approach considers first-order dynamics in the motors.

The dynamics of the DC motor is given by the difference equation shown in (1.5), where $u(k)$ represents the input to the system (k is the sample number) and $\omega(k)$ is the output that simulates a first-order closed-loop control of the motor's angular speed as a servo-controller is driving it. The parameters in this equation are sampling time, T, and time constant, τ. The linear speed of the wheels is given by (1.6).

$$\omega(k) = \left(\frac{\tau}{\tau + T}\right)\omega(k-1) + \left(\frac{T}{\tau + T}\right)u(k) \tag{1.5}$$

$$v(k) = r\omega(k) \tag{1.6}$$

Equations (1.5) and (1.6) can be used to express equations for the left and right wheels as follows:

$$\omega_l(k) = \left(\frac{\tau}{\tau + T}\right)\omega_l(k-1) + \left(\frac{T}{\tau + T}\right)u_l(k) \tag{1.7}$$

$$v_l(k) = r\omega_l(k) \tag{1.8}$$

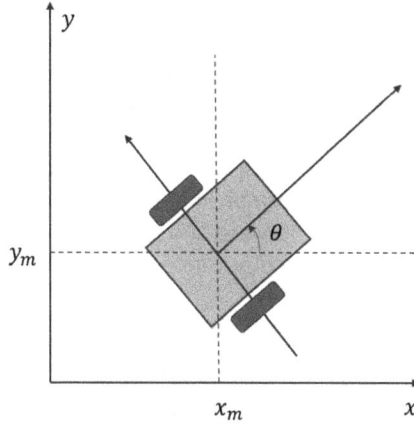

Fig. 1.8. Cartesian coordinates and robot variables.

$$\omega_r(k) = \left(\frac{\tau}{\tau+T}\right)\omega_r(k-1) + \left(\frac{T}{\tau+T}\right)u_r(k) \tag{1.9}$$

$$v_r(k) = r\omega_r(k) \tag{1.10}$$

A set of equations can be generated for the angular and linear speeds of the robot by incorporating the previously proposed simple dynamical response of the motors driving the left and right wheels:

$$\omega(k) = \frac{v_r(k) - v_l(k)}{L} \tag{1.11}$$

$$v(k) = \frac{v_r(k) + v_l(k)}{2} \tag{1.12}$$

Finally, the linear displacement produced in the robot can be calculated by numerical integration as in (1.13), where the sampling time, T, is used to integrate the robot's speed.

$$d(k) = d(k-1) + Tv(k) \tag{1.13}$$

5.3 Robot Control Algorithms

The basic control of the robot can be accomplished at two levels: (1) control of the locomotion by commanding the wheels, and (2) kinematic control of the robot's speed and position. From the equation of the previous section, 5.2, the angular speeds that must be applied to the left and right wheels must be found; this is done by solving the simultaneous system of equations expressed by (1.1) and (1.2). The control of the wheels can be performed in closed-loop with quadrature encoders; in this case, a more precise position and orientation of the robot can be calculated in the microcontroller by computing the robot kinematics to determine the orientation and position of the robot, by applying directly equations (1.11) to (1.13). It is also possible to control the robot without feedback sensors by calculating in real time the estimated angular and linear velocities of the wheels, given by equations (1.7) to (1.10), with these calculations, to estimate the orientation and position of the robot. More control functions can be implemented to follow consecutive linear trajectories by executing motion segments composed of a longitudinal displacement and a rotation or by more advanced trajectories considering curvature radio or polynomial trajectories. Other advanced control functions can be implemented by adding proximity sensors to avoid obstacles, for example, by considering only the three simulated proximity sensors and rotating 90° to the direction where no obstacle is detected.

5.4 Graphical 3D Model and Animation

In the mechatronics design process, a 3D model of the system is conventionally carried out to validate and specify the mechanical components. Any 3D CAD software may be used. SolidWorks® is a computer-aided design (CAD) software for product and machine parts and mechanisms. Optionally, the 3D model can be enhanced to incorporate an animation of the mechatronic system. Blender® and Unity® are multiplatform engines for the animation of graphic models. For the mobile robot, the 3D model of each part of the virtual robot can be designed in SolidWorks®, color, and texture details added in Blender®, then each part of the robot can be imported into Unity®.

Since the virtual robot's objective is to validate the control system's design, the computational simulation does not require many details of the real robot, so only the parts that directly affect the simulation of the robot's locomotion and kinematics were considered. The primary actuators are the two DC motors of the wheels. For the numerical simulation, all equations of the previous subsection 5.2 can be programmed in Unity® within scripts developed in C# and incorporated into the 3D model. This allows the simulation of the rotation of the wheels and the animation of the robot's locomotion.

5.5 Real-Time HiL Simulation

The virtual robot considering a numerical simulation of the robot wheels' dynamic and robot kinematics, can be connected to the Arduino® microcontroller in the HiL simulation scheme that is depicted in Fig. 1.9. The bidirectional communication between the application developed in Unity® and Arduino® was carried out through the Modbus RTU communication protocol. Modbus is an open industrial communication protocol widely used by many manufacturers that allow communications over RS-485, RS-232, or USB with the Modbus RTU (Remote Terminal Unit) protocol, and there is also a Modbus implementation over TCP/IP in case an Ethernet link is available in the microcontroller.

The implementation of the HiL scheme was done, as previously described, with the Modbus protocol. Since the protocol uses a UART connection, the communication speed between Unity® and Arduino® is defined to be 100 ms; this happens because there are independent functions for reading and writing registers and coils. The control manipulation values to the wheels are determined by the

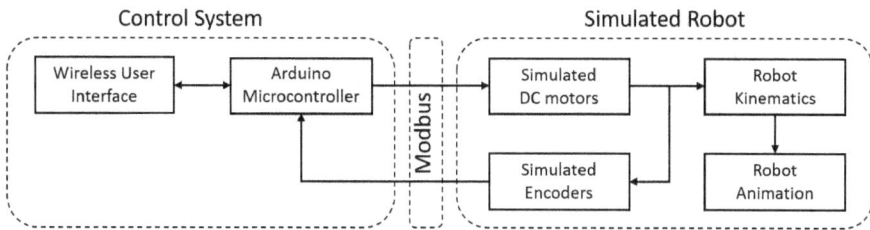

Fig. 1.9. HiL simulation for control verification of the mobile robot.

control algorithms programmed in the external control unit, the microcontroller. Therefore, only the dynamics of the actuators and sensors are programmed in the virtual robot. Two binary signals are used on the Modbus communication to control each wheel's rotation direction. The wheel's angular position is the feedback to the external microcontroller through the virtual encoder. The virtual encoder has 1000 pulses per revolution and a virtual counter of fast pulses through the holding registers of the Modbus protocol towards the microcontroller.

The HiL scheme allows the verification of the control algorithms and the connectivity to the user interface. The response with the Modbus link is slower than the embedded controller in the virtual robot, but this does not imply a disadvantage because the delay in the communication link can represent an additional and particular effect of the dynamical response of the mobile robot and, therefore, it results in more realistic locomotion and displacement that is adequate for the remote visualization of the mobile robot. Once the software integration is performed, which implies the control system verification, the software function can be tested with the actual hardware components making the appropriate wiring connections and required software changes to control the real robot. If the control functions are verified in the simulated robot, they are expected to perform correctly in the real mobile robot.

6. Conclusions

Mechatronics is an interdisciplinary field in which the following disciplines act together: (1) mechanical systems, (2) electronic systems, and (3) information technology. Mechatronics deals with the application of computational intelligence, complex decision-making, and the operation of physical systems. The integration of mechanisms with electronic devices and information systems is performed with the hardware components (mechanisms, sensors, actuators) and the software functions (control and connectivity).

The mechatronics design process uses an interdisciplinary approach instead of a multidisciplinary one, based on a concurrent instead of sequential design approach. In designing and fabricating mechatronics systems, employing good design practices, such as 3D CAD-based design, model-based design, and hardware-in-the-loop simulations, is convenient.

Mathematical modeling and computational real-time simulation are required for the concurrent interdisciplinary design of the mechatronic system and its integration and verification stages. The model is required at the mechatronic design process's design stage to better understand the systems and their process and control variables. Later at the system integration, the model can be converted to discrete-time to be partially or fully implemented in a "hardware in the loop" real-time simulation useful for verifying the control algorithms and other control functions like the correct operation of the user interface or the connectivity functions. In this chapter, a mechatronic design methodology was reinterpreted and defined in several stages with clearly identified inputs and outputs for each system design and integration stage.

References

Bishop, R.H. 2002. *The Mechatronics Handbook*. First Edition. CRC Press.

Cardenas, M.M., Villalobos, J., Santacruz, A.F., Paredes, C.A., Sanchez, I.Y. and Martell, F. 2021. Mobile robot with modbus RTU connectivity to arduino microcontroller for remote-online education in control systems. *2021 International Conference on Electrical, Computer and Energy Technologies (ICECET)*, pp. 1–6. DOI: 10.1109/ICECET52533.2021.9698696.

Harashima, F., Tomizuka, M. and Fukuda, T. 1996. Mechatronics: "What Is It, Why, and How? An Editorial". *IEEE/ASME Transactions on Mechatronics* 1(1): 1–4. DOI: 10.1109/TMECH.1996.7827930.

Isermann, R., Schaffnit, J. and Sinsel, S. 1999. Hardware-in-the-loop simulation for the design and testing of engine-control systems. *Control Engineering Practice*.

Isermann, R. 2005. *Mechatronic Systems: Fundamentals*. Springer-Verlag, Berlin, Heidelberg.

Martell, F., Uribe, J.M., Sarabia, J., Ruiz A., Martínez, Á.E. and Licurgo, E. 2022. Mechatronic design methodology for fast-prototyping of a pressure controlled mechanical ventilator. *In:* Flores Rodríguez, K.L., Ramos Alvarado, R., Barati, M., Segovia Tagle, V. and Velázquez González, R.S. (eds.). *Recent Trends in Sustainable Engineering. ICASAT 2021. Lecture Notes in Networks and Systems*, 297. Springer, Cham. https://doi.org/10.1007/978-3-030-82064-0_15.

Shetty, D. and Kolk, R.A. 2011. *Mechatronics System Design*. Second Edition. Cengage Learning.

Chapter 2

Industrial Automation and Control

1. Introduction

Automation is "the application of technologies by which a process or procedure is accomplished without human assistance" (Groover, 2015). Some reasons for automating are to achieve the following: increase labor productivity, reduce labor cost, improve product quality, improve worker safety, and reduce manufacturing lead time. Productivity and quality control are obtained by continuous innovation of the control systems governing the automated systems. Automating industrial processes and plants certainly reduces the personnel required by the overall production systems. However, it also offers employment opportunities for more specialized work functions and activities needed to design, implement, operate, and maintain automated and digitalized production and manufacturing systems (Groover, 2015). Industrial automation requires computer hardware and software combined with electronics, electromechanical, electro-pneumatic, and electro-hydraulic devices.

System Integrators and automation engineers in the automation industry must have specialized knowledge and skills in applying good design practices in developing and supporting industrial control systems. This chapter reviews the tasks for developing the control system required in automation projects from the systems integrator's perspective. The models and their computational simulation can be helpful for system control verification and system integration, particularly for performing a software acceptance test where the control system can be verified and all the automation functions defined in the functional specification can be validated.

In this chapter, Section 2, Automation Technologies, reviews the automation pyramid and briefly describes the most popular automation technologies, such as PLC, DCS, and SCADA. Section 3, Industrial Control Systems, describes industrial control types and the control system design process. Section 4, Control Software Development, explains how control software, control logic, and algorithms can be developed concurrently if a model and HiL simulation are used to verify the control software. Design and integration of industrial automation projects, particularly activities dealing with control software development are discussed in

Section 5. A hardware-in-the-loop simulation scheme for control software verification is depicted in Section 6, and Section 7 reviews the conclusions.

2. Automation Technologies

The term 'automation' includes all traditionally identified topics such as instrumentation and control, process control, process automation, control systems, automation and control, manufacturing control, manufacturing automation, and system integration (Sands and Verhappen, 2018). Automation includes control systems and other automation functions such as measurement, instrumentation, data acquisition and logging, data communications, and software applications to monitor and control the industrial processes and the overall plant operation in real time. A control system conventionally includes sensors, actuators, the electronic control unit, and the user interface.

There is a main differentiation among the technologies employed for complete plant automation. The lower levels implement operational technologies (OT), and the middle to upper levels are automated with information technologies (IT). The main difference between OT and IT devices is that OT devices control the physical world. At the same time, IT systems manage data (Brandl, 2012) and other automation technologies such as industrial robots and computerized numerical control (CNC).

2.1 Automation Pyramid

Today an industrial plant with a higher degree of automation can be conceptualized as a distributed computer network made out of several industrial communication protocols, ranging from field buses to industrial Ethernet networks that interconnect all the machines, processes, and systems. The plant automation can be depicted as a pyramid, as shown in Fig. 2.1; the machines and processes that use sensors and actuators are on the plant floor levels. At an intermediate level, the supervisory control and manufacturing execution systems are found. The production and quality

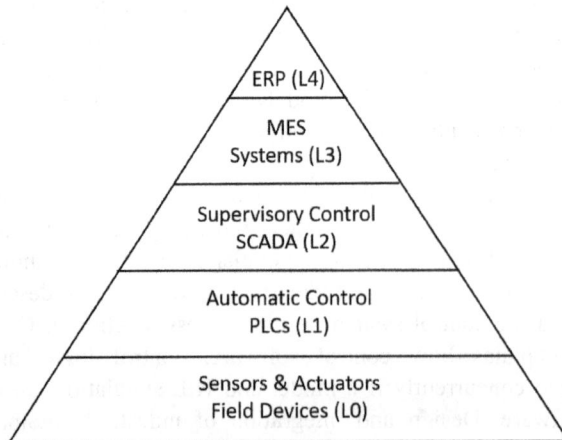

Fig. 2.1. The automation pyramid.

Table 2.1. Automation levels.

Automation Levels	Description
Sensors and Actuators Layer	The pyramid's base layer is called the 'level 0' layer. This layer is closest to the processes and machines used to convert and collect signals from processes for analysis and control. Hence, control signals can be applied to the processes.
Automatic Control Layer	This layer consists of automatic control and monitoring systems, which drive the actuators using the process information given by sensors. This is called the 'level 1' layer.
Supervisory Control Layer	Supervisory Control looks over the equipment, which may consist of several control loops. This is named the 'level 2' layer. This layer drives the automatic control system by setting targets or goals for the controller.
Production Control Layer	Manufacturing execution systems (MES) characterize this layer and support decisions about production targets, resource and task allocation to machines, maintenance management, etc. This is the 'level 3' layer.
Enterprise Control layer	Less technical and commercial activities like supply, demand, cash flow, product marketing, etc., are carried out with special enterprise resource planning software (ERP). This is called the 'level 4' layer.

management functions and administrative and business systems are considered at the upper levels. Table 2.1 details the functions implemented at each plant automation level or layer.

2.2 Automation Levels in Manufacturing Industries

In the discrete manufacturing industries, production operations are performed on quantities of materials: parts or product units, such as in automobile and electronics manufacturing. Basic operations in manufacturing include processing and assembly operations, material handling, inspection and testing, and coordination and control. In the manufacturing industries, the automation concept can be applied at different levels that implement progressive groupings (Groover, 2015). Sensors and actuators are operated in control loops at level one or device level. At level 2, or machine level, the operation of various devices is integrated. At level 3 or system or cell level, a manufacturing cell is defined and operated as a group of machines or workstations interconnected by a material handling system, including production and assembly lines. The 4th level, or plant level, produces operational plans such as processing plans, inventory control, purchasing, quality control, etc. The 5th level is the enterprise level which handles corporate information related to marketing, sales, accounting, research and design, production scheduling, etc. Computer Integrated Manufacturing (CIM) technologies refer to computer-controlled machinery and automation systems in manufacturing products.

2.3 Automation Levels in Processes Industries

In the process industries, the production operations are performed on amounts of materials: liquids, gases, powders, etc. Process industries include chemicals, petrochemicals, oil and mineral refining, food processing, pharmaceuticals, power generation, etc. Continuous processes and fluid processing characterize these industries. Process Control typically implies the automation of process industries. In the process industries, automation level 1 includes process controls conventionally implemented in distributed control systems (DCS) and programmable logic controllers (PLC). DCS of industrial grade are more robust and fault-tolerant control platforms. The level 2 process control can be implemented in workstations with higher computational capabilities suitable for non-critical, supervisory control applications. PLC, DCS, and SCADA have become synonymous in process instrumentation and control (Dey et al., 2020).

2.4 Operational Technologies

Programmable Logic Controllers. The automation of discrete event processes in the manufacturing industries has been based mainly on the use of PLC, which are programmable microprocessor-based computer hardware suitable to control assembly lines and machinery on the shop floor as well as many other types of mechanical, electrical, and electronic equipment in a plant. PLC are designed for real-time use in rugged industrial environments. Connected to sensors and actuators, PLC are categorized by the number and type of I/O ports they provide and their I/O scan rate. In the late 1960s, PLC were first used to replace the hardwired networks of relays and timers in automobile assembly lines, which were partially automated then. The programmability of the PLC enabled changes to be effected considerably faster.

Programmable Automation Controllers (PAC). PAC are programmable microprocessor-based devices for discrete manufacturing, process control, and remote monitoring applications. PAC combines the functions of a programmable logic controller (PLC) with the greater flexibility of a PC. A PAC is a modern term for a PLC with higher computing and communications capabilities and performance, and, like PLC, typically has a RISC-based processor that can be programmed in the IEC 61131 programming languages.

Human Machine Interface and Operators Panels. The user interface in a manufacturing or process control system provides a graphics-based visualization of an industrial control and monitoring system. Previously called an MMI (man-machine interface), an HMI typically resides in an office-based Windows computer that communicates with a specialized computer (DCS, PLC, or PAC) in the plant. An operator panel is a low-level graphical interface for operating processes and machines on the plant floor. Typically built into a ruggedized terminal for shop floor use, it provides a remote interface to the push buttons, switches, and gauges found on the equipment itself. An operator interface does not offer the rich graphics and flexibility of a Windows-based interface.

Distributed Control System (DCS). A DCS is a process control system that uses an industrial network or field buses to interconnect sensors, controllers, operator terminals, and actuators. A DCS is widely used in the process industries because it is robust hardware capable of connecting thousands of input/output points from digital and analog sensors and actuators. A DCS typically contains several computers for control and uses proprietary interconnections. Generally, it is a very large and costly system.

Numerical Control. Computer numerical control (CNC) is the automated control of machining tools using computer hardware. A CNC machine processes a piece of metal, plastic, wood, ceramic, or composite materials to meet manufacturing specifications by following a program with instructions to operate automatically. A list of instructions forms a sequential G-code program executed in a CNC machine. The program can be written by a specialized person or generated by graphical computer-aided design (CAD) or computer-aided manufacturing (CAM) software. Machining processes include drills, mills, grinders, routers, and 3D printers.

Industrial Robots. An industrial robot is a robot system used for manufacturing. Industrial robots can move on three or more axes for positioning and orientation in Cartesian coordinates. Articulated robots are the most common industrial robots. They have 5, 6, or 7 degrees of freedom, called robotic or manipulator arms. SCARA is an acronym for selective compliance assembly robot arm. SCARA robots are recognized by their two parallel joints, which provide faster movement in the X-Y plane than Cartesian CNC robots. SCARA is considered ideal for assembly tasks. Other applications of robots include product inspection and testing, material handling, welding, painting, pick and place, packaging, labeling, and other manufacturing processes that require speed and precision.

2.5 Information Technologies

Supervisory Control and Data Acquisition. The supervisory control and data acquisition (SCADA) system is an automation technology that collects data from sensors and controllers on the production systems or remote locations and logs them in computer databases for management and control. A SCADA is conventionally a networked system composed of supervisory control computers connected to PLC serving as Remote Terminal Units (RTU). The operator interfaces within the SCADA enable monitoring and issuing process commands, like set points to the controllers of machines on the shop floor or to other controllers connected to sensors and actuators. SCADA systems typically use a tag database, which contains data elements called tags corresponding to the process's digital and analog inputs and outputs related to specific instrumentation (sensors or actuators).

Manufacturing Execution Systems (MES). An MES tracks product and order details on the plant floor, collects transactions for reporting to financial and planning systems, and electronically dispatches orders and manufacturing instructions to shop floor personnel. ISA 95 standards are used from shop floor to top floor integration. The standards provide a formal model for exchanging data between business and manufacturing systems. The models also include a definition of manufacturing operations management and the activities on the shop floor to follow production schedules and perform the work required to manufacture products.

3. Industrial Control Systems

The industrial control system is a term used to describe different types of control systems and the required instrumentation, which includes the devices, systems, networks, and controls used to automate industrial processes. The control system causes the process to accomplish its defined function, to carry out some manufacturing operation. A control system comprises hardware such as controller units, sensors, and actuators (see Fig. 2.2). The controller unit executes a program of instructions and performs the control calculation (or control laws) and the logic sequences. The programmed instructions are executed and activate the proper commands or signals to the actuating devices interfaced by output modules. Actuators are electromechanical devices such as relays, valves, and motors. Process variables and conditions are sensed and monitored through input modules. The program of instructions is the set of commands that specifies the sequence of steps in the work cycle and the details of each stage (Groover, 2015). During each step, one or more activities involve changes in process variables. Programs of instructions are, for example, CNC part programs, robot programs, and PLC programs. Developing industrial control systems can benefit from good mechatronic design practices such as model-based design, software-in-the-loop, and hardware-in-the-loop simulations.

All control systems can typically be defined as having inputs, outputs, and some form of computational intelligence to implement control functions so that the outputs are controlled based on the status of the inputs. The decision-making functions can be performed by a PLC or any other Electronic Control Unit (ECU), which has inputs, outputs, and a central processing unit that uses logic programming to make decisions based on input status and the logical conditions set in the program. The controller processes continuous and binary variables. Continuous variables and parameters are continuous as time proceeds. They are also called analog variables since they can take on any value within a specific range. An ADC converts a continuous analog signal from a transducer into digital code for a computer. A DAC converts the computer's digital output into a continuous analog signal to drive an analog actuator (or another

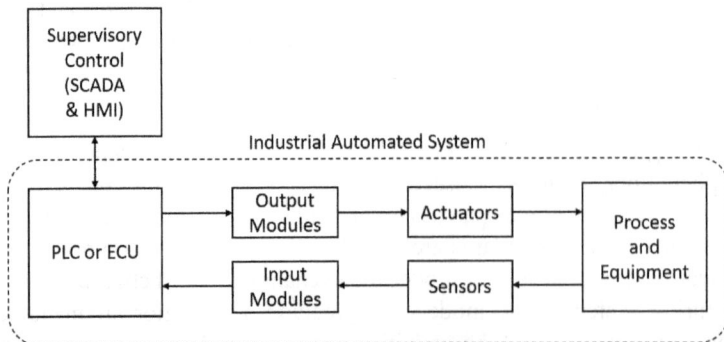

Fig. 2.2. Industrial automated system.

analog device). The reference parameter or set point represents the desired value of the output process variable.

3.1 Feedback Regulatory Control and Servo-Control

Control variables and parameters are continuous in time and analog or discrete in their domain. Examples of continuous systems are those for controlling the temperature in a furnace, the tank level, the pressure or flow in a water treatment system, etc. A usual objective of continuous control is to maintain the value of an output variable at a desired level; this is possible through implementing a feedback control system. In the closed loop (feedback) control system, the output variable is compared with an input parameter (set point), and any difference between the two is used to drive the output into agreement with the input. The open loop control system operates without the feedback and is more straightforward and less expensive but with the risk that the actuator could not have the intended effect.

In many industrial processes, feedback control loops, as shown in Fig. 2.3, are used to keep the process variables at specific values despite disturbances (***regulatory control***), and other process variables must be adjusted to required set points in a stable manner (***servo-control or set-point tracking***).

For example, the temperature in an industrial furnace is set to a certain value for controlling a heating process. For the temperature to be controlled, it first has to be measured through sensors. The disturbance variables can be the temperature of the raw materials entering the furnace or even the ambient temperature. This measured temperature, which corresponds to the current value of the process variable, is then compared with the desired value or set point. The difference between the actual value and the set point is the error or deviation. When the control deviation is calculated, it is possible to perform control actions. In the case of temperature control with burners, the pressure or flow control valves can be opened slightly more when the temperature measured value is less than the desired set point, or they can be slightly closed when the measured temperature is higher than the required set point. For the process to react automatically, a control loop is required. The Proportional Integral Derivative (PID) is a popular controller that operates in a closed loop and calculates the manipulated variable based solely on the current deviation. PID controllers remain preferred in industrial process control. They can be used for a wide variety of processes and tuned with few parameters.

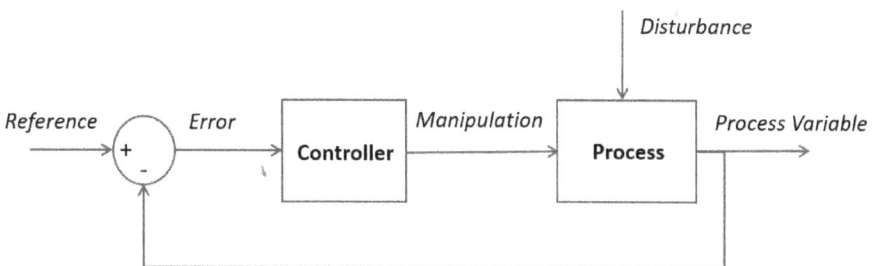

Fig. 2.3. Control loop.

3.2 Discrete Event Systems and Sequential Logic Control

Discrete event systems are systems where variables and parameters are mostly binary discrete. Digital variables and parameters can take on only specific values within a given range. Binary variables can take on only two different values: on or off, 1 or 0, etc. Discrete manipulations or actions are taken when the system's state has changed (event-driven changes) or a certain amount of time has elapsed (time-driven changes). The controller executes event-driven changes in response to some event that has altered the system's state. Some examples are as follows: (1) a robot loads a work part into a fixture, and a limit switch in the fixture senses the part; (2) the diminishing level of plastic in the hopper of an injection molding machine triggers a low-level switch, which opens a valve to start the flow of more plastic into the hopper; (3) counting parts moving along a conveyor past an optical sensor; control of the position of a cutting tool relative to a work part in a CNC machine tool.

Time-driven systems are operated by controlling the activation of the actuators in the proper order; for example, a thermal treating operation may be applied for a certain length of time. The controller executes time-driven events at specific times; for example, a factory shop clock sounds a bell to indicate the start of shift, break start and stop times, and end of shift. Systems may require both event and time-driven manipulations, as in the case of a washing machine: the agitation is set to operate for a certain length of time; by contrast, filling the tub is event-driven because it should stop when a high water level is detected.

Combinational logic controls the execution of event-driven changes, also known as logic control. In combinational logic control, the output at any moment depends on the values of the inputs. Sequential logic is applied to control the execution of time-driven changes and uses internal timing devices to determine when to initiate changes in output variables. Sequential logic control refers to discontinuous time and event-based automation sequences within continuous processes. These may be implemented as a collection of time and logic function blocks, a custom algorithm, or a formal sequential function chart methodology.

4. Control Systems Design

The development of industrial control systems follows design and implementation stages as depicted in Fig. 2.4. The design of the control system requires preparing documentation where the design can be described adequately to be easily understood. Anyone must be able to interpret the information and eventually update it. Industrial organizations use applicable standards and codes in their production systems, particularly on the plant floor. In the engineering drawings, particularly the process and instrumentation schematics, it is important to use standard symbols such as those defined by the ISA or EIC. The symbols assigned to the sensors and actuators and documented in the instruments list can become the input and output (IO) tags programmed in the PLC.

The system integrators assemble and test the automation system. Automation engineers and support personnel perform the design tasks using different methods and tools. In addition to the system design, the system integrator makes test and

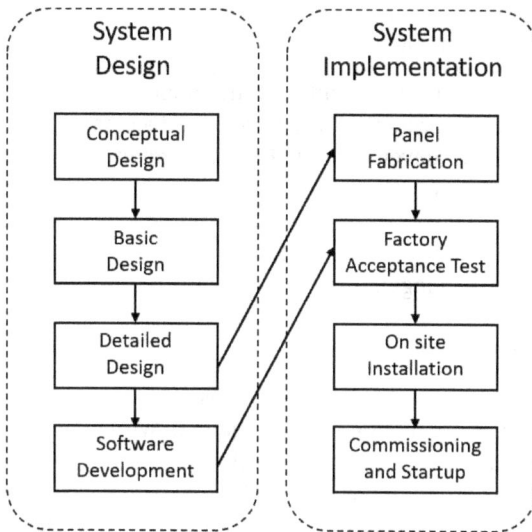

Fig. 2.4. Automation project development process.

installation plans. An approved factory acceptance test (FAT) is a requirement to comply before the system is delivered to the customer and implies that the system is ready for on-site installation.

4.1 System Design

Requirements and Conceptual Design. The conceptual design of a cost-effective automation solution should fulfill the needs of a particular process or application. During the conceptual design stage, the customer and the system integrator describe the automation system functions for agreement and specific design and execution. The system integrators or automation engineers define the design documents required to be prepared for system execution at the system design phase. In this stage, it is preferable to elaborate a preliminary functional specification of the system.

Basic Design. At this stage, automation engineers elaborate schematics of the automated control, including the system configurations. At this stage, the control schemes and the functionality of the user interface should be defined. Control schematics are usually drawn as flow charts or state diagrams, showing the various control and process variables. The functional specification of the system must be completed at this stage.

Detailed Design and Procurement. A complete set of process and instrumentation drawings must be completed in this stage, including a final instrumentation list. The control panel design includes panel layouts and drawings showing the logical wiring of the inputs and outputs modules. Required instruments and control hardware (controller, interface, input/output modules, and communication options) are selected and purchased. The bill of materials (BOM) should list each component in the automated control system, its quantity, designation to identify it on the schematics, a brief description, and part number.

Software Development. Control software development involves designing and programming control algorithms and the user interface application (typically PLC programming and HMI development). PLC programming should preferably be performed using technics and methods such as finite state machines (FSM) for discrete event systems and model-based PID. HMI development includes designing and programming operation and maintenance screens, systems configuration, alarms, historical trends, and data logging.

4.2 System Implementation

System Construction or Panel Fabrication. This stage concerns the control panel's fabrication, including the mounting and internal wiring of the control system components. The fabrication of the control panel often considers the use of a removable subpanel for easy replacement of any failed device or component in the future. Terminal blocks mounted on the subpanel connect all the external devices. Another terminating method is mating connectors so the field wiring can be plugged into connectors mounted on the panel.

Factory Acceptance Test (FAT). The FAT is done to verify the correct fabrication of the control panel and correct wiring of the input/output modules to the terminal blocks; this implies the powering and configuration of the controller units and, if possible, a connection of the networked devices, including the HMI. In this stage, following good system integration practices, the integration of control hardware and software functions can be performed and verified in a hardware-in-the-loop simulation; this can help to achieve a functional verification test to comply with the specifications.

On-Site Installation. At the installation phase, the automation system with all its components and software is delivered and installed at the destination. On-site system installation includes control panel mounting and wiring to machines and field instruments. The system is tested and checked to fulfill the design descriptions and requirements. A complete I/O checkout will ensure that the point-to-point wiring between the I/O module terminals and the field wiring terminal blocks has been done correctly.

System Commissioning and Startup. The startup of our automated control system begins once the control system is installed and all wiring from control panel I/O to field sensors and actuators is completed. Commissioning is the on-site test of manual and automatic control functions in the real process and implies a system operational validation.

5. Control Software Development

Automating many production systems, manufacturing or industrial processes, is the automation of mechatronic processes. In practice, mechatronic design methodologies and practices can be extended to the design and integration of industrial automation projects, particularly the activities related to control software development and verification.

5.1 Concurrent Engineering

Concurrent engineering is possible to apply by splitting the control system development activities into hardware and software, this is, into control panel fabrication (including the panel verification) and control software development (programming and verification) as depicted in Fig. 2.5. Once control panel fabrication and verification are completed, the FAT can be done.

In most cases, for event-driven processes, the first step in designing the control algorithms will be to define the sequence of operations and control actions derived from interpreting the functional specification. The logical sequence should list the operation steps and the actuators and sensors involved in each step. State diagrams or finite states machines derived from automata theory are beneficial for designing sequential control logic.

Real-time simulations like software-in-the-loop and hardware-in-the-loop are helpful for software verification and factory acceptance tests. For automation projects, it is desirable to use the final physical control unit, in most cases a PLC, as the hardware component and simulate the process. The HiL is required to verify the correct functioning of the control software with regards to the operation sequence but also to verify some other automation functions related to the production data that can be generated, such as the cycle work time, the number of mixing work cycles (Fig. 2.6). This configuration allows to test not only level 1 control function but also connectivity to HMI and other SCADA systems, even connection to data basses collecting relevant production process data.

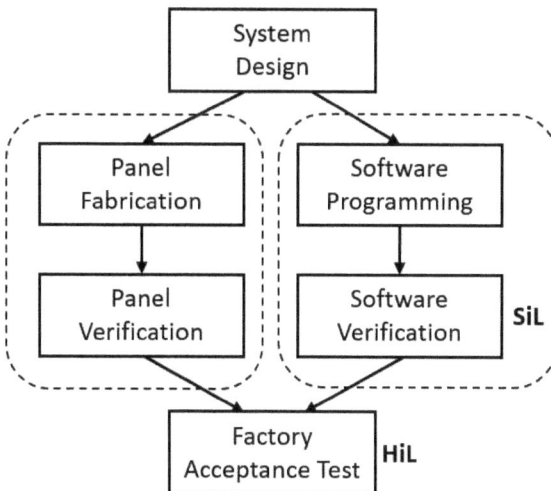

Fig. 2.5. Industrial control system development.

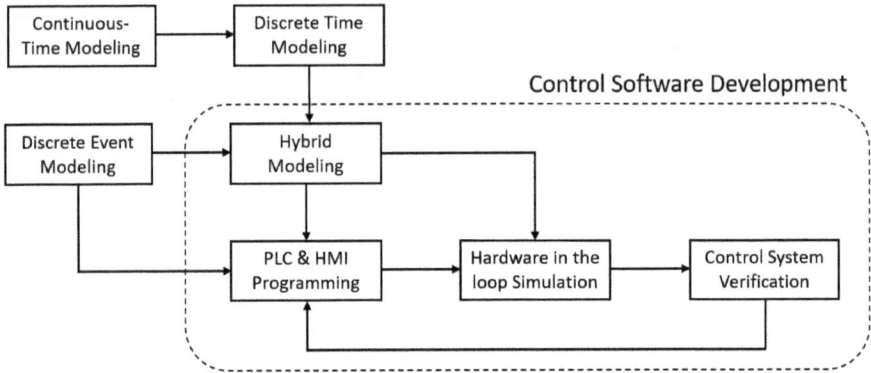

Fig. 2.6. Industrial control system design and implementation.

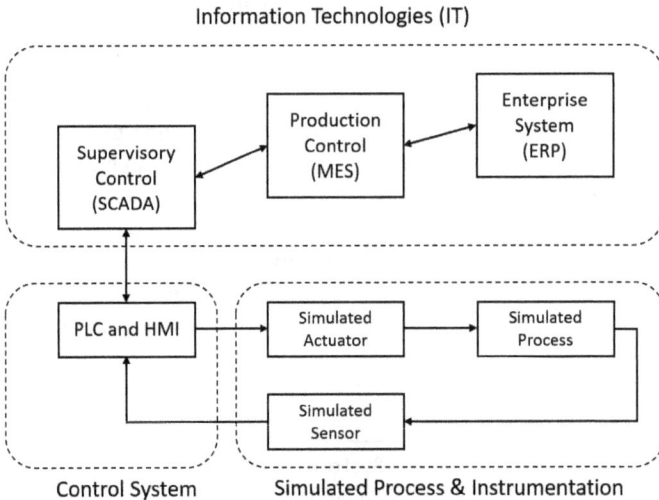

Fig. 2.7. Interaction of high level automation systems for the industrial control system design and implementation.

Performing the I/O check of the control panel implies that the control hardware (PLC in most cases) is energized and configured; the process simulation can be used to verify the control algorithms to a certain level, which can be done in HiL scheme. This way, the FAT can cover the control panel hardware and the control software. If the automation application requires keeping data records for reference, traceability, history, trending, meeting regulations, etc., then the control system will have some functionality of a SCADA system. Under this scenario, the FAT test performed with a HiL can also be used to test some SCADA data logging and supervisory functions (Fig. 2.7).

5.2 Safety Considerations

Some important considerations are necessary for developing and implementing industrial control systems and are related to safety issues (Whitt, 2015).

Automated Fail Safe. The control system must follow a fail-safe procedure to ensure proper shutdown of the control system when feedback signals are lost, preventing injury to personnel or damage to the equipment. A good selection of sensors, actuators, and protection devices is necessary to implement automated fail-safe procedures.

Control Power Distribution. The control of power distribution in the electrical and electronic circuits of the control system should be designed appropriately. All circuits must be protected with proper fusing, thermomagnetic breakers, or other circuit-interrupting devices. All enclosures and cabinets with energized circuits must be secured to prevent access from unauthorized or untrained personnel.

Grounding. Proper grounding and the correct use of shielding in wires reduce electromagnetic interference and induced noise to instrumentation and communications signals and other sensitive electronic devices. A good grounding practice is required to prevent electrical shock and must be incorporated into the industrial control system's design, construction, and installation.

6. Case Study

A case study shows the simplicity of developing a hardware-in-the-loop simulation scheme for control software verification. The mixing and heating process of Fig. 2.8 is automated using a Simatic® S7-1200 PLC and LabVIEW® as the user interface. The process consists of a tank, two feeding valves SVA and SVB, one discharge valve SVC, and three binary level switches: LS1 to sense high level, LS2 for intermediate level, and LS3 for low level or empty tank. The motor to drive the mixer is energized with contactor M1. There is a temperature control loop with an analog temperature indicator and transmitter, TIT, and a flow control valve, FCV, to set the temperature in the heating process (Fig. 2.8).

There are two possible approaches to process simulation. The first approach is to simulate the mixer as a fluidic process; the tank can be considered as an integrator where the constant volumetric flow of each ingredient is integrated in discrete time to obtain a volume process variable form which the level switches can be generated

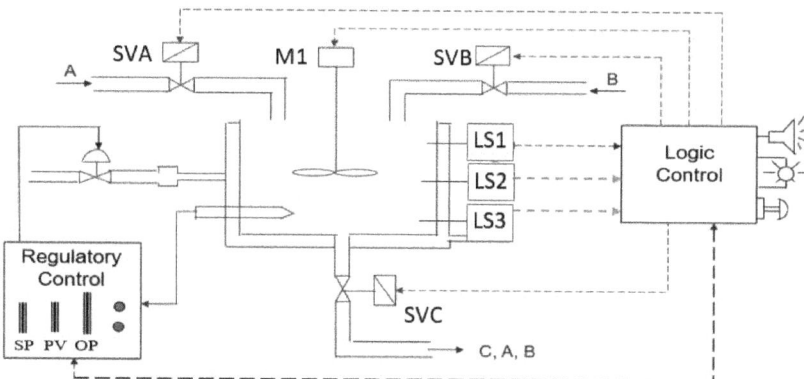

Fig. 2.8. Mixing and heating process diagram.

as simple comparisons against specific level references or thresholds. When the volume is discharged, a first-order process can be simulated to decrease the level and eventually empty the tank. The logic to simulate the flows of substances A, B, and C, given by f_A, f_B, and f_C, based upon the opening status of the actuator valves are as follows:

$$if\,(SVA = TRUE)\ then f_A = 1 \quad else\ f_A = 0 \tag{2.1}$$

$$if\,(SVB = TRUE)\ then f_B = 1 \quad else\ f_B = 0 \tag{2.2}$$

$$if\,(SVC = TRUE)\ then f_C = 1 \quad else\ f_C = 0 \tag{2.3}$$

SVA, SVB, and SVC are the solenoid valves for ingredients A and B and product C, respectively.

The logic to fill or discharge the tank is a function that integrates the volumetric flows f_A, f_B, and f_C with a sample time, T_s:

$$if\,(V_k < V_{max}\ and\ V_k > V_{min})\ then\ V_k = V_{k-1} + T_s\,(f_A + f_B - f_C) \tag{2.4}$$

where V is the tank's volume, k is the current discrete time, V_{min} and V_{max} are minimum and maximum values.

The logic to activate the level sensors based on the tank volume is as follows:

$$if\,(V_k \geq V_{max})\ then\ LS1 = TRUE\ else\ LS1 = FALSE \tag{2.5}$$

$$if\,(V_k \geq V_{med})\ then\ LS2 = TRUE\ else\ LS2 = FALSE \tag{2.6}$$

$$if\,(V_k \geq V_{min})\ then\ LS3 = TRUE\ else\ LS3 = FALSE \tag{2.7}$$

For the temperature regulation, a temperature dynamical response can be simulated with a first-order difference equation considering the percentage of the opening of the FCV as input, and the temperature T as output in the range between ambient temperature and maximum temperature, $T_{amb} \geq T_k \geq T_{max}$.

$$T_n = a\,T_{k-1} + k_p\,(1-a)\,FCV_k \tag{2.8}$$

$$if\,(T_k \geq T_{max})\ then\ T_k = T_{max} \tag{2.9}$$

$$if\,(T_k \leq T_{amb})\ then\ T_k = T_{amb} \tag{2.10}$$

where a is an adjustable parameter.

The HiL scheme can be prepared for the FAT connecting the LabVIEW computer to the PLC with an Ethernet Connection (Fig. 2.9). Two possibilities for the implementation of the simulation can be considered: (1) program the simulation in the PLC, and (2) program the simulation in LabVIEW. The modeling and simulation in an HiL scheme require more effort. However, they can serve to gain knowledge about the process, verify the correct functioning of the control software regarding the operation sequence, and also to verify some other automation functions related to the

Mixer Process Computer Simulation

Programmable Controller

Ethernet Network

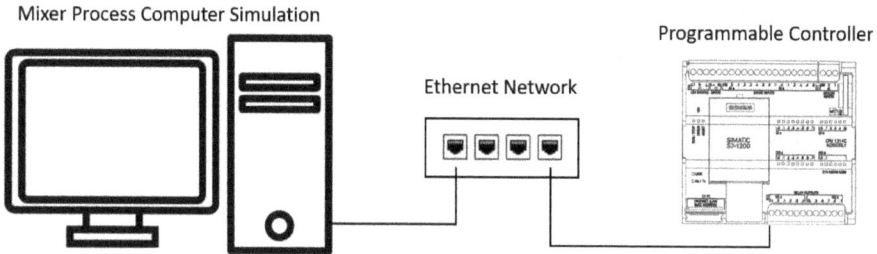

Fig. 2.9. HiL simulation with the PLC.

production data that can be generated, such as the cycle work time and the number of mixing work cycles.

7. Conclusions

Industrial automation also refers to diverse technologies implemented to control industrial processes. In production systems, operational technologies (OT) refer to the hardware and software used to change, monitor, and control physical devices, processes, and events on the plant floor. The automation pyramid represents various levels of automation. Levels 0 and 1 focus on hardware, sensors, actuators, and controllers, essential components for industrial automation. Levels 2 and 3 of the pyramid represent the SCADA and MES systems critical to maximizing the efficiency and resources of a modern, highly automated facility. The top level (Level 4) of the pyramid, Enterprise Resource Planning, is governed by business processes and decisions. There are slight differences in the interpretation of the automation levels between process and discrete manufacturing industries and the preponderant automation technologies and information systems. In industrial automation, OT refers to the hardware and software needed to monitor and control. Advanced process controls may reside in the DCS or the supervisory computer, depending on the application. Critical process controls are conventionally implemented in the DCS and PLC.

The process is the operation or function being controlled; the process output values or variables are measured with appropriate sensors and read by the controller. The controller closes the loop between the input references or set points and the process variables. The controller compares the process variable with the set point and makes the required adjustment. The adjustment is accomplished using one or more actuators, the hardware devices that carry out the control actions. Most continuous industrial processes have multiple feedback loops and combine continuous control and logic control. Low-order linear systems can be models for most continuous loops to design model-based control strategies. Basic event-driven systems modeled by finite state automata or ordinary Petri networks can be used as computational models to program sequential logic control. Combining continuous process and automata theories can be used to model hybrid dynamical systems for complete modeling and simulation of control systems. When a FAT is planned as a key activity of an automation project, it can split the design and development activities from the installation and commission phase. However, it is often considered as a development completion stage.

References

Brandl, D. 2012. *Plant IT: Integrating Information Technology into Automated Manufacturing.* Momentum Press.

Dey, C. and Sen, S.K. 2020. *Industrial Automation Technologies.* First Edition. CRC Press.

Groover, M.P. 2008. *Automation, Production Systems, and Computer-Integrated Manufacturing.* Third Edition. Prentice Hall.

Sands, N. and Verhappen, I. 2018. *A Guide to the Automation Body of Knowledge.* Third Edition. ISA.

Whitt, M.D. 2003. *Successful Instrumentation and Control Systems Design.* Second Edition. ISA.

Chapter 3
Linear Systems Analysis and Modeling

1. Introduction

Signal and systems theory is fundamental in designing and developing engineered systems such as electronic systems for communications, signal processing of audio and video, control systems, and all types of instrumentation, including sensors and actuators. Their application ranges over various consumer products, medical devices, industrial production processes, and many others. Control systems are designed using signals and systems analysis tools in mechatronics systems and industrial automation areas. Examples of these control systems include motion control (control of speed and position); motor control (variable frequency drives, soft-starters); positioning control of hydraulic and pneumatic cylinders, linear actuators, control valves, etc., regulatory and set point tracking control of thermal processes and machines, fluidic and chemical processes.

The mathematical representation of signals and systems applied in linear systems theory has been a key tool for developing control systems with classical and modern control theory. The tools and techniques employed are the following: time-domain solution of differential equations, Laplace Transform and transfer functions in the frequency domain; z transform, stability analysis, pulse width, and amplitude modulation (PWM and PAM) used in Power Electronics; and control of actuators. The Laplace and Fourier transforms are mathematical tools for analyzing the system response (in time and frequency domains). An alternative representation of ordinary differential equations (ODEs) is a rational function in the frequency domain. The Laplace Transform is a mathematical tool to convert a time-domain ODE into a transfer function in the frequency domain. Real and complex exponentials are solutions for ordinary differential equations. Complex exponentials can represent exponentially decaying signals with oscillations that represent the real response of some physical systems. Many physical systems can be modeled with first- and second-order linear systems. Transfer functions are also used in system identification (when experimental data can be obtained), where a first- or second-order system response can be proposed.

This chapter reviews signals and systems concepts, Laplace and Fourier Transforms, and their application in analyzing first- and second-order linear time invariant (LTI) systems. Section 2 is for a brief review of basic definitions of signals and systems. Section 3 presents the impulse and step response of basic linear time invariant systems. Section 4 introduces complex exponentials involved in the definition of the Laplace transform and its application in the analysis of the step response of first- and second-order systems. Section 5 on frequency response is for a basic analysis in the frequency domain of the response of first- and second-order LTI systems and reviews concepts of bandwidth gain and phase margin. Section 6 presents some examples of the use of LTI systems in modeling electrical circuits, spring-mass-damper mechanical systems, and electromechanical systems. Conclusions are in Section 7.

2. Signals and Systems

The theory of signals and systems analyzes the mechatronic systems components and the design of control systems.

Signals may describe a wide variety of physical phenomena. Although signals can be represented in many ways, the information in a signal is contained in a pattern of variations of some form (Oppenheim et al., 1997). Signals encode information and can be processed by electronic systems. The information is not purposefully encoded in physical and biological systems, where signals occur naturally. The data is always encoded in instrumentation, control, and communication applications.

A system combines elements that manipulate one or more signals to accomplish a function and produce some output. A system takes a signal as an input and transforms it into another signal. The transformation may be as simple as a microphone converting a sound pressure wave into an electrical waveform. Linearity and time invariance are fundamental in signal and system analysis for two primary reasons. First, many physical processes possess these properties and thus can be modeled as LTI systems. In addition, LTI systems can be analyzed in considerable detail, providing insight into their properties and a set of powerful tools that form the core of signal and system analysis (Oppenheim et al., 1997).

2.1 Signal Representation and Classification

Mathematically, signals are represented as a function of one or more independent variables. The independent variable is time, t, but it could be another variable of interest, such as position. A speech waveform is an example of a one-dimensional signal with time as the independent variable. An image is an example of a two-dimensional signal, a function of two spatial variables. If the idea is put into motion, as in a movie or video, a third independent variable (in discrete time) is required.

Signals are classified depending on their time existence, continuous time and discrete time; their value range, analogical and digital signals; periodicity, periodic, and non-periodic signals. An exponential decaying signal is an example of a continuous, analogical, non-periodic signal. Sinusoidal signals are examples of continuous, analogical, and periodic signals where $x(t + T) = x(t)$. Signals can also be deterministic if they can be mathematically represented; therefore, these signals' behavior is predictable in time. If there is no uncertainty of its value, these signals can be expressed mathematically. Random signals have an unpredictable behavior in time, that is, their value is uncertain at any time. These signals cannot be expressed mathematically.

Most signals in the real world are continuous in time, as the scale is infinitesimal, for example, voltage, current, velocity, and acceleration. Continuous signals are denoted by $x(t)$, where the time interval may be bounded (finite) or infinite. Some real-world and digital signals are discrete in time, as they are sampled, like anything a digital computer processes. Discrete signals are denoted by $x_n = x[n]$, where n is an integer value that varies discretely. Sampled continuous signals have a sampling period, T_s, and can also be represented as $x_n = x(nT_s)$ (Fig. 3.1).

Another helpful classification of the signals is if they are considered to be power signals or energy signals. A signal is regarded as a power signal if its average power P is finite, i.e., $0 < P < \infty$. The periodic signals are examples of power signals. An energy signal is a signal whose energy is finite, and power is zero, whereas a power signal has a finite power and an infinite energy. Periodic signals are power signals; non-periodic signals (pulses) are energy signals. When both power and energy are infinite, the signal is neither a power nor an energy signal. A true power signal cannot exist in the real world because it would require a power source that operates for an infinite amount of time.

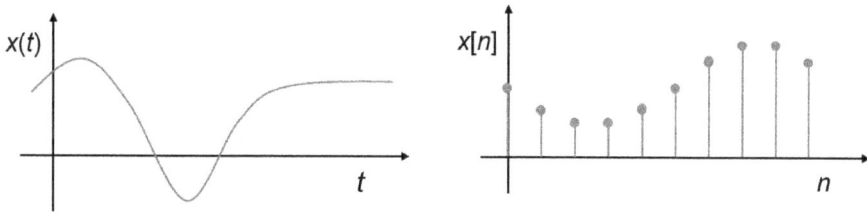

Fig. 3.1. Continuous-time signal (left) and discrete-time signal (right).

2.2 Systems Classifications and Properties

The systems can be classified depending on their properties into: (a) linear or nonlinear systems; (b) continuous-time and discrete-time systems; (c) time-variant or time-invariant systems; (d) systems with or without memory; (e) invertible and noninvertible systems; (f) stable and unstable systems; and (g) causal or non-casual systems. Primary systems classifications and their properties are briefly described below (Smith, 1997).

Linear Systems

The system response can be analyzed as the sum of more straightforward responses. A linear system must satisfy two properties: homogeneity (scaling) and superposition (addition). The scaling property applies the same factor that scales or multiplies the input to scale the output:

If the input is $ax_1(t)$ then the output is $ay_1(t)$ (3.1)

The superposition property consists of adding the individual responses to different inputs when the summation of those inputs is entered into the system.

If the input is $x_1(t) + x_2(t)$ then the output is $y_1(t) + y_2(t)$ (3.2)

If we can analyze how simple signals affect the output response, we can analyze more complex inputs as a linear combination of simple signals. Linear systems are relevant for engineering analysis since they help realize certain physical systems' mathematical analysis. More complex nonlinear systems can be considered linear for some operational ranges and some perturbations.

Continuous-Time and Discrete-Time Systems

A continuous-time system takes a continuous-time signal as input and generates a continuous-time signal as output. A discrete-time system takes a discrete-time signal as input and generates a discrete-time signal as output. The scaling and superposition properties can be combined and are valid for both linear continuous-time and discrete-time systems:

if the input is $ax_1(t) + bx_2(t)$ then the output is $ay_1(t) + by_2(t)$ (3.3)

if the input is $ax_1[n] + bx_2[n]$ then the output is $ay_1[n] + by_2[n]$ (3.4)

Figure 3.2 shows a block representation of continuous-time and discrete-time systems with the representation of signals as input and output arrows.

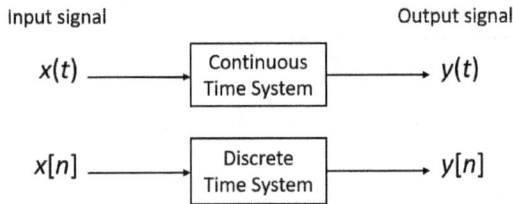

Fig. 3.2. Continuous-time systems and discrete-time systems.

Time-Invariant Systems

A system is classified as time-invariant if a time delay or a time advance in the input signal leads to an identical time shift in the output signal. The following continuous-time system is time-invariant:

$y(t) = ax(t)$ (3.5)

$y(t - \tau) = ax(t - \tau)$ (3.6)

Systems with and without Memory

A memory corresponds to a mechanism in the system that retains information about previous input values. In a memory system, the output signal depends on past or future values of the input signal. A system is said to be memoryless if its output signal depends only on the present value of the input signal; for example, a resistor is a memoryless system where the current flowing through it is instantly affected by the applied voltage. An inductor has memory since the current flowing through it is proportional to the time integral of the voltage across it. A discrete-time system with memory can be an accumulator (integrator). A moving-average system has memory.

Stable and Unstable Systems

A system is stable only if every bounded input results in a bounded output (BIBO). From an engineering perspective, the system of interest must remain stable under all possible operating conditions.

Invertible and Non-invertible Systems

A system is invertible if the input of the system can be recovered from the output. For example, the following continuous-time system is invertible:

$$y(t) = ax(t) \tag{3.7}$$

since the input value can be recovered:

$$x(t) = \frac{1}{a}y(t) \tag{3.8}$$

This continuous-time system is not invertible:

$$y(t) = x^2(t) \tag{3.9}$$

Because two possible input signals can generate the same output signal.

$$x(t) = \pm\sqrt{y(t)} \tag{3.10}$$

Casual and Non-casual Systems

A system is causal if the present value of the output signal depends only on the current or past values of the input signal. This is an example of a casual system:

$$y[n] = ax[n] + bx[n-1] \tag{3.11}$$

A non-casual system cannot operate in real time since its response depends on future values of the signal:

$$y[n] = ax[n] + bx[n+1] \tag{3.12}$$

$$y(t) = ax(t + \tau) \tag{3.13}$$

3. Continuous Time Linear Systems

From a systems perspective, the impulse response, $h(t)$, of a linear system, H, is defined as the response to a unitary impulse, $\delta(t)$:

$$h(t) = H\{\delta(t)\} \tag{3.14}$$

Any continuous LTI system can be determined entirely (characterized) by measuring its impulse response, $h(t)$. The step response, $s(t)$, of a system is defined as the convolution in the time domain of the impulse response with a unit step, $u(t)$:

$$s(t) = h(t) * u(t) = \int_{-\infty}^{\infty} h(t)u(t-\tau)d\tau = \int_{-\infty}^{t} h(\tau)d\tau \tag{3.15}$$

The system response to an arbitrary input signal, $x(t)$, is the convolution in the time domain of the impulse response, $h(t)$, with the input signal:

$$y(t) = h(t) * x(t) = \int_{-\infty}^{\infty} h(t)x(t-\tau)d\tau = \int_{-\infty}^{\infty} x(t)h(t-\tau)d\tau = x(t) * h(t) \tag{3.16}$$

Figure 3.3 shows the impulse response, the step response, and the response to an arbitrary input.

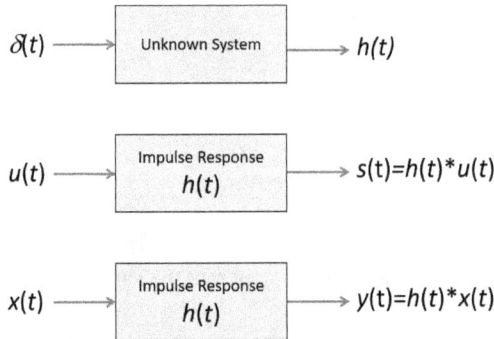

Fig. 3.3. System response in time domain.

3.1 Ordinary Differential Equations

In many engineered systems applications, linear differential equations can model a system's input/output behavior. Linear systems with memory are typically described by ordinary differential equations (ODE) that relate the input signal, $x(t)$, to the output signal, $y(t)$. ODE can have first, second, or even higher-order derivatives:

$$a_1 \frac{dy(t)}{dt} + a_0 y(t) = x(t) \tag{3.17}$$

$$a_2 \frac{d^2 y(t)}{dt^2} + a_1 \frac{dy(t)}{dt} + a_0 y(t) = x(t) \tag{3.18}$$

$$a_n \frac{d^n y(t)}{dt^n} + \cdots + a_2 \frac{d^2 y(t)}{dt^2} + a_1 \frac{dy(t)}{dt} + a_0 y(t) = x(t) \tag{3.19}$$

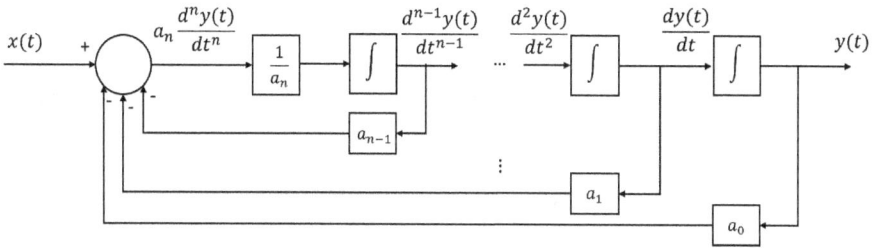

Fig. 3.4. Block diagram of ordinary differential equation.

The response of systems with memory requires the application of at least a onetime integration function and, therefore, has particular dynamic characteristics. The ODE can be depicted using block diagrams as shown in Fig. 3.4. Integration of one or more derivatives is feedback and subtracted from the input. Therefore, systems represented by ODE are feedback systems with the corresponding properties of such systems.

The dynamical response of a system modeled by an ODE can be obtained analytically by solving the differential equation. Before we analyze the response to an input, let us consider the input is zero, which is called a homogeneous differential equation:

$$a\frac{dy(t)}{dt} + by(t) = 0 \tag{3.20}$$

$$\frac{dy(t)}{dt} = -\frac{b}{a}y(t) \tag{3.21}$$

The derivative of $y(t)$ is the same function affected by a $-(b/a)$ factor that is the rate of change of the decaying exponential; the higher the value of this factor, the faster the decaying of the response. Real exponential functions are candidates to be the solution of first order homogeneous differential equations since the derivative of a real exponential is the same real exponential.

$$\frac{d\left(e^{-(b/a)t}u(t)\right)}{dt} = \left(-\frac{b}{a}\right)e^{-(b/a)t}u(t) \tag{3.22}$$

Real exponentials can represent exponentially decaying signals for the natural response of stable first-order linear systems.

4. Laplace Transform

The Laplace Transform is a mathematical transformation obtained by the integral operation over the multiplication of a complex exponential function by a time-domain signal. The time-domain signal can be either an input signal to the system or the system's impulse response. The coefficients of the exponential functions in the time domain are converted to poles (roots) of the characteristic polynomial of the complex rational function in the frequency domain.

4.1 Complex Exponentials

Basic trigonometric signals in the time domain, such as sine and cosine, can be represented in a complex plane by complex exponentials. Before reviewing basic concepts and applications of the Laplace Transform, recalling some properties of the complex exponential functions is convenient. The Euler formula establishes the fundamental relationship between complex exponential and trigonometric functions:

$$e^{j\omega t} = cos(\omega t) + jsin(\omega t) \tag{3.23}$$

If we add one pair of complex conjugates exponentials, we can find an equivalence for a cosine signal:

$$e^{j\omega t} + e^{-j\omega t} = cos(\omega t) + jsin(\omega t) + cos(\omega t) - jsin(\omega t) \tag{3.24}$$

$$cos(\omega t) = \frac{e^{j\omega t} + e^{-j\omega t}}{2} \tag{3.25}$$

We can also subtract one pair of complex conjugates exponentials to find an equivalence for a sine signal:

$$e^{j\omega t} - e^{-j\omega t} = cos(\omega t) + jsin(\omega t) - cos(\omega t) + jsin(\omega t) \tag{3.26}$$

$$sin(\omega t) = \frac{e^{j\omega t} - e^{-j\omega t}}{2j} \tag{3.27}$$

Complex exponentials provide a convenient way to combine sine and cosine terms with the same frequency in the complex plane (while the cosine function is at its maximum, the sine function is at its minimum, and vice versa). The complex plane in Fig. 3.5 shows a complex exponential's Euler relation and vectorial components. The real part is the cosine component, and the imaginary part is the sine component.

Complex exponentials are particularly interesting for the analysis of LTI systems since they are used for time domain to frequency domain transformations. Sine and cosine trigonometric functions are a linear combination of complex exponentials. Moreover, we can say that a large class of signals can be represented as linear combinations of complex exponentials, for example, all the periodic signals that can be represented with complex Fourier series. Pairs of complex exponential conjugates

Complex plane

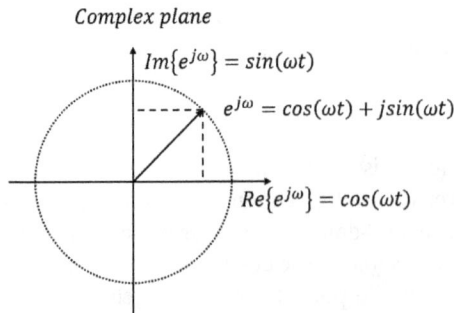

Fig. 3.5. Complex plane with Euler relation.

can also represent exponentially decaying signals with oscillations representing the natural response of stable second-order linear systems.

4.2 Introduction to Laplace Transform

A decaying complex exponential signal, e^{-st}, can multiply any real signal, $x(t)$. This multiplication results in a bounded function in the time domain that can be integrated to obtain $X(s)$; that is, the integration result will be a function in the complex frequency $s = (\sigma + j\omega)$ domain. This is the concept of the Laplace Transform $X(s)$ of a general signal $x(t)$:

$$X(s) = \int_{-\infty}^{\infty} x(t)e^{-st}dt \tag{3.28}$$

The Laplace Transform of common functions is presented in Fig. 3.6.

The Laplace Transform is a mathematical tool for the analysis of linear systems. There is a particular case where the input is:

$$x(t) = e^{st} \tag{3.29}$$

The output response replicates the input behavior and is only affected by a scalar value. For this reason, complex exponentials are called eigen (proper) functions of LTI systems; when they are used as inputs to the system, they exhibit characteristic scalar values that are called the eigen values:

$$y(t) = \lambda\, e^{st} \tag{3.30}$$

The region of convergence (ROC) of the Laplace Transform is the set of values for $s = (\sigma + j\omega)$ for which the Fourier Transform of $x(t)\, e^{st}$ exists. The ROC is generally displayed in the complex plane. In general, roots of the characteristic polynomial of the denominator of the rational functions obtained by the Laplace Transform must be located at the left part of the complex plane for the system to be stable and causal (Fig. 3.7).

Function	$f(t)$		$F(s)$
Impulse	$f(t) = \begin{cases} 1 & t = 0 \\ 0 & t < 0, t > 0 \end{cases}$		1
Step	$f(t) = u(t)$		$\dfrac{1}{s}$
Ramp	$f(t) = tu(t)$		$\dfrac{1}{s^2}$
Exponential	$f(t) = e^{-at}u(t)$		$\dfrac{1}{s+a}$
Sine	$f(t) = \sin(\omega t)$		$\dfrac{1}{\omega^2 + s^2}$

Fig. 3.6. Laplace Transform of common signals.

$$s - plane$$

$$\uparrow j\omega$$

Causal & stable systems	Non causal & unstable systems

$$\sigma$$

Fourier ....◥
Transform
$s = j\omega$

$$s = \sigma + j\omega$$

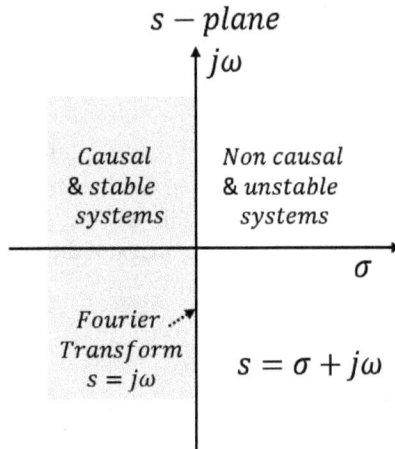

Fig. 3.7. Complex Plane of $s = (\sigma + j\omega)$.

4.3 Transfer Functions

A differential equation in the time domain has an equivalent transfer function in the frequency domain. The Laplace Transform of the impulse response $h(t)$ of an LTI system is called the transfer function of the system, $H(s)$:

$$H(s) = \int_{-\infty}^{\infty} h(t)e^{-st}\, dt \tag{3.31}$$

Suppose an input signal $X(s)$ is passed through the LTI system $H(s)$ in the complex frequency domain. In that case, the output response results in a multiplication in the frequency domain that is equivalent to a convolution in the time domain:

$$Y(s) = H(s)X(s) \tag{3.32}$$

The transfer function can also be expressed as the relation of the output over the input signal as follows:

$$H(s) = \frac{Y(s)}{X(s)} \tag{3.33}$$

The Laplace Transform is used to obtain the transfer function of a system, $H(s)$, analytically, and its response, $Y(s)$, to a transformed input, $X(s)$. The transfer function of the system $H(s)$ is then a useful means to calculate the system response $y(t)$ to an arbitrary input x but in the domain of the variable s. The inverse Laplace Transform of $Y(s)$ will give the time response y (see Fig. 3.8).

A causal system with a rational system function, $H(s)$, is stable if and only if all of the poles of $H(s)$ lie in the left half of the s-plane; that is, all poles have negative real.

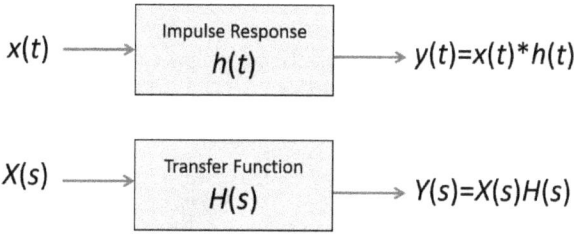

$$x(t) \longrightarrow \boxed{\begin{array}{c} \text{Impulse Response} \\ h(t) \end{array}} \longrightarrow y(t)=x(t)*h(t)$$

$$X(s) \longrightarrow \boxed{\begin{array}{c} \text{Transfer Function} \\ H(s) \end{array}} \longrightarrow Y(s)=X(s)H(s)$$

Fig. 3.8. System response in time domain.

4.4 Impulse Response of First-Order Systems

Consider an LTI first-order system described by the ODE:

$$\frac{dy(t)}{dt} + ay(t) = x(t) \tag{3.34}$$

The equivalent differential equation to express the impulse response is:

$$\frac{dh(t)}{dt} + ah(t) = \delta(t) \tag{3.35}$$

The Laplace Transform is applied to obtain the transfer function:

$$L\left\{ \frac{dh(t)}{dt} + ah(t) = \delta(t) \right\} = L\{h(t)\}(s+a) = 1 \tag{3.36}$$

$$H(s) = \frac{1}{s+a} \tag{3.37}$$

The analytical expression for the system output, in this case, the impulse response, can be found by taking the inverse Laplace Transform:

$$h(t) = e^{-at} u(t) \tag{3.38}$$

The exponential parameter, a, is the rate of the decaying exponential; this parameter also affects the system gain. If a unit gain process is desired, the transfer function needs to be multiplied by a:

$$\frac{Y(s)}{X(s)} = \frac{a}{s+a} \tag{3.39}$$

Then, for a unit gain process, the impulse response is given by:

$$h(t) = ae^{-at} u(t) \tag{3.40}$$

The time constant, $\tau = 1/a$, represents a more comprehensive parameter to understand the impulse response:

$$h(t) = \frac{1}{\tau} e^{-t/\tau} u(t) \tag{3.41}$$

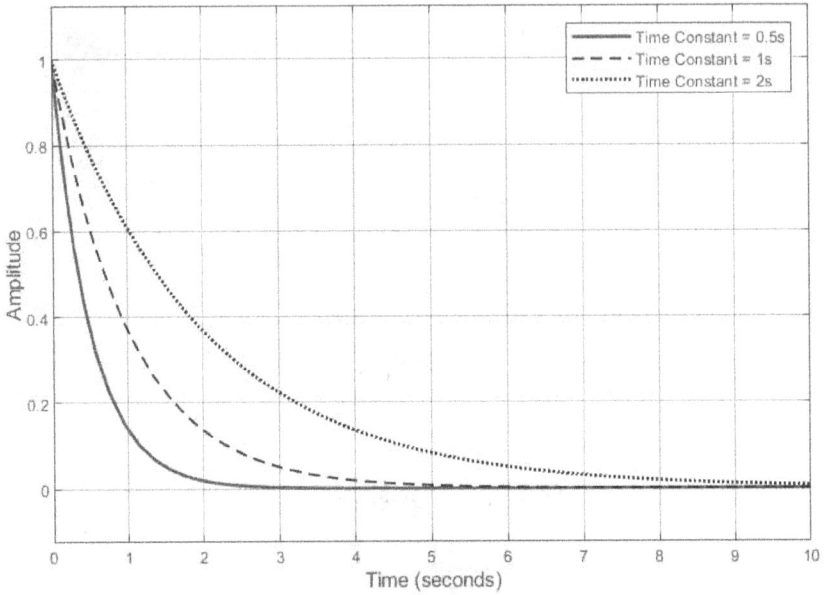

Fig. 3.9. First-Order impulse response with unit gain and three different time constants (0.5, 1.0, 2.0 s).

Figure 3.9 shows the impulse response of several first-order LTI systems with different time constants. Let us consider a time constant, $\tau = 1$. If the time constant is increased to $\tau = 2$, the signal decays more slowly, and if the time constant is reduced to $\tau = 0.5$, the signal decays faster.

4.5 Impulse Response of Second-Order Systems

Consider a general second-order LTI system modeled with a second-order differential equation:

$$a\frac{dy^2(t)}{dt^2} + b\frac{dy(t)}{dt} + cy(t) = x(t) \tag{3.42}$$

The ODE can be expressed with the impulse response and the impulse signal as input:

$$a\frac{dh^2(t)}{dt^2} + b\frac{dh(t)}{dt} + ch(t) = \delta(t) \tag{3.43}$$

The Laplace Transform is applied to obtain the impulse response or transfer function:

$$L\left\{a\frac{dh^2(t)}{dt^2} + b\frac{dh(t)}{dt} + ch(t) = \delta(t)\right\} = L\{h(t)\}(as^2 + bs + c) = 1 \tag{3.44}$$

$$H(s) = \frac{1}{as^2 + bs + c} \tag{3.45}$$

Depending on the roots, whether they are a pair of real roots or complex conjugates, the impulse response will have three different dynamical responses: overdamped, critically damped, and underdamped. In the case of real roots, the behavior of the system is overdamped or critically damped:

$$H(s) = \frac{1}{(as^2 + bs + c)} = \frac{k}{(\tau_1 s + 1)(\tau_2 s + 1)} \tag{3.46}$$

For this case of two real roots, the system output can be found by taking the inverse Laplace Transform and is the sum of the real exponentials taking the form of the analytical expression:

$$h(t) = k(A_1 e^{-t/\tau_1} + A_2 e^{-t/\tau_2})u(t) \tag{3.47}$$

For the case of complex roots, the system output presents oscillations, and the system behavior is said to be underdamped, where the damping coefficient, ζ, defines the oscillation that is produced with the natural frequency, ω_n, which is explicit in the transfer function as:

$$H(s) = \frac{1}{(as^2 + bs + c)} = \frac{k\omega_n^2}{s^2 + 2\omega_n \zeta s + \omega_n^2} \tag{3.48}$$

For the case of complex roots, the inverse Laplace Transform is the sum of the complex conjugates exponentials and gives the following expression that incorporates a sinusoidal function:

$$h(t) = k A_1 e^{-\zeta \omega_n t} \cos(\omega_n t) u(t) \tag{3.49}$$

Figure 3.10 shows the impulse response of second-order LTI systems with the three possible scenarios: overdamped, ($\zeta = 2$, $\omega_n = 2$); critically damped, ($\zeta = 1$, $\omega_n = 2$); and underdamped ($\zeta = 0.5$, $\omega_n = 2$).

A brief explanation of the partial fraction expansion method is presented below before explaining how to obtain analytically the step response of first- and second-order systems.

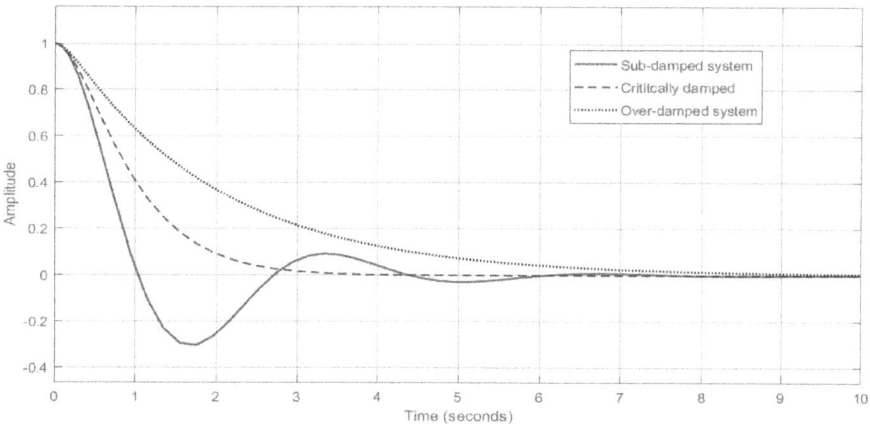

Fig. 3.10. Second-order impulse response: underdamped, overdamped, and critically damped.

4.6 Partial Fractions Expansion Method

The partial fractions expansion method is applied to ease the procedure to obtain the inverse Laplace transfer function of second- or higher-order systems. The procedure decomposes the higher order rational function as a sum of first-order functions. This method helps obtain the response to arbitrary signals and is often applied to get the step response of LTI systems.

Consider a rational function in a complex frequency domain (also called Laplace domain) with different poles, which can be expressed as a sum of partial fractions:

$$F(s) = \sum_{i=1}^{n} \frac{A_i}{s + s_i} \tag{3.50}$$

Where:

$$A_i = (s + s_i)F(s)\big|_{s = s_i} \tag{3.51}$$

The inverse Laplace Transform is the sum of the real or complex exponentials of complex roots.

$$f(t) = \sum_{i=1}^{n} A_i e^{s_i t} \tag{3.52}$$

Example 1. Find the inverse Laplace Transform of the following function, $F(s)$:

$$F(s) = \frac{2}{s^2 + 3s + 2} \tag{3.53}$$

$$F(s) = \frac{2}{(s+1)(s+2)} \tag{3.54}$$

$$F(s) = \sum_{i=1}^{n} \frac{A_i}{s + s_i} = \frac{A_1}{s+1} + \frac{A_2}{s+2} \tag{3.55}$$

$$A_1 = (s+1)\frac{2}{(s+1)(s+2)}\bigg|_{s=-1} = \frac{2}{(s+2)}\bigg|_{s=-1} = \frac{2}{(-1+2)} = 2 \tag{3.56}$$

$$A_2 = (s+2)\frac{2}{(s+1)(s+2)}\bigg|_{s=-2} = \frac{2}{(s+1)}\bigg|_{s=-2} = \frac{2}{(-2+1)} = -2 \tag{3.57}$$

$$F(s) = \frac{2}{(s+1)(s+2)} = \frac{2}{s+1} + \frac{-2}{s+2} \tag{3.58}$$

$$f(t) = \sum_{i=1}^{n} A_i e^{s_i t} = A_1 e^{s_1 t} + A_2 e^{s_2 t} = 2(e^{-t} - e^{-2t})u(t) \tag{3.59}$$

Example 2. Find the inverse Laplace Transform of the following function, $F(s)$:

$$F(s) = \frac{2}{s^2 + 4s + 5} \tag{3.60}$$

$$F(s) = \frac{2}{(s+2+j)(s+2-j)} \tag{3.61}$$

Apply partial fractions expansion:

$$F(s) = \sum_{i=1}^{n} \frac{A_i}{s+s_i} = \frac{A_1}{(s+2+j)} + \frac{A_2}{(s+2-j)} \tag{3.62}$$

$$A_1 = (s+2+j)\frac{2}{(s+2+j)(s+2-j)}\bigg|_{(s=-2-j)} = \frac{2}{(s+2-j)}\bigg|_{(s=-2-j)} = \frac{2}{-2i} = j \tag{3.63}$$

$$A_2 = (s+2-j)\frac{2}{(s+2+j)(s+2-j)}\bigg|_{(s=-2+j)} = \frac{2}{(s+2+j)}\bigg|_{(s=-2+j)} = \frac{2}{2i} = -j \tag{3.64}$$

$$F(s) = \frac{i}{(s+2+j)} - \frac{i}{(s+2-j)} \tag{3.65}$$

Obtain inverse Laplace Transform:

$$f(t) = \sum_{i=1}^{n} A_i\ e^{s_i t} = (je^{-(2+j)t} - je^{-(2-j)t})u(t) \tag{3.66}$$

$$f(t) = e^{-2t}(je^{-jt} - je^{jt})u(t) = e^{-2t}(-j)(e^{jt} - e^{-jt})u(t) \tag{3.67}$$

$$f(t) = e^{-2t}\frac{2}{2j}(e^{jt} - e^{-jt})u(t) = 2e^{-2t}\sin(t)u(t) \tag{3.68}$$

4.7 Step Response of First-Order Systems

For a practical analysis, consider a transfer function with gain k and time constant τ:

$$H(s) = \frac{Y(s)}{X(s)} = \frac{k}{\tau s+1} \tag{3.69}$$

$$X(s) = U(s) = \frac{1}{s} \tag{3.70}$$

$$Y(s) = H(s)X(s) = \frac{k}{\tau s+1}\left(\frac{1}{s}\right) = \frac{k/\tau}{s+1/\tau}\left(\frac{1}{s}\right) \tag{3.71}$$

Apply partial fractions expansion:

$$Y(s) = \sum_{i=1}^{n} \frac{A_i}{s+s_i} = \frac{A_1}{s} + \frac{A_2}{s+1/\tau} \tag{3.72}$$

$$A_1 = (s)\frac{k/\tau}{s(s+1/\tau)}\bigg|_{s=0} = \frac{k/\tau}{(s+1/\tau)}\bigg|_{s=0} = k \tag{3.73}$$

$$A_2 = (s+1/\tau)\frac{k/\tau}{s(s+1/\tau)}\bigg|_{s=-1/\tau} = \frac{k/\tau}{s}\bigg|_{s=-1/\tau} = -k \tag{3.74}$$

$$Y(s) = \left(\frac{1}{s}\right)\frac{k/\tau}{s+1/\tau} = \frac{k}{s} + \frac{-k}{s+1/\tau} \tag{3.75}$$

Obtain inverse Laplace Transform:

$$y(t) = \sum_{i=1}^{n} A_i e^{s_i t} = \left(ke^0 - ke^{-\frac{t}{\tau}}\right)u(t) = k(1-e^{-t/\tau})u(t) \tag{3.76}$$

Figure 3.11 shows the step responses of three different first-order LTI systems. The initial response corresponds to exponential functions with a rate of change defined by the time constant, τ. If the time constant increases, the transitory is slower, and if the time constant is reduced, the step response is faster.

The steady state time, t_{ss}, is the time to achieve 98% of the total response change, four times the time constant, $t_{ss} = 4\tau$.

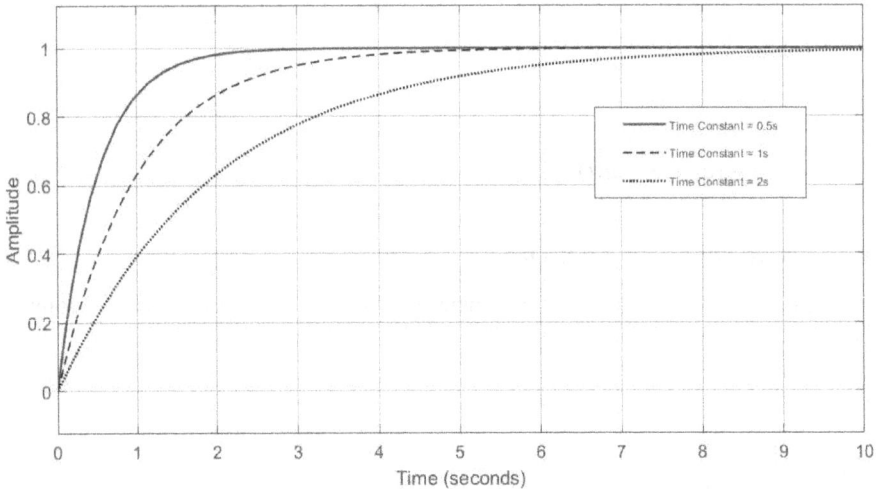

Fig. 3.11. First-order step response with three different time constants (0.5, 1.0, 2.0).

Example 3. Find the step response of first-order system with the following transfer function, $H(s)$:

$$H(s) = \frac{Y(s)}{U(s)} = \frac{3}{2s+1} \tag{3.77}$$

Express $Y(s)$:

$$Y(s) = \frac{3}{2s+1}U(s) = \frac{1.5}{s+0.5}\left(\frac{1}{s}\right) \tag{3.78}$$

$$Y(s) = \frac{1.5}{s(s+0.5)} \tag{3.79}$$

Apply partial fraction expansion:

$$Y(s) = \sum_{i=1}^{n} \frac{A_i}{s+s_i} = \frac{A_1}{s} + \frac{A_2}{s+0.5} \tag{3.80}$$

$$A_1 = (s)\frac{1.5}{(s)(s+0.5)}\bigg|_{s=0} = \frac{1.5}{(s+0.5)}\bigg|_{s=0} = \frac{1.5}{0.5} = 3 \tag{3.81}$$

$$A_2 = (s+0.5)\frac{1.5}{(s)(s+0.5)}\bigg|_{s=-0.5} = \frac{1.5}{s}\bigg|_{s=-0.5} = \frac{1.5}{-0.5} = -3 \tag{3.82}$$

$$Y(s) = \frac{3}{s} - \frac{3}{s+0.5} \tag{3.83}$$

Obtain inverse Laplace Transform:

$$y(t) = A_1 e^{s_1 t} + A_2 e^{s_2 t} = 3(e^0 - e^{-0.5t})u(t) = 3(1 - e^{-0.5t})u(t) \tag{3.84}$$

4.8 Step Response of Second-Order Systems

For practical analysis, consider a second-order system with parameters: gain, k, damping factor, ζ, and natural frequency, ω_n:

$$G(s) = \frac{k\omega_n^2}{s^2 + 2\omega_n\zeta s + \omega_n^2} \tag{3.85}$$

Consider a step input, $U(s)$, and the output response, $Y(s)$:

$$X(s) = U(s) = \frac{1}{s} \tag{3.86}$$

$$Y(s) = H(s)X(s) = \frac{k\omega_n^2}{s^2 + 2\omega_n\zeta s + \omega_n^2}\left(\frac{1}{s}\right) \tag{3.87}$$

The inverse Laplace Transform gives a general form for the step response:

$$y(t) = k\left[1 - e^{-\zeta\omega_n t}\left(k_1\cos\left(\omega_n\sqrt{1-\zeta^2}t\right) + k_2\sin\left(\omega_n\sqrt{1-\zeta^2}t\right)\right)\right]u(t) \tag{3.88}$$

For second-order systems with two real roots or poles, the transfer function can be expressed as:

$$G(s) = \frac{k}{(\tau_1 s + 1)(\tau_2 s + 1)} \tag{3.89}$$

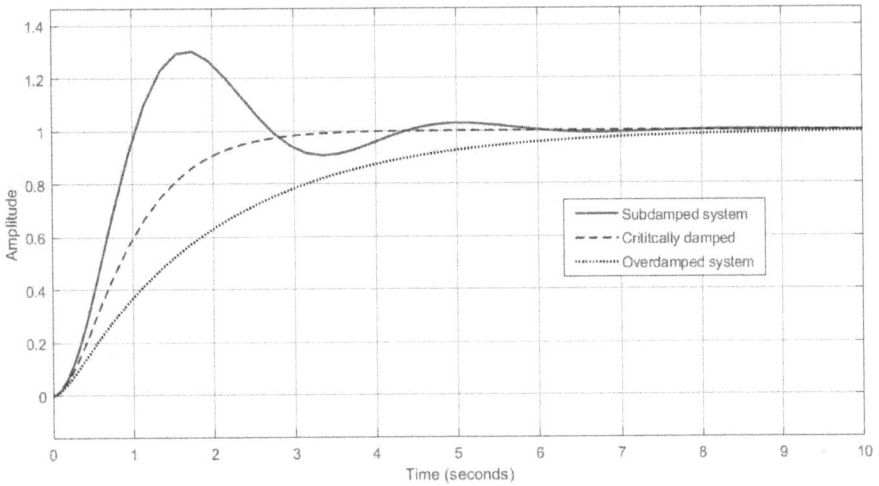

Fig. 3.12. Step response of overdamped, critically damped, and underdamped systems.

If the two poles are different, the inverse Laplace Transform gives a form for the step response with only real exponentials:

$$y(t) = k(1 - A_1 e^{-t/\tau_1} - A_2 e^{-t/\tau_2})u(t) \tag{3.90}$$

Figure 3.12 shows the impulse response of second-order LTI systems with the three possible scenarios: overdamped with $\zeta = 2$, $\omega_n = 2$, that is equivalent to a system with $\tau_1 = 1.866025$, $\tau_2 = 0.1339746$; critically damped with $\zeta = 1$, $\omega_n = 2$, that is equivalent to a system with $\tau_1 = 0.5$, $\tau_2 = 0.5$; and underdamped with $\zeta = 0.5$, $\omega_n = 2$.

Example 4. Find the step response of a second-order system with the following transfer function, $G(s)$:

$$G(s) = \frac{4}{s^2 + 5s + 6} \tag{3.91}$$

$$\frac{Y(s)}{U(s)} = \frac{4}{s^2 + 5s + 6} \tag{3.92}$$

$$Y(s) = \frac{4}{s^2 + 5s + 6}U(s) = \frac{4}{(s+2)(s+3)}\left(\frac{1}{s}\right) \tag{3.93}$$

Apply partial fraction expansion:

$$Y(s) = \sum_{i=1}^{n} \frac{A_i}{s + s_i} = \frac{A_1}{s} + \frac{A_2}{s+2} + \frac{A_3}{s+3} \tag{3.94}$$

$$A_1 = (s)\frac{4}{s(s+2)(s+3)}\bigg|_{s=0} = \frac{4}{(s+2)(s+3)}\bigg|_{s=0} = \frac{4}{(2)(3)} = \frac{2}{3} \tag{3.95}$$

$$A_2 = (s+2)\frac{4}{s(s+2)(s+3)}\bigg|_{s=-2} = \frac{4}{s(s+3)}\bigg|_{s=-2} = \frac{4}{-2(-2+3)} = -2 \qquad (3.96)$$

$$A_3 = (s+3)\frac{4}{s(s+2)(s+3)}\bigg|_{s=-3} = \frac{4}{s(s+2)}\bigg|_{s=-3} = \frac{4}{-3(-3+2)} = \frac{4}{3} \qquad (3.97)$$

$$Y(s) = \frac{4}{s(s+2)(s+3)} = \frac{2}{3} - \frac{2}{s+2} + \left(\frac{4}{3}\right)\frac{1}{s+3} \qquad (3.98)$$

Obtain inverse Laplace Transform:

$$y(t) = A_1 e^{s_1 t} + A_2 e^{s_2 t} + A_3 e^{s_3 t} = \left(\frac{2}{3} - 2e^{-2t} + \frac{4}{3}e^{-3t}\right)u(t) \qquad (3.99)$$

5. Frequency Response Analysis

The frequency response of a system is an analysis performed on the magnitude and phase of the output signal by varying the input signal frequency. The frequency response is widely used in the modeling and analysis of physical systems and control systems design and analysis. The differential equations in continuous time can be converted into algebraic equations in the frequency domain. The frequency response can be designed to have a specific bandwidth. Stability can also be analyzed using the concepts of gain and phase margin. In control systems like a vehicle's cruise control, it may be used to assess system stability, often through Bode plots. Systems with a specific frequency response can be designed using analog and digital filters (Hsu, 2011).

The frequency response characterizes systems in the frequency domain, just as the impulse response characterizes systems in the time domain. In linear systems, either response completely describes the system, and thus there is a one-to-one correspondence: the frequency response is the Fourier Transform of the impulse response. The frequency response allows a more straightforward analysis of cascade systems as the response of the overall system can be found through the multiplication of the individual stages' frequency responses (as opposed to the convolution of the impulse response in the time domain).

5.1 Fourier Transform

The frequency response is closely related to the transfer function in linear systems since it is the Laplace Transform of the impulse response.

The Fourier Transform is the Laplace Transform when s is purely imaginary:

$$X(s)\big|_{s=j\omega} = H(j\omega) = \int_{-\infty}^{\infty} h(t)e^{-j\omega t}\,dt \qquad (3.100)$$

Bode plots are used for systems analysis. A sinusoid signal can be applied to a system, sensor, or actuator to detect the attenuation and the phase shift experimentally; they can even be used to identify the order or the LTI that can be used to model a physical system. Other concepts like gain and phase margins are relevant for design control systems. The Bode diagram for a system consists simply of the curves of magnitude and phase in the logarithmic scale against frequency. Log magnitude and phase effects are now additive.

5.2 Bode Plot of First-Order Systems

Consider a LTI first-order system described by:

$$H(s) = \frac{1}{\tau s + 1} \tag{3.101}$$

The Fourier transfer function is:

$$H(j\omega) = H(s)\big|_{s=j\omega} = \frac{1}{j\omega\tau + 1} \tag{3.102}$$

Now consider $\tau = 1/\omega_c$,

$$H(j\omega) = \frac{1}{j\left(\dfrac{\omega}{\omega_c}\right) + 1} \tag{3.103}$$

The magnitude and the angle of $H(j\omega)$ are:

$$|H(j\omega)| = \frac{1}{\sqrt{\left(\dfrac{\omega}{\omega_c}\right)^2 + 1}}, \quad \angle H(j\omega) = \angle 1 - \angle\left(j\left(\frac{\omega}{\omega_c}\right) + 1\right) = -tan^{-1}\left(\frac{\omega}{\omega_c}\right) \tag{3.104}$$

If we evaluate at the low frequency of one-tenth of the cutting frequency, $\omega = \omega_c/10$, the magnitude tends to unity; this means the output signal is not attenuated.

$$|H(j\omega)| = \frac{1}{\sqrt{\left(\dfrac{\omega_c}{10\omega_c}\right)^2 + 1}} = \frac{1}{\sqrt{\left(\dfrac{1}{10}\right)^2 + 1}} = \frac{1}{\sqrt{\dfrac{1}{100} + 1}} \approx 1 \tag{3.105}$$

At the cutting frequency, $\omega = \omega_c$, the amplitude of the output signal has a 0.7071 fraction of the input signal:

$$|H(j\omega)| = \frac{1}{\sqrt{\left(\dfrac{\omega_c}{\omega_c}\right)^2 + 1}} = \frac{1}{\sqrt{1+1}} = \frac{1}{\sqrt{2}} = 0.7071 \tag{3.106}$$

At higher frequencies, like ten times the cutting frequency, $\omega = 10\,\omega_c$, the magnitude tends to zero, meaning the output signal is completely attenuated.

Bode Diagram

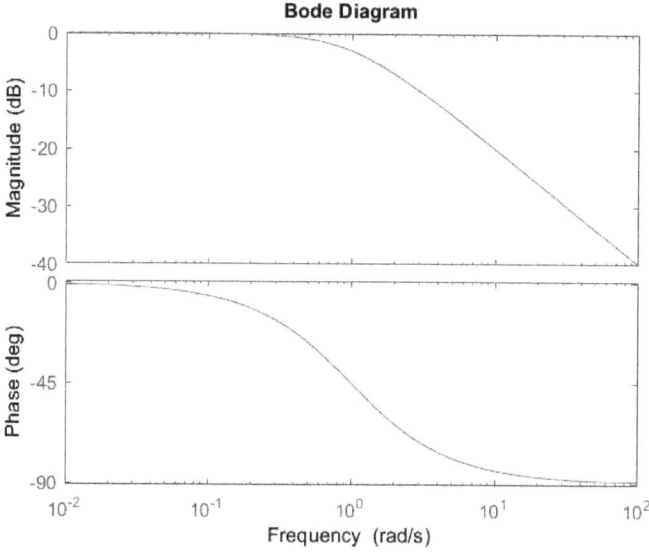

Fig. 3.13. Bode plots of first-order system.

$$|H(j\omega)| = \frac{1}{\sqrt{\left(\dfrac{10\omega_c}{\omega_c}\right)^2 + 1}} = \frac{1}{\sqrt{(10)^2 + 1}} = \frac{1}{\sqrt{101}} \approx 0 \tag{3.107}$$

Figure 3.13 shows the Bode diagrams as logarithmic plots for a system with cutting frequency $\omega_c = 1$, at which the amplitude is 0.7071, and the phase is shifted 45°.

5.3 Bode Plot of Second-Order Systems

Consider an LTI second-order system with damping coefficient, ζ, and natural frequency, ω_n:

$$H(s) = \frac{\omega_n^2}{s^2 + 2\omega_n \zeta s + \omega_n^2} \tag{3.108}$$

The Fourier transfer function is:

$$H(j\omega) = H(s)\big|_{s=j\omega} = \frac{\omega_n^2}{(j\omega)^2 + 2\omega_n \zeta (j\omega) + \omega_n^2} \tag{3.109}$$

The magnitude of $H(j\omega)$ is:

$$|H(j\omega)| = \frac{\omega_n^2}{\sqrt{\left(\omega_n^2 - \omega^2\right)^2 + (2\omega_n \zeta \omega)^2}} \tag{3.110}$$

Bode Diagram

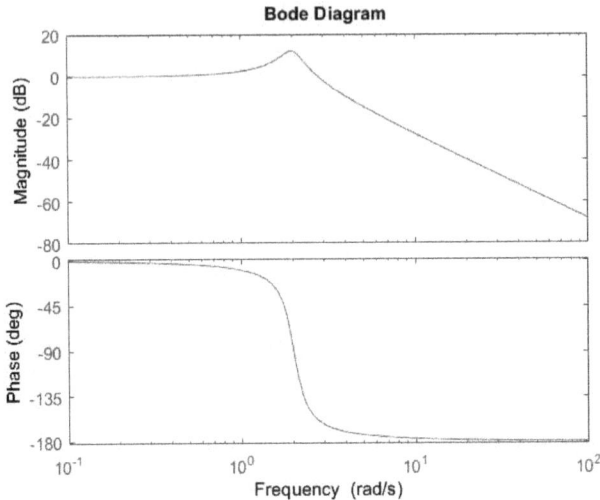

Fig. 3.14. Bode diagram of second order underdamped system.

As for the first-order system, evaluating the magnitude in frequencies lower than the cutting frequency, the output signal is not attenuated, and assessing at a higher frequency than the cutting frequency, the magnitude of the output signal is attenuated. Underdamped second-order systems present the phenomenon of resonance that occurs at the natural frequency $\omega = \omega_n$, where the output signal may be amplified, and the magnitude of the amplification is inversely proportional to the damping ratio:

$$|H(j\omega)| = \frac{\omega_n^2}{\sqrt{\left(\omega_n^2 - \omega^2\right)^2 + (2\omega_n \zeta \omega)^2}} = \frac{\omega_n^2}{2\omega_n^2 \zeta} = \frac{1}{2\zeta} \qquad (3.111)$$

Figure 3.14 shows the Bode diagrams for an underdamped second-order system.

6. Modeling of Physical Systems

Many physical systems have dynamical responses that can be modeled as linear systems with memory; this is, physical variables cannot be suddenly changed, such as the voltage in a capacitor, the displacement of a spring mass system, the speed of a car, the temperature furnace, the pressure of fluidic systems, etc. The theory of signals and systems and the response of dynamical systems are helpful for a basic approach to modeling and simulating physical systems; in particular, they are widely used in designing electronic and mechatronic systems and industrial control systems.

6.1 Modeling First-Order Systems

Consider the circuit of Fig. 3.15 with a resistor and a capacitor in series (Feucht, 1990), with resistance $R = 5\ k\Omega$ and capacitance $C = 100\ \mu F$, respectively. The input voltage is $V_i(t)$, and the output voltage is $V_o(t)$.

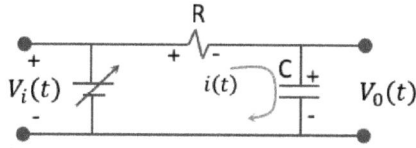

Fig. 3.15. Resistor–Capacitor circuit.

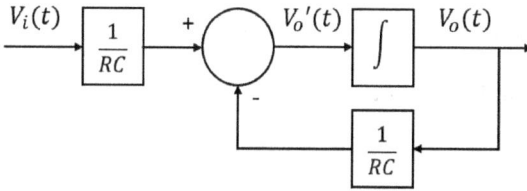

Fig. 3.16. Resistor–Capacitor circuit ODE block diagram.

The model of the circuit is given by:

$$V_i(t) - Ri(t) - \frac{1}{C}\int i(t)dt = 0 \qquad (3.112)$$

The output voltage through the capacitor is

$$V_o(t) = \frac{1}{C}\int i(t)dt \qquad (3.113)$$

The current then can be expressed in terms of the output voltage as:

$$i(t) = C\frac{dV_o(t)}{dt} \qquad (3.114)$$

Substitution of equation (3.114) in equation (3.112) gives the relation between the input and output voltages:

$$V_i(t) - RC\frac{dV_o(t)}{dt} - V_o(t) = 0 \qquad (3.115)$$

Figure 3.16 shows the block diagram of the differential equation.

The transfer function, given the specific resistance and capacitance values, is:

$$G_p(s) = \frac{V_o(s)}{V_i(t)} = \frac{1}{RCs+1} = \frac{1}{0.5s+1} \qquad (3.116)$$

The response to a step input of amplitude 5V is:

$$V_o(t) = 5(1 - e^{-t/(RC)})u(t) \qquad (3.117)$$

6.2 Modeling of Second-Order Systems

Consider the example of a spring-damper-mass system shown in Fig. 3.17.

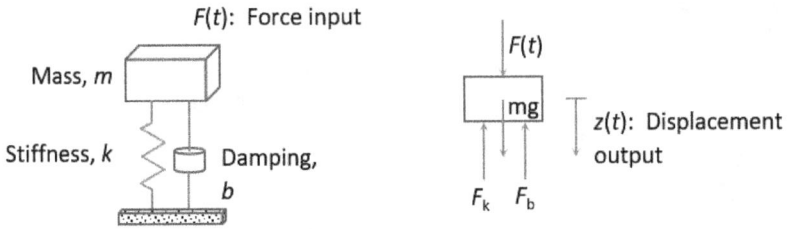

Fig. 3.17. Spring-damper-mass system and its free-body diagram.

The second-order differential equation for this system is given by:

$$m\frac{d^2Z(t)}{dt^2} + b\frac{dZ(t)}{dt} + kZ(t) = F(t) + mg \tag{3.118}$$

where Z is the displacement, F is the force input, g is gravity, m is mass, k is the spring stiffness, and b is the damping coefficient of the damper.

The evaluation at the initial steady state is:

$$kZ(0) = F(0) + mg \tag{3.119}$$

The subtraction of the initial steady state evaluation from the differential equation can be expressed in terms of deviation variables $z(t)$ and $f(t)$:

$$\frac{d^2z(t)}{dt^2} + \frac{b}{m}\frac{dz(t)}{dt} + \frac{k}{m}z(t) = \frac{1}{m}f(t) \tag{3.120}$$

The block diagram of Fig. 3.18 represents this differential equation.

Let's assume the physical parameters of the system determine the coefficients of the differential equation as shown next:

$$\frac{d^2z(t)}{dt^2} + 1.5\frac{dz(t)}{dt} + 0.5z(t) = f(t) \tag{3.121}$$

where $z(t)$ is the process variable (PV), and $f(t)$ is the control variable (CV).

In the domain of the complex variable, s, the transfer function is:

$$G_p(s) = \frac{Z(s)}{F(s)} = \frac{1}{s^2 + 1.5s + 0.5} = \frac{2}{(2s+1)(s+1)} \tag{3.122}$$

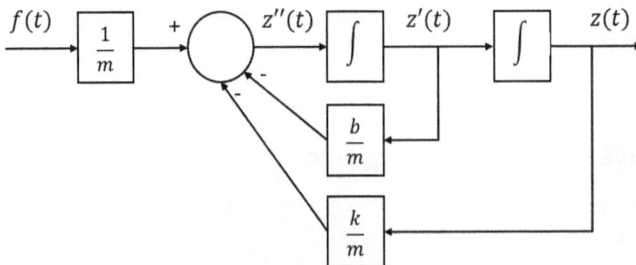

Fig. 3.18. Block diagram of second-order differential equation of the spring-damper-mass system.

The displacement response to a unit step in the force input of 1N is:

$$z(t) = 2(1 - 2e^{-t/2} + e^{-t})u(t) \tag{3.123}$$

6.3 Modeling of a DC Motor

DC motors are well-known devices that can be used as actuators of many mechanisms where the speed and position of the axis of movement need to be controlled. The mathematical model of the DC motor requires two equations: the first to model the electric circuit's dynamic response, and the second to model the mechanical device's dynamic response. These equations are based on Kirchhoff's circuit laws and Newton's motion laws. Figure 3.19 shows the electromechanical model of a DC motor.

The model illustrated in Fig. 3.19 has the following electrical and mechanical parameters; V = voltage, ω = angular speed, $i(t)$ = armature current, L = armature inductance, R = armature resistance, J = rotor inertia, V_b = back electromagnetic field (EMF), B = friction coefficient, τ_M = DC motor load torque.

Applying Kirchhoff's laws, the first-order differential equation (3.124) can be expressed for the electric circuit:

$$V(t) = Ri(t) + L\frac{di(t)}{dt} + K_b\omega(t) \tag{3.124}$$

The back emf is:

$$V_b(t) = K_b\omega(t) \tag{3.125}$$

Equation (3.126) is obtained by substituting Eq. (3.125) into Eq. (3.124);

$$V(t) = Ri(t) + L\frac{di(t)}{dt} + V_b(t) \tag{3.126}$$

For the mechanical part of the DC motor, the first-order differential equation (3.127) is considered, where τ_L is the disturbance variable:

$$\tau_M - \tau_L = J\frac{d\omega(t)}{dt} + B\omega(t) \tag{3.127}$$

The motor generated torque is:

$$\tau_M(t) = K_t\, i(t) \tag{3.128}$$

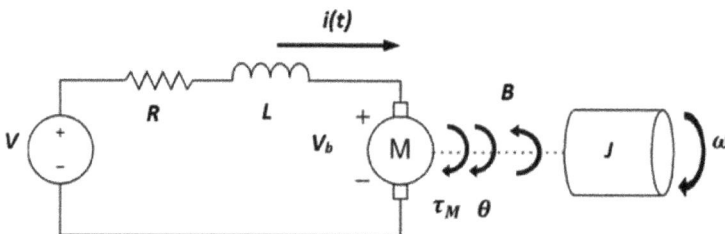

Fig. 3.19. Circuit of the DC Motor.

Substituting in Eq. (3.127) and considering null load torque $\tau_L = 0$:

$$K_t i(t) = J\frac{d\omega(t)}{dt} + B\omega(t) \tag{3.129}$$

Equation (3.126) and Eq. (3.129) can also be represented in the frequency domain by applying the Laplace Transform:

$$V(s) = R\, i(s) + Ls\, i(s) + V_b(s) \tag{3.130}$$

$$K_t\, i(s) = Js\, \omega(s) + B\, \omega(s) \tag{3.131}$$

where:

$$i(s) = \frac{V(s) - V_b(s)}{R + Ls} \tag{3.132}$$

substituting in Eq. (3.131):

$$K_t \frac{V(s) - V_b(s)}{R + Ls} = Js\omega(s) + B\omega(s) \tag{3.133}$$

Figure 3.20 shows a typical block diagram of the mathematical model of a DC motor, representing on the left side the electromagnetic process of the stator and on the right side the mechanical rotation of the rotor, where the electric circuit current is converted to torque, the load torque is subtracted, the mechanical circuit is integrated, having feedback of the induced voltage.

From the block diagram of Fig. 3.20, where K_t = torque constant and K_b = emf constant, the transfer function of interest that relates the motor developed speed (output signal) with the voltage input signal is:

$$G(s) = \frac{\omega(s)}{V(s)} = \frac{K_t}{JLs^2 + JRs + LBs + BR + K_t \cdot K_b} \tag{3.134}$$

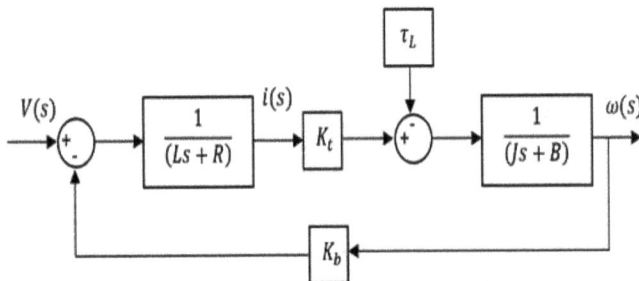

Fig. 3.20. Block diagram of the DC motor model.

7. Conclusions

This chapter reviewed mathematical and engineering tools required for model-based design, such as signals and systems concepts. Signals represent physical phenomena and convey information from one point to another. They can be generated by natural phenomena or by electronic means. Ordinary differential equations can model LTI continuous-time systems in the time domain or transfer functions in the frequency domain. A differential equation relates an output variable with a dynamical response to an input variable. The mathematical operation is a convolution in the time domain that is an integration process in which the output signal is infinitesimally passed through systems a converted into an output signal. A first-order differential equation's solution is typically exponential; the second-order system can exhibit oscillations.

Many physical systems have dynamical responses that can be modeled as linear systems with memory; this is, physical variables cannot be suddenly changed, such as the voltage in a capacitor, the displacement of a spring mass system, the speed of a car, the temperature of a furnace, the pressure of a fluidic system, etc. An ODE in the time domain has an equivalent transfer function in the frequency domain. Laplace Transform is a mathematical tool to convert a time domain ODE into a rational function in the frequency domain. Partial fractions expansion methods allow splitting a rational function into simpler additive components. The transfer function of a system is used to obtain its response by multiplying by the Laplace Transform of the input signal. Partial fractions ease the application of the inverse Laplace Transform to obtain the continuous-time domain response of first and second-order LTI systems. The theory of signals and systems and the analysis of the response of dynamical systems are necessary tools for process modeling and simulation, as well as for control system design. Laplace and Fourier Transforms are mathematical tools used for: (1) analysis of the system response in both time and frequency domains; (2) system identification, when experimental data can be obtained and a first- or second-order system can be proposed as a model; and (3) model-based design of control systems.

References

Feucht, D.L. 1990. *Handbook of Analog Circuit Design*. Elsevier Science. p. 192.

Hsu, H.P. 2011. *Signals and Systems*. Second Edition. McGraw Hill.

Oppenheim, A.V., Willsky, A.S. and Nawab, S.H. 1997. *Signals and Systems*. Second Edition. Prentice Hall.

Smith, S.W. 1997. *The Scientist and Engineer's Guide to Digital Signal Processing*. California Technical Pub. pp. 177–180.

Chapter 4
Discretization of Linear Systems

1. Introduction

The signals which are defined only at discrete instants of time are known as discrete-time signals. A discrete time system operates on a discrete time signal input and produces a discrete-time signal output. Discrete time linear time invariant (LTI) systems are of particular interest since the properties of linearity and time invariance together with the application of difference equations are used to computationally: (1) implement models of dynamical systems, (2) design digital controllers, and (3) implement digital filters. Computer simulation of physical systems relies on using difference equations that are expressed as recursive equations for modeling discrete-time systems. Difference equations are easy to program in any computer platform, such as microcontrollers, programmable controllers, personal computers, or more powerful workstations. Simple or complex control laws in discrete time can be expressed as difference equations. Difference equations are also mathematical tools for representing linear or logarithmic ramps used to ramp manipulation variables, for example.

Digital filters are discrete-time systems used in digital signal processing for filtering images or sounds (Oppenheim et al., 1997). In mechatronics and control systems, digital filters are used for filtering the noise induced in sensor signals (Bishop, 2007). Discrete transfer functions in z-domain are very practical in the design of digital filters and digital controllers (Hsu, 2011). Discrete transfer functions can be directly specified in the z-domain obtained by analog to digital conversion methods. Discretization methods are important since many analysis and design tools are for continuous-time systems or s-domain. However, the computational implementation requires either difference equations or transfer functions in the z-domain from which difference equations can be obtained. Two popular conversion methods are Euler backward differences in the time domain, and bilinear transform in the frequency domain.

This chapter reviews linear systems' representation in discrete time and discrete frequency domains. Section 2 presents the basic representation of discrete-time signals and a system's impulse and step responses in discrete time. Section 3 gives an introduction to z-transform and discrete frequency transfer functions. Section 4 presents the backward Euler method with the approach to obtain difference equations for computational implementation. Section 5 reviews the bilinear transform and its practical application for discretizing continuous time transfer functions. The last Section 6 reviews methods for converting basic analog filters to digital filters often used for filtering noise in sensor signals or implementing logarithmic ramps for manipulation variables. Conclusions are in Section 7.

2. Discrete-Time Signals and Systems

The signals defined only at discrete-time instants are known as discrete-time signals. To convert a signal from continuous time to discrete time, a process called sampling is used. The value of the signal is measured at certain intervals in time. Each measurement is referred to as a sample. The time instants at which the signal is defined are the signal's sample times, and the associated signal values are the signal's samples:

$$x[n] = x(t = nT) \tag{4.1}$$

Sampling has two important effects on the signal. The first effect is the selection of specific signal values that become the samples in discrete time. The second effect is known as spectral replication, which means that a sampled signal's frequency spectrum is periodically repeated in the frequency domain, with a period equal to the sampling frequency. The discrete-time signal can be represented as a series of samples distributed by a train of impulses; in this sense, the impulse signal can also be used to express a generic signal, $x[n]$ mathematically:

$$x[n] = \sum_{k=-\infty}^{\infty} x[n]\delta[n-k] \tag{4.2}$$

2.1 Impulse Response

A discrete-time system takes a discrete-time signal as input and generates a discrete-time signal as output. Similarly, to continuous-time systems, any discrete-time linear system can be characterized by measuring its impulse response.

The impulse response, $h[n]$, of a linear system H, is defined as the response to a unit impulse:

$$h[n] = H\{\delta[n]\} \tag{4.3}$$

If the system is time invariant (LTI), a delay in time in the input signal is passed to the output signal:

$$h[n-k] = H\{\delta[n-k]\} \tag{4.4}$$

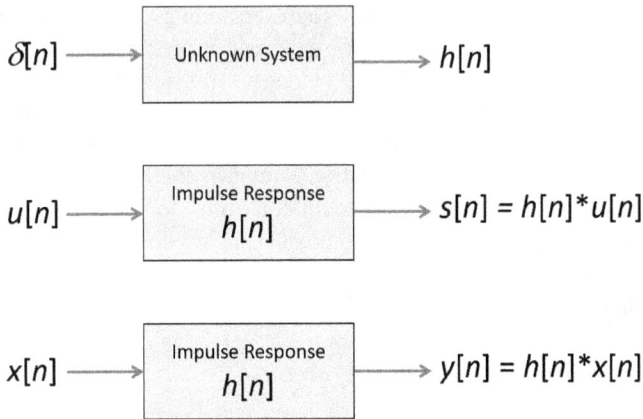

Fig. 4.1. Discrete-time system response.

The discrete unit sample function is typically used as an input to a discrete system to find its dynamical response. The step response of a discrete-time LTI system can be expressed with the convolution sum:

$$s[n] = h[n] * u[n] = \sum_{k=-\infty}^{\infty} h[n]u[n-k] = \sum_{k=-\infty}^{\infty} h[n] \tag{4.5}$$

The output of any discrete-time LTI system is the convolution of any input with the impulse response. The convolution sum is conventionally used to find the discrete-time system response to an arbitrary input:

$$y[n] = h[n] * x[n] = \sum_{k=0}^{\infty} h[n]x[n-k] \tag{4.6}$$

Figure 4.1 shows the impulse response obtained with a unit impulse and their use to obtain the discrete-time system response to arbitrary signals.

If the input $x[n] = \delta[n]$, the system output can be represented with a power series. If $a < 1$, the system is stable since its impulse response corresponds to a decaying power function of time. If $a > 1$, the system is unstable because its impulse response represents a growing power function of time:

$$y[n] = a^n u[n] \tag{4.7}$$

If the input $x[n] = Au[n]$, then the output is the step response. If the system is stable ($a < 1$), the step response corresponds to a series that converges to a constant value. For a unit gain system, this steady state value is the step input amplitude A, with a decaying rate of change given by the power series a^n:

$$y[n] = A(1 - a^n)u[n] \tag{4.8}$$

Figure 4.2 shows the impulse and step responses of a first order discrete time linear system. The power series of the impulse response is equivalent to a sampled exponentially decaying signal of the impulse response in continuous time. The step response discrete-time signal is equivalent to a sampled signal of the step response in continuous time.

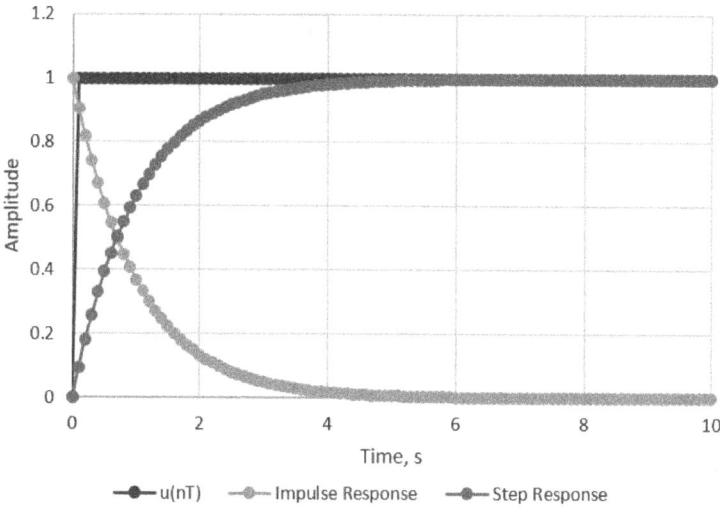

Fig. 4.2. First-order discrete-time system impulse and step response.

2.2 Difference Equations

Discrete-time systems can be described in terms of rates of change through linear constant coefficient difference equations, a type of recurrence relation. Linear constant coefficient difference equations are discrete-time equivalents of differential equations in continuous-time and reproduce the response of dynamical systems.

A general *N-th* order LTI difference equation can be expressed in the form:

$$\sum_{k=0}^{N} a_k y[n-k] = \sum_{k=0}^{M} b_k x[n-k] \tag{4.9}$$

The system has memory if the equation involves time shifted terms of $y[n]$ or $x[n]$. An example of a first order difference equation is:

$$y[n] - a_1 y[n-1] = b_1 x[n] \tag{4.10}$$

$$y[n] = a_1 y[n-1] + b_1 x[n] \tag{4.11}$$

The notation of the difference equations can be simplified assuming $y_n = y[n]$:

$$y_n = a_1 y_{n-1} + b_1 x_n \tag{4.12}$$

A second order differences equation requires two delayed samples, the output function; a *n*-th order differences equation requires delayed samples. An example of a second order difference equation is:

$$y[n] = a_1 y[n-1] + a_2 y[n-2] + b_1 x[n] \tag{4.13}$$

Expressed in simplified notation:

$$y_n = a_1 y_{n-1} + a_2 y_{n-2} + b_1 x_n \tag{4.14}$$

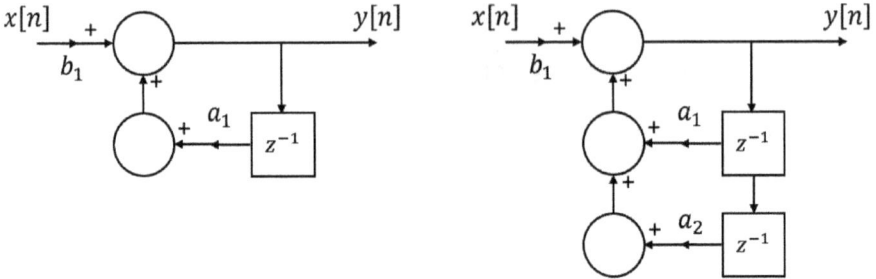

Fig. 4.3. Block diagrams of first (left) and second order (right) difference equations.

Basic first and second order difference equations are represented with block diagrams in Fig. 4.3. Notice how difference equations are feedback systems.

One great advantage of differences equations is that they can be directly implemented computationally; for this reason, difference equations are widely used and are an important and powerful tool in digital control and signal processing.

3. Introduction to Z-Transform

The Laplace and z-transforms are the most important methods to transform calculus problems into algebraic operations. The Laplace Transform transforms ordinary differential equations into the frequency domain, which can be solved by ordinary algebra. As LTI continuous-time systems are analyzed using the Laplace Transform, discrete-time systems are analyzed with the z-transform. The Laplace Transform deals with differential equations in the s-domain and transforms the time-domain ODE into complex rational functions with poles and zeros in the s-plane. The z-transform deals with difference equations converted into rational functions in the z-domain with poles and zeros in the z-plane.

The main objective of these two transforms is to obtain transfer functions in the frequency domain useful for system analysis to find the system's poles and zeros and to obtain the system response to input signals such as steps, ramps, sinusoids, and exponentials. However, the two techniques are not equivalent since the s-plane employs a rectangular coordinate system, while the z-plane uses a polar coordinate system. The z-transform is the most general concept for the transformation of discrete-time series:

$$X(z) = \sum_{n=0}^{\infty} x[n]z^{-n} \tag{4.15}$$

For example, the following is the z-transform of the unit step expressed with positive and negative powers of z:

$$X(z) = \sum_{n=0}^{\infty} u[n]z^{-n} = \frac{z}{z-1} = \frac{1}{1-z^{-1}} \tag{4.16}$$

Another example is the z-transform function of a decaying power series ($a < 1$):

$$X(z) = \sum_{n=0}^{\infty} a^{n}z^{-n} = \frac{z}{z-a} = \frac{1}{1-az^{-1}} \tag{4.17}$$

3.1 Discrete Transfer Functions

Discrete transfer functions in the z-domain are widely applied in digital signal processing and control systems. Digital controllers can be directly designed in discrete time as difference equations or as a transfer function in the complex frequency (Ogata, 1996). Discrete transfer functions are very practical in the design of digital filters, particularly for analogic to digital conversion methods to obtain difference equations that can be implemented computationally:

$$H(z) = \frac{Y(z)}{X(z)} = \frac{a_0 z^N + a_1 z^{N-1} + a_2 z^{N-2} + \cdots + a_N}{b_0 z^N + b_1 z^{N-1} + b_2 z^{N-2} + \cdots + b_N} \tag{4.18}$$

$$H(z) = \frac{Y(z)}{X(z)} = \frac{\sum_{k=0}^{N} a_k y[n-k]}{\sum_{k=0}^{N} a_k y[n-k]} \tag{4.19}$$

Transfer functions in the z-domain can be directly transformed into difference equations required for computational implementation. The exponent of the z complex variable represents a shift in discrete time; to ease the inverse transform process, it is helpful to express the transfer function in negative powers of z so that the time delays can be identified. This algebraic manipulation can be done by multiplying by the inverse of the maximum power of z:

$$H(z) = \frac{Y(z)*z^{-N}}{X(z)*z^{-N}} = \frac{a_0 + a_1 z^{-1} + a_2 z^{-2} + \cdots + a_N z^{-N}}{b_0 + b_1 z^{-1} + b_2 z^{-2} + \cdots + b_N z^{-N}} \tag{4.20}$$

Rewriting the equation by manipulating the denominators as a simple two polynomial side equation:

$$[b_0 + b_1 z^{-1} + b_2 z^{-2} + \cdots + b_N z^{-N}]Y(z) = [a_0 + a_1 z^{-1} + a_2 z^{-2} + \cdots + a_N z^{-N}]X(z) \tag{4.21}$$

Solving for $Y(z)$, moving the delayed terms to the right side:

$$Y(z) = \frac{1}{b_0}[b_1 z^{-1} + b_2 z^{-2} + \cdots + b_N z^{-N}]Y(z)$$
$$+ \frac{1}{b_0}[a_0 + a_1 z^{-1} + a_2 z^{-2} + \cdots + a_N z^{-N}]X(z) \tag{4.22}$$

The process to apply the inverse z-transform is simplified by replacing the negative powers of z for corresponding delays of the input and output signals in discrete time:

$$y[n] = \frac{1}{b_0}(b_1 y[n-1] + b_2 y[n-2] + \cdots + b_N y[n-N])$$
$$+ \frac{1}{b_0}(a_0 x[n] + a_1 x[n-1] + a_2 x[n-2] + \cdots + a_N x[n-N]) \tag{4.23}$$

Simplifying the expression and assuming $b_0 = 1$, as is often the case:

$$y_n = b_1 y_{n-1} + b_2 y_{n-2} + \cdots + b_N y_{n-N} + a_0 x_n + a_1 x_{n-1} + a_2 x_{n-2} + \cdots + a_N x_{n-N} \qquad (4.24)$$

Transfer functions are conventionally expressed in positive powers of z. To ease the procedure to obtain the inverse z-transform, it is convenient to express the transfer function in negative powers of z; this can be done at any step of the algebraic procedure.

As a first example, obtain the difference equation of an integrator with gain k:

$$H(z) = \frac{Y(z)}{X(z)} = \frac{kz}{z-1} \qquad (4.25)$$

Multiply by z^{-1}:

$$H(z) = \frac{Y(z)z^{-1}}{X(z)z^{-1}} = \frac{k}{1-z^{-1}} \qquad (4.26)$$

Solve for $Y(z)$:

$$(1 - z^{-1})Y(z) = k\,X(z) \qquad (4.27)$$

$$Y(z) = z^{-1}\,Y(z) + k\,X(z) \qquad (4.28)$$

Apply inverse z-transform and notice that the gain k only affects the input; this serves to integrate at a given rate of change defined by the gain:

$$y_n = y_{n-1} + k\,x_n \qquad (4.29)$$

If the system has unit gain, $k = 1$, a basic integrator is obtained:

$$y_n = y_{n-1} + x_n \qquad (4.30)$$

The obtained expression is very useful for the numerical integration of many processes, for example, to integrate speed and obtain a position, integrate flow and obtain volume, simulate a hydraulic or pneumatic cylinder, etc. Some considerations are necessary for computational implementation, such as the maximum and minimum values to limit the output and the shifting of input and output variables; see below the pseudocode to be included in an execution loop:

```
While true, do:
    yn=yn1+k*xn
    if yn>ymax then yn=ymax
    if yn<ymin then yn=ymin
    yn1=yn
```

Consider a variation of the previous example, and obtain the difference equation of the first-order process with gain k. The $(1 - a)$ factor in the numerator is required to compensate for the gain originated by the pole from $(z - a)$. For stability of the difference equation, $a < 1$:

$$H(z) = \frac{k(1-a)z}{z-a} \tag{4.31}$$

Multiply by z^{-1}:

$$H(z) = \frac{Y(z)z^{-1}}{X(z)z^{-1}} = \frac{k(1-a)}{1-az^{-1}} \tag{4.32}$$

Algebraic manipulation for a simple two polynomial side equation:

$$(1 - az^{-1})Y(z) = k(1-a)X(z) \tag{4.33}$$

$$Y(z) = az^{-1}Y(z) + k(1-a)X(z) \tag{4.34}$$

Finally, apply inverse z-transform to obtain the recursive equation:

$$y_n = a\,y_{n-1} + k(1-a)\,x_n \tag{4.35}$$

For additional practice, consider the transfer function, $G(z)$, obtain the difference equation, and implement an algorithm to obtain the response to a pulse $x_n = u_n - u_{n-10}$ from $n = 0$ to 20:

$$G(z) = \frac{1.2z}{z-0.7} \tag{4.36}$$

Multiply by z^{-1}:

$$G(z) = \frac{1.2z}{z-0.7}\left(\frac{z^{-1}}{z^{-1}}\right) = \frac{1.2}{1-0.7z^{-1}} \tag{4.37}$$

Express as two-side equation:

$$(1 - 0.7\,z^{-1})Y(z) = 1.2\,X(z) \tag{4.38}$$

Solve for $Y(z)$:

$$Y(z) = 0.7\,z^{-1}\,Y(z) + 1.2\,X(z) \tag{4.39}$$

Apply inverse z-transform

$$y_n = 0.7\,y_{n-1} + 1.2\,x_n \tag{4.40}$$

Please notice that the real gain of the system can be calculated from the factor multiplying the input signal x_n: $k(1-a) = k(1-0.7) = 1.2$; in this case, the gain value is 4:

$$k = \frac{1.2}{0.3} = 4.0 \tag{4.41}$$

Programming difference equations requires floating point variables, and bounding the input and output variables in an appropriate range is important. The pseudocode for its implementation is as follows:

```
float un=0.0, yn=0.0, a=0.7, k=4.0
float ymax=100.0, ymin=0.0
While (0<n<21) do:
        If (n>0) and (n<10) xn=1
                else xn=0
        yn=a*yn1+k*(1–a)*xn
        if (yn>ymax) then yn=ymax
        if (yn<ymin) then yn=ymin
        yn1=yn
```

4. Euler Backward Differences

In numerical analysis in scientific and engineering computing, the Euler's backward differences also called Backward Euler Method (BEM) is one of the basic tools for solving ordinary differential equations in discrete time. Although the BEM leads to a discretization with certain small numerical errors, its use is justified for most applications since any model already has a certain deviation from the real process (Arenas-Rosales et al., 2022). Euler backward differences can be applied directly in discrete time or by conversion methods in the frequency domain.

4.1 BEM Digitalization in Time-Domain

The expression for the discrete time equivalent of first-order derivative, that is, the first backward difference, considering a sampling time T, is given by:

$$\frac{dx(t)}{dt} \approx \frac{x_n - x_{n-1}}{\Delta T} = \frac{x_n - x_{n-1}}{T} \tag{4.42}$$

An ordinary differential equation (ODE) to model a first order dynamical system is:

$$\tau \frac{dy(t)}{dt} + y(t) = ku(t) \tag{4.43}$$

A straightforward procedure to obtain the difference equation is to apply the discrete first Euler backward difference approximation to the time derivative and to express variables in discrete time:

$$\tau \cdot \left(\frac{y_n - y_{n-1}}{T} \right) + y_n = ku_n \tag{4.44}$$

$$y_n - y_{n-1} + y_n \left(\frac{T}{\tau} \right) = ku_n \left(\frac{T}{\tau} \right) \tag{4.45}$$

Rearranging:

$$y_n = y_{n-1}\left(\frac{\tau}{\tau+T}\right) = ku_n\left(\frac{T}{\tau+T}\right) \tag{4.46}$$

The difference equation obtained directly with the BEM discretization can have the parameters α, τ, and T in discrete time, defining the following relations:

$$\alpha = \frac{\tau}{\tau+T} \tag{4.47}$$

$$(1-\alpha) = \frac{T}{T+\tau} \tag{4.48}$$

The discrete model can be expressed in a more compact form as:

$$y(n) = \alpha\, y_{k-1} + (1-\alpha)\ k\, u_n \tag{4.49}$$

As an example, consider an ODE with a time constant $\tau = 0.5$ s and gain $k = 1.2$:

$$0.5\frac{dy(t)}{dt} + y(t) = 1.2x(t) \tag{4.50}$$

The corresponding transfer function is:

$$G(s) = \frac{1.2}{0.5s+1} \tag{4.51}$$

The differential equation can be converted with the BEM using a sampling period of $T = 0.05$ s. The obtained difference equation is:

$$y_n = 0.9091y_{n-1} + (0.1091)1.5x_n \tag{4.52}$$

The response of the discrete-time system modeled with the previous difference equation to a pulse defined by $x_n = u_n -$ is shown in Fig. 4.4.

Consider an LTI second-order system described by the ODE:

$$\frac{dy^2(t)}{dt^2} + 2\zeta\omega_n\frac{dy(t)}{dt} + \omega_n^2 y(t) = k\omega_n^2 x(t) \tag{4.53}$$

To simplify the procedure, redefine the parameters of the equation as $a = 1$, $b = 2\zeta$, $c = \omega_n^2$:

$$a\frac{dy^2(t)}{dt^2} + b\frac{dy(t)}{dt} + cy(t) = kcx(t) \tag{4.54}$$

The expression for the discrete time equivalent of the second-order derivative, that is, the second backward difference, considering a sampling time T, is given by:

$$\frac{d^2x(t)}{dt^2} \approx \frac{x_n - 2x_{n-1} + x_{n-2}}{T^2} \tag{4.55}$$

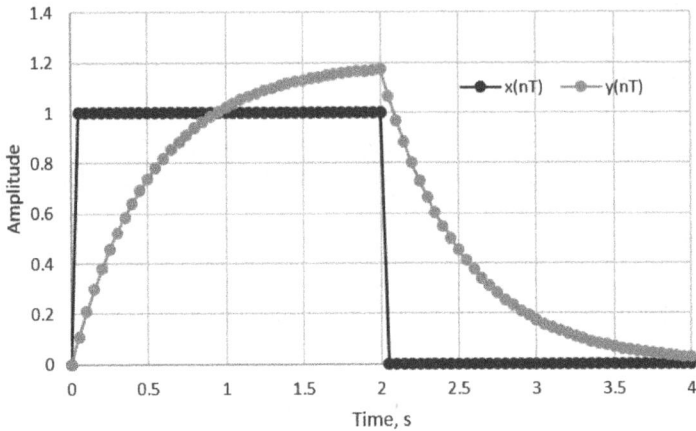

Fig. 4.4. First-order system's response to a pulse signal.

Substitution of BEM approximations for both first- and second-order derivatives into the differential equation results in:

$$a\frac{y_n - 2y_{n-1} + y_{n-2}}{T^2} + b\frac{y_n - y_{n-1}}{T} + cy_n = kcx_n \tag{4.56}$$

Multiplying by T^2:

$$a(y_n - 2y_{n-1} + y_{n-2}) + bT(y_n - y_{n-1}) + cT^2 y_n = k\,cT^2\,x_n \tag{4.57}$$

Grouping common terms:

$$(a + bT + cT^2)\,y_n - (2a + bT)\,y_{n-1} + ay_{n-2} = k\,cT^2\,x_n \tag{4.58}$$

Solving for y_n:

$$y_n = \frac{1}{a + bT + cT^2}[(2a + bT)y_{n-1} - ay_{n-2} + kcT^2x_n] \tag{4.59}$$

Substituting back the original parameters:

$$y_n = \frac{1}{1 + 2\zeta\omega_n T + \omega_n^2 T^2}[(2 + 2\zeta\omega_n 1T)y_{n-1} - y_{n-2} + k\omega_n^2 T^2 x_n] \tag{4.60}$$

The obtained difference equation can be programmed and tested for different damping and natural frequency parameters to obtain overdamped, critically damped, or underdamped (oscillatory) systems.

For example, consider a second-order ODE of a critically damped system:

$$\frac{d^2 y(t)}{dt^2} + 1.5\frac{dy(t)}{dt} + 0.5y(t) = x(t) \tag{4.61}$$

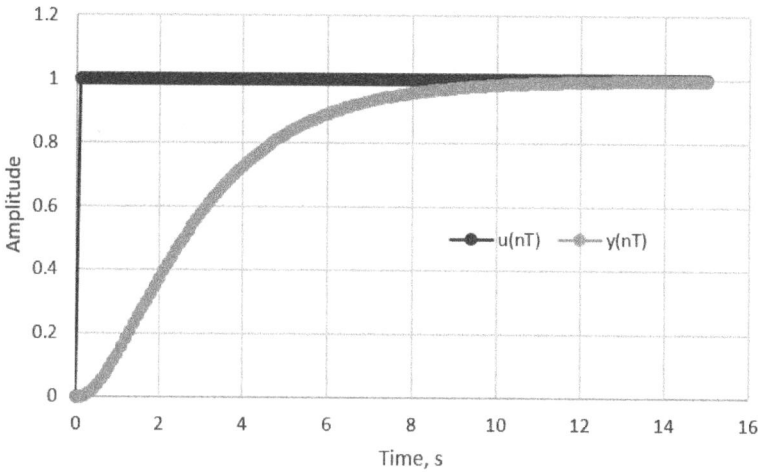

Fig. 4.5. Discrete time critically damped second order system response to a unit step.

The second-order ODE can be converted with the BEM using Eq. (4.60), identifying the parameters properly and considering a sampling period of $T = 0.1$ s. The obtained difference equation is:

$$y_n = \frac{1}{1+1.5(0.1)+0.5(0.1)^2}[(2+1.5(0.1))y_{n-1} - y_{n-2} + 0.5(0.1)^2 x_n] \tag{4.62}$$

$$y_n = \frac{1}{1.155}[2.15y_{n-1} - y_{n-2} + 0.005x_n] \tag{4.63}$$

The response of the discrete-time system modeled with the previous difference equation to a step signal, defined by $x_n = u_n$, is shown in Fig. 4.5.

Another example consists of an underdamped second-order system:

$$\frac{d^2 y(t)}{dt^2} + 0.5\frac{dy(t)}{dt} + 0.5y(t) = x(t) \tag{4.64}$$

The second ordinary differential equation can be converted as in the previous example by identifying the parameters properly and considering a sampling period of $T = 0.1$ s:

$$y_n = \frac{1}{1+0.5(0.1)+0.5(0.1)^2}[(2+0.5(0.1))y_{n-1} - y_{n-2} + 0.5(0.1)^2 x_n] \tag{4.65}$$

$$y_n = \frac{1}{1.055}[2.05y_{n-1} - y_{n-2} + 0.005x_n] \tag{4.66}$$

The response of the discrete-time system modeled with the previous difference equation to the same step signal, defined by $x_n = u_n$, is shown in Fig. 4.6.

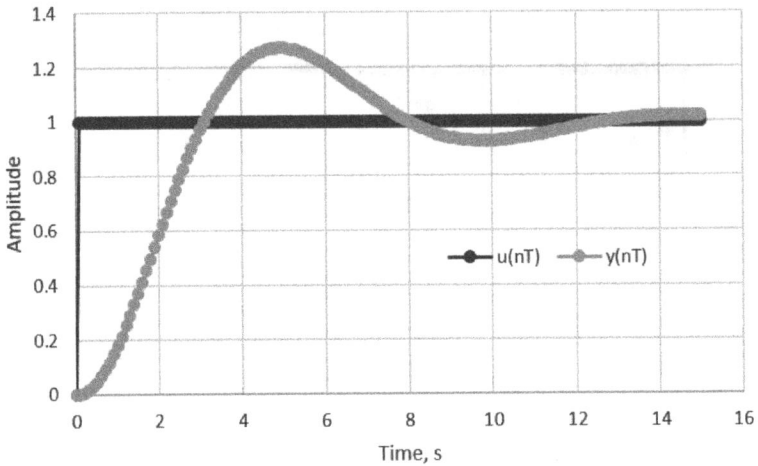

Fig. 4.6. Discrete time underdamped second order system response to a unit step.

4.2 BEM Digitalization in Frequency Domain

An alternative procedure to obtain a difference equation that can be implemented computationally is to apply the conversion in the frequency domain.

Apply z-transform to the difference equation obtained by the Euler backward difference approximation:

$$\frac{dy(t)}{dt} \approx \left(\frac{y_n - y_{n-1}}{T} \right) \tag{4.67}$$

Applying the Laplace and the z-transforms on the expressions on each side of the above equation separately produces:

$$\frac{dy(t)}{dt} \overset{L}{\leftrightarrow} sY(z) \tag{4.68}$$

$$\frac{y_n - y_{n-1}}{T} \overset{z}{\leftrightarrow} \left(\frac{Y(z) - z^{-1}Y(z)}{T} \right) = \left(\frac{1 - z^{-1}}{T} \right) Y(z) \tag{4.69}$$

The factors of the right-hand side members of the previous two equations can be matched. The BEM transformation from s-domain to z-domain leads to:

$$s \approx \frac{(1 - z^{-1})}{T} \tag{4.70}$$

The transfer function in the z-domain can be algebraically obtained by substituting Eq. (4.70) in the transfer function in the s-domain:

$$H(z) = H(s)\Big|_{s = \frac{(1 - z^{-1})}{T}} \tag{4.71}$$

As an example, consider the transfer function in the s-domain (analog frequency):

$$H(s) = \frac{k}{\tau s + 1} \tag{4.72}$$

Apply the BEM equivalence:

$$H(z) = \frac{k}{\tau\left(\dfrac{1-z^{-1}}{T}\right)+1} = \frac{kT}{\tau(1-z^{-1})+T} = \frac{kT}{(\tau+T)-\tau z^{-1}} \tag{4.73}$$

Divide by $(\tau + T)$:

$$H(z) = \frac{Y(z)}{X(z)} = \frac{k\left(\dfrac{T}{\tau+T}\right)}{1-\left(\dfrac{\tau}{\tau+T}\right)z^{-1}} \tag{4.74}$$

Rearrange and solve for $Y(z)$:

$$\left[1-\left(\frac{\tau}{\tau+T}\right)z^{-1}\right]Y(z) = k\left(\frac{T}{\tau+T}\right)X(z) \tag{4.75}$$

$$Y(z) = \left(\frac{\tau}{\tau+T}\right)z^{-1}Y(z) + k\left(\frac{T}{\tau+T}\right)X(z) \tag{4.76}$$

Apply the inverse z-transform:

$$y_n = \left(\frac{\tau}{\tau+T}\right)y_{n-1} + k\left(\frac{T}{\tau+T}\right)x_n \tag{4.77}$$

The result is equivalent to the direct conversion presented in the time domain, Eq. (4.46). Gain, time constant, and sampling period must be appropriately selected for a stable difference equation.

Now consider the transfer function:

$$G(s) = \frac{3}{s+2} \tag{4.78}$$

Express the gain and time constant parameters explicitly:

$$G(s) = \left(\frac{3}{s+2}\right)\Big/2 = \frac{1.5}{0.5s+1} \tag{4.79}$$

Use Eq. (4.46) with the desired sample time of $T = 0.1$ s:

$$y_n = \frac{0.5}{(0.5+0.1)}y_{n-1} + 1.5\frac{0.1}{(0.5+0.1)}x_n \tag{4.80}$$

$$y_n = 0.8333y_{n-1} + 0.25x_n \tag{4.81}$$

In difference equations, using at least four precision digits is convenient to obtain more reliable calculations and favor convergence to steady state numerical values.

5. Bilinear Transform

The bilinear transform method (BLT) is commonly used in digital signal processing and digital control. This mathematical method is often used to transform the continuous-time representation of signals or systems into discrete time and vice versa.

The BLT performs a mapping between the s-plane and z-plane according to Eq. (4.82), where T is the sampling period of the discrete time control system under consideration:

$$s = \frac{2}{T}\frac{(z-1)}{(z+1)} = \frac{2}{T}\frac{(1-z^{-1})}{(1+z^{-1})} \tag{4.82}$$

5.1 BLT Mapping

Transfer functions conversion from $H(s)$ to $H(z)$ by the BLT method is the most common transform; the converted transfer function in (z-domain) can be algebraically obtained by:

$$H(z) = H(s)\big|_{s=\frac{2(z-1)}{T(z+1)}} \tag{4.83}$$

The selection of the sample time is important in the BLT. Theoretically, the sampling frequency must be at least double the cutoff frequency for first-order systems, $\omega_s = 2\omega_c$, which implies a sample period $T = \tau/2$. Similarly, for a second-order frequency, double the natural frequency is the minimum sampling frequency, $\omega_s = 2\omega_n$. In practice, it is desirable to have a sample time with at least one order of magnitude less than the time constant, that is, $T = \tau/10$, which means that ten samples would be collected before reaching a time equal to τ, for instance, in the step response of the system.

The order of the transfer function in the analog complex frequency domain will be the order of the transfer function in the digital complex frequency domain. BLT is a nonlinear transformation that maps the entire s-plane into the unit circle in the z-plane, see Fig. 4.7 (Ogata, 1996). The stability of the discrete-time systems

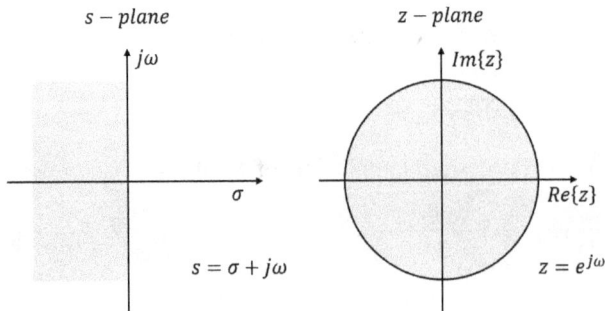

Fig. 4.7. BLT s to z mapping.

z – plane

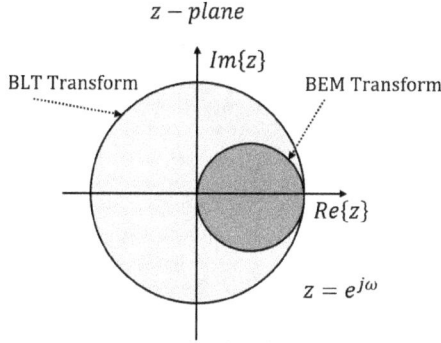

Fig. 4.8. BLT and BEM mapping areas.

is related to the roots of the polynomial denominator of *H*; *all* the roots need to be inside the unit circle.

In the complex plane, the unit circle mapping achieved by BLT is reduced or shrunk to a "half size" circle by the BEM-based transform, as shown in Fig. 4.8.

5.2 BLT Conversion Procedure

Consider the conversion of a first order transfer function with gain, *k*, and time-constant parameter *τ*,

$$H(s) = \frac{k}{\tau s + 1} \tag{4.84}$$

Substitute $s = \frac{2}{T} \frac{(z-1)}{(z+1)}$ into $H(s)$ to obtain $H(z)$:

$$H(z) = \frac{k}{\tau \left(\frac{2(z-1)}{T(z+1)} \right) + 1} \tag{4.85}$$

Multiply by $T(z + 1)$:

$$H(z) = \frac{kT(z+1)}{2\tau(z-1) + T(z+1)} = \frac{kT(z+1)}{(T+2\tau)z + (T-2\tau)} \tag{4.86}$$

Multiply by z^{-1}:

$$H(z) = \frac{Y(z)}{X(z)} = \frac{kT(1+z^{-1})}{(T+2\tau) + (T-2\tau)z^{-1}} \tag{4.87}$$

Express the result as a two polynomial side equation:

$$[(T+2\tau) + (T-2\tau)z^{-1}]Y(z) = k\,T(1+z^{-1})X(z) \tag{4.88}$$

Rearrange:

$$(T+2\tau)Y(z) = (2\tau - T)z^{-1}\,Y(z) + k\,T(1+z^{-1})X(z) \tag{4.89}$$

$$Y(z) = \frac{(2\tau - T)}{(2\tau + T)} z^{-1} Y(z) + \frac{kT}{(2\tau + T)}(1 + z^{-1})X(z) \tag{4.90}$$

Apply the inverse z-transform:

$$y_n = \frac{(2\tau - T)}{(2\tau + T)} y_{n-1} + \frac{kT}{(2\tau + T)}(x_n + x_{n-1}) \tag{4.91}$$

Notice that the BLT requires two input signal samples, gain time constant, and sampling period parameters. The sample time must be appropriately selected to obtain a stable difference equation.

This previous example can be reconsidered for the application of the BLT. The transfer function is:

$$G(s) = \frac{3}{s+2} \tag{4.92}$$

The parameters can be distinguished by a simple algebraic manipulation:

$$G(s) = \left(\frac{3}{s+2} \right) \Big/ 2 = \frac{1.5}{0.5s + 2} \tag{4.93}$$

Use Eq. (4.91) with the desired sample time $T = 0.1$:

$$y_n = \frac{(2(0.5) - 0.1)}{(2(0.5) + 0.1)} y_{n-1} + \frac{1.5(0.1)}{(2(0.5) + 0.1)}(x_n + x_{n-1}) \tag{4.94}$$

$$y_n = \frac{0.9}{1.1} y_{n-1} + \frac{0.15}{1.1}(x_n + x_{n-1}) \tag{4.95}$$

$$y_n = 0.8182 y_{n-1} + 0.1367(x_n + x_{n-1}) \tag{4.96}$$

Compare the obtained difference equations with both methods: the numerical coefficients obtained by BLT, Eq. (4.96), and with the previous result of BEM, Eq. (4.81) are close; both of them can reproduce numerically the behavior of a discretized first-order system corresponding to Eq. (4.92). The slight difference is compared below with the study of an alternative transformation.

5.3 Hybrid Transformation

The Hybrid transformation is a combination of BEM and BLT that is possible with the introduction of the α parameter to 'weight' the contribution of the BEM or BLT:

$$s = \frac{(1 + \alpha)}{T} \frac{(1 - z^{-1})}{(1 + \alpha z^{-1})} \tag{4.97}$$

With $\alpha = 1$, the BLT is obtained:

$$s = \frac{2}{T} \frac{1 - z^{-1}}{1 + z^{-1}} \tag{4.98}$$

With $\alpha = 0$, the BEM is obtained:

$$s = \frac{1 - z^{-1}}{T} \tag{4.99}$$

Let us consider the system:

$$H(s) = \frac{k}{\tau s + 1} \tag{4.100}$$

Obtain the transfer function in the z domain:

$$H(z) = H(s)\Big|_{s = \frac{(1+\alpha)(1-z^{-1})}{T(1+\alpha z^{-1})}} \tag{4.101}$$

$$H(z) = \frac{k}{\tau \left(\dfrac{(1+\alpha)(1-z^{-1})}{T(1+\alpha z^{-1})} \right) + 1} \tag{4.102}$$

Perform algebraic operations:

$$H(z) = \frac{kT(1+\alpha z^{-1})}{\tau(1+\alpha)(1-z^{-1}) + T(1+\alpha z^{-1})} \tag{4.103}$$

$$H(z) = \frac{Y(z)}{X(z)} = \frac{kT(1+\alpha z^{-1})}{\tau(1+\alpha)(1-z^{-1}) + T(1+\alpha z^{-1})} \tag{4.104}$$

Rearrange the equation:

$$[\tau(1 + \alpha)(1 - z^{-1}) + T(1 + \alpha z^{-1})]Y(z) = k\,T(1 + \alpha z^{-1})X(z) \tag{4.105}$$

$$[(\tau(1 + \alpha) + T) + (-\tau(1 + \alpha) + \alpha T)\,z^{-1}]Y(z) = k\,T(1 + \alpha z^{-1})X(z) \tag{4.106}$$

$$[(\tau(1 + \alpha) + T)]Y(z) = (\tau(1 + \alpha) - \alpha T)\,z^{-1}\,Y(z) + k\,T(1 + \alpha z^{-1})X(z) \tag{4.107}$$

Obtain inverse z-transform:

$$[(\tau(1 + \alpha) + T)]\,y_n = [(\tau(1 + \alpha) - \alpha T)]\,y_{n-1} + k\,T(x_n + \alpha x_{n-1}) \tag{4.108}$$

Solve for y_n:

$$y_n = \frac{\tau(1+\alpha) - \alpha T}{\tau(1+\alpha) + T}\,y_{n-1} + \frac{kT}{\tau(1+\alpha) + T}(x_n + \alpha x_{n-1}) \tag{4.109}$$

Now, evaluate in $\alpha = 0$ to verify equivalence with BEM:

$$y_n = \frac{\tau(1+0) - 0T}{\tau(1+0) + T}\,y_{n-1} + \frac{kT}{\tau(1+0) + T}(x_n + 0x_{n-1}) \tag{4.110}$$

$$y_n = \frac{\tau}{\tau + T}\,y_{n-1} + \frac{kT}{\tau + T}\,x_n \tag{4.111}$$

The previous result coincides with the difference equation obtained with BEM.

Finally, to verify equivalence with BLT, evaluate in $\alpha = 1$:

$$y_n = \frac{\tau(1+1)-1T}{\tau(1+1)+T}y_{n-1} + \frac{kT}{\tau(1+1)+T}(x_n + 1x_{n-1}) \qquad (4.112)$$

$$y_n = \frac{2\tau - T}{2\tau + T}y_{n-1} + \frac{kT}{2\tau + T}(x_n + x_{n-1}) \qquad (4.113)$$

The previous result coincides with the differences equation obtained with the BLT method.

5.4 Comparison of Discretization Methods

The BEM and BLT lead to slightly different difference equations (Arenas-Rosales et al., 2022). To study more in detail, the discretization methods, a comparison can be made between them and the Hybrid transformation (HT). Conversions of a first-order system with unit gain and time constant ($k = 1$, $\tau = 1$) are performed using a sampling period $T = \tau/10$, for $n = 1$ to 20 and compared with the step response in discrete time. The exact solution is:

$$y(n) = 1 - e^{-nT/\tau} \qquad (4.114)$$

Table 4.1 shows the obtained values for the numerical evaluation of the exact solution, the BEM, BLT, and HT with $\alpha = 0.5$ conversion methods.

The error obtained with the three conversion methods (BEM, BLT, and HT) versus the exact solution are shown in Fig. 4.9. For time values $t < 1$, the BEM has a smaller error compared to the BLT method, but for time values $t > 1$, the BLT method gets the lower error.

The cumulative error obtained with the three conversion methods (BEM, BLT, and HT) versus the exact solution is shown in Fig. 4.10. It can be seen that in the transient response, the BLT accumulates a greater value than the BEM but in the long term, cumulative errors tend to converge.

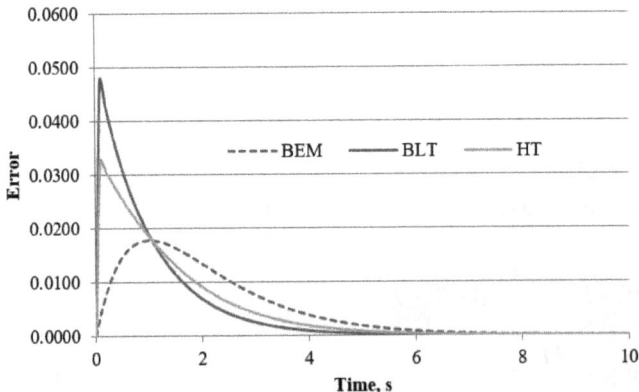

Fig. 4.9. Error comparison.

Table 4.1. Numerical step response obtained with different conversion methods.

nT	BEM	BLT	HT
0	0	0	0
0.1	0.0909	0.0476	0.0625
0.2	0.1736	0.1383	0.1504
0.3	0.2487	0.2204	0.2300
0.4	0.3170	0.2946	0.3022
0.5	0.3791	0.3618	0.3676
0.6	0.4355	0.4226	0.4269
0.7	0.4868	0.4776	0.4807
0.8	0.5335	0.5273	0.5293
0.9	0.5759	0.5724	0.5735
1.0	0.6145	0.6131	0.6135
1.1	0.6495	0.6499	0.6497
1.2	0.6814	0.6833	0.6825
1.3	0.7103	0.7134	0.7123
1.4	0.7367	0.7407	0.7393
1.5	0.7606	0.7654	0.7637
1.6	0.7824	0.7878	0.7859
1.7	0.8022	0.8080	0.8059
1.8	0.8201	0.8263	0.8241
1.9	0.8365	0.8428	0.8406
2.0	0.8514	0.8578	0.8556
2.1	0.8649	0.8713	0.8691
2.2	0.8772	0.8836	0.8814
2.3	0.8883	0.8947	0.8925
2.4	0.8985	0.9047	0.9026
2.5	0.9077	0.9138	0.9117
2.6	0.9161	0.9220	0.9200
2.7	0.9237	0.9294	0.9275
2.8	0.9307	0.9361	0.9343
2.9	0.9370	0.9422	0.9404
3.0	0.9427	0.9477	0.9460
3.1	0.9479	0.9527	0.9511
3.2	0.9526	0.9572	0.9557
3.3	0.9569	0.9613	0.9598
3.4	0.9609	0.9650	0.9636
3.5	0.9644	0.9683	0.9670

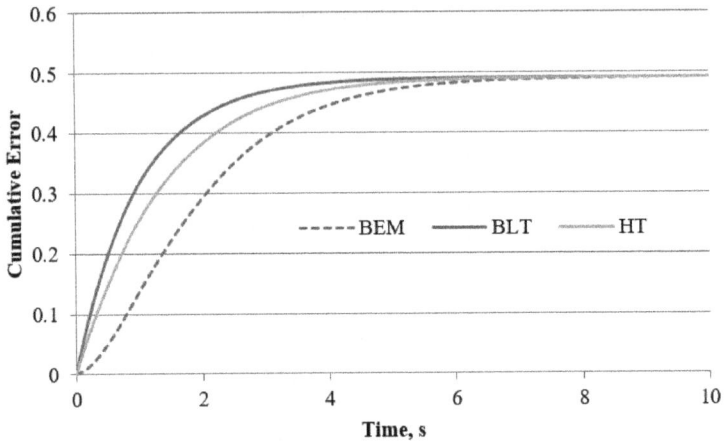

Fig. 4.10. Cumulative error comparison.

In summary, the several conversion methods have slight differences but are helpful as long as stable difference equations are obtained.

6. Practical Digital Filtering

Digital filters are a very important part of modern instrumentation and control systems. In instrumentation systems, a signal needs to be filtered when a measurement from a sensor has been contaminated with electromagnetic interference or any other type of noise. In digital signal processing, signals may also require to be filtered if they have been distorted somehow. In control systems, more straightforward control techniques can be implemented if a digital filter is specified; this is, complex control techniques may be avoided by using good filtering practices in both signals measured from sensors and signals sent to actuators or final control elements.

Filters allow to pass some frequencies while completely suppressing other frequencies. The filter bandwidth refers to those frequencies that are passed. Low-pass filters are the most widely used and practical type of filters. Real signals have finite bandwidth, and, in most cases, noise with higher frequencies is induced in the measurement system, and digital filtering is required to be implemented in the computational platform. Computational implementation of digital filters requires difference equations that can be obtained from transfer functions directly defined in the z-domain or from the transformation of transfer functions in the Laplace domain.

6.1 Finite and Infinite Impulse Response Filters

A filter can be represented by block diagrams, which can be used as references to program a computational algorithm in a computer. A filter may also be described as a difference equation or transfer function with defined zeros and poles or can also be specified with the impulse response or step response. The Finite Impulse Response (FIR) filters are linear discrete-time systems that are always stable. There are no

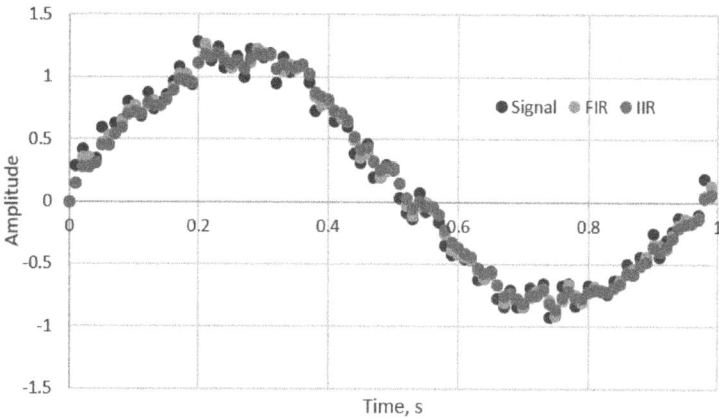

Fig. 4.11. FIR and IIR comparison.

poles in these systems. Each sample in the output is calculated by weighting the samples in the input and adding them together; an example is the moving average:

$$y_n = \frac{1}{3}(x_{n-1} + x_{n-2} + x_{n-3}) \tag{4.115}$$

The Infinite Impulse Response (IIR) filters are a type of linear discrete-time systems. This type of recursive filters uses previously calculated values from the outputs and one or more samples from the input. An IIR is only stable if poles are inside the unit circle or an unstable pole is canceled by a zero of equal value.

$$y_n = b_1\, y_{n-1} + b_2\, y_{n-2} + \cdots + b_N\, y_{n-N} + a_0\, x_n + a_1\, x_{n-1} + \cdots + a_N\, x_{n-N} \tag{4.116}$$

Consider as an example a sine signal with random noise. Then two filters are designed. An FIR filter of two terms is given next and weighs the actual and previous inputs equally (50%):

$$y_n = \frac{1}{2}(x_n + x_{n-1}) \tag{4.117}$$

An IIR filter that is recursive considers a 50% weight for the previous output and a 50% weight for the input as follows:

$$y_n = 0.5\, y_{n-1} + 0.5\, x_n \tag{4.118}$$

Figure 4.11 shows the sine signal with added noise and the performance of the FIR and IRR filters. Both filters suppress some noise and, in some way, reconstruct the sine wave.

6.2 Basic Digital Filter Design

A practical method to design an IIR filter uses transfer functions of continuous-time systems as a model or prototype filter, called the analog filter. The natural or cutoff frequency of the analog filter is selected and represented in the first or second order corresponding s-function. The next step to digitalize that filter is to apply some transformations: first from the s-domain to the z-domain and then from the z-domain to the discrete-time domain. To discretize the continuous transfer functions, several methods can be used. The bilinear transform is the discretization method often used in designing a digital filter to convert a transfer function in the s-domain to the z-domain, from which the difference equation can be obtained.

The bilinear transform is an approximation and therefore introduces some errors in the discretization process. The magnitude of the errors depends upon the sampling frequency. Very high sampling frequency reduces the discretization error. In real systems, where the sampling frequency depends on the processing unit or CPU and the analog to digital signal conversion process, high sampling frequencies cannot be achieved (unless expensive DSP or DAQ are used). This is mainly a problem in PLC and other control units where implementing the digital filter introduces a significant error. In the case of the bilinear transform, the most important introduced error is due to frequency warping. This effect produces a shift in the natural frequency of the filter due to the sampling process. This effect may be negligible in most cases but can be critical for some applications.

The frequency warping occurs because all the analog frequency range ($-\infty$ to $+\infty$) is mapped to the discrete frequency range ($-\pi$ to $+\pi$) due to sampling frequency. The units of the analog frequency, ω_a, are radians per second, and the units of the discrete frequency, ω_d, are radians per sample. Therefore, the sample time needs to be used in the following expression to obtain an analog cutoff frequency equivalent to a digital cutoff frequency:

$$\omega_a = \frac{2}{T} tan\left(\frac{\omega_d T}{2}\right) \tag{4.119}$$

And vice versa, the expression to obtain a digital cutoff frequency equivalent to an analog cutoff frequency is:

$$\omega_d = \frac{2}{T} arctan\left(\frac{\omega_a T}{2}\right) \tag{4.120}$$

Figure 4.12 shows the frequency warping. There is a fairly linear range from ($-\pi/2$ to $+\pi/2$), but a highly nonlinear behavior from $-\pi$ to $\pi/2$ and from $\pi/2$ to π.

The frequency response of digital filters obtained from converted transfer functions in the s-domain can be corrected by a procedure called frequency pre-warping. This procedure adjusts the analog filter frequency before the discretization through the following design steps:

1. Determine the digital cutoff frequency specification
2. Pre-warp the cutoff frequency (or the cutoff frequencies)
3. Design the necessary analog filter, that is, perform the filter conversion with a low pass filter prototype
4. Convert the transfer function using bilinear transform
5. Obtain the difference equations from the digital filter transfer function.

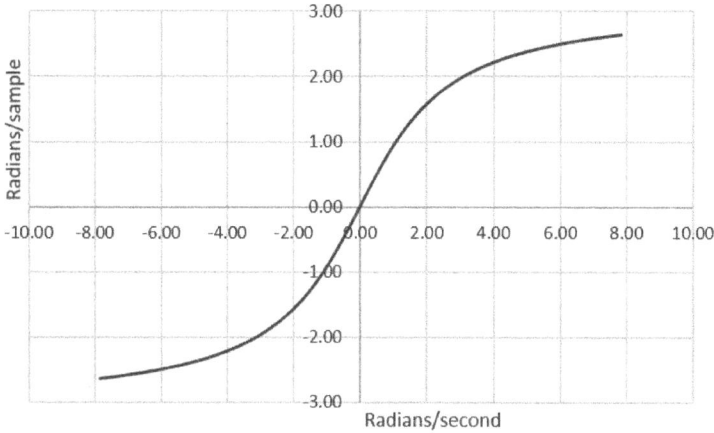

Fig. 4.12. Frequency warping.

To illustrate this procedure, assume that a digital filter is required to suppress the noise of a sensor signal. The sensor has a frequency response with a cutoff frequency of 50 rad/s. The sampling period that can be implemented in the electronic control unit is 10 ms. Following the design procedure:

1. *Specify digital cutoff frequency.* The digital filter cutoff frequency, ω_d, must be higher than the cutoff frequency of the sensor, i.e., two times: $\omega_c = 2 * 50 = 100$ rad/s.

2. *Pre-warp frequency:* Calculate the analog frequency corresponding to the desired digital frequency by applying the frequency warping with $T = 0.01$s:

$$\omega_a = \frac{2}{T} tan\left(\frac{\omega_d T}{2}\right) = \frac{2}{0.01} tan\left(\frac{100 * 0.01}{2}\right) = 200 tan(0.5) = 109.26 \qquad (4.121)$$

3. *Analog Filter Design:* Use a low pass analog filter prototype with a cutoff frequency of 1 rad/s, and convert the filter transfer function to the desired cutoff frequency $\omega_a = 109.26$:

$$H_p(s) = \frac{1}{s+1} \qquad (4.122)$$

$$H_a(s) = H_p(s)\big|_{s=s/\omega_a} \qquad (4.123)$$

$$H_a(s) = \frac{1}{\dfrac{s}{109.26}+1} = \frac{109.26}{s+109.26} \qquad (4.124)$$

4. Transform the analog filter to a digital filter by applying the bilinear transform:

$$H_d(z) = H_a(s)\big|_{s=\frac{2(z-1)}{T(z+1)}} \qquad (4.125)$$

$$H_d(z) = \frac{109.26}{\frac{2(z-1)}{T(z+1)}+109.26} = \frac{109.26T(z+1)}{2(z-1)+109.26T(z+1)} \tag{4.126}$$

Multiply by z^{-1}:

$$H_d(z) = \frac{109.26(0.01)(1+z^{-1})}{2(1-z^{-1})+109.26(0.01)(1+z^{-1})} = \frac{1.0926(1+z^{-1})}{2(1-z^{-1})+1.0926(1+z^{-1})} \tag{4.127}$$

$$H_d(z) = \frac{Y(z)}{X(z)} = \frac{1.0926(1+z^{-1})}{(2+1.0926)-(2-1.0926)z^{-1}} \tag{4.128}$$

$$H_d(z) = \frac{Y(z)}{X(z)} = \frac{1.0926(1+z^{-1})}{3.0926-0.9074z^{-1}} \tag{4.129}$$

5. Obtain the difference equation from the digital filter transfer function:

$$[3.0926 - 0.9074z^{-1}]Y(z) = 1.0926\,(1+z^{-1})X(z) \tag{4.130}$$

Solve for $Y(z)$:

$$3.0926\,Y(z) = 0.9074\,z^{-1}\,Y(z) + 1.0926\,(1+z^{-1})X(z) \tag{4.131}$$

$$Y(z) = \frac{0.9074}{3.0926}z^{-1}Y(z) + \frac{1.0926}{3.0926}(1+z^{-1})X(z) \tag{4.132}$$

Apply inverse z-transform:

$$y_n = \frac{0.9074}{3.0926}y_{n-1} + \frac{1.0926}{3.0926}(x_n + x_{n-1}) \tag{4.133}$$

Finally, the difference equation of the digital filter with pre-warping is:

$$y_n = 0.2934\,y_{n-1} + 0.3533\,(x_n + x_{n-1}) \tag{4.134}$$

Without the frequency pre-warping, the digital filter, if obtained with the analog frequency of 100 rad/s:

$$H_d(z) = \frac{100}{\frac{2(z-1)}{T(z+1)}+100} = \frac{100T(z+1)}{2(z-1)+100T(z+1)} \tag{4.135}$$

$$H_d(z) = \frac{Y(z)}{X(z)} = \frac{(1+z^{-1})}{(2+1)-(2-1)z^{-1}} = \frac{(1+z^{-1})}{3-z^{-1}} \tag{4.136}$$

$$(3 - z^{-1})Y(z) = (1 + z^{-1})X(z) \tag{4.137}$$

Solve for $Y(z)$:

$$Y(z) = \frac{1}{3}z^{-1}Y(z) + \frac{1}{3}(1+z^{-1})X(z) \tag{4.138}$$

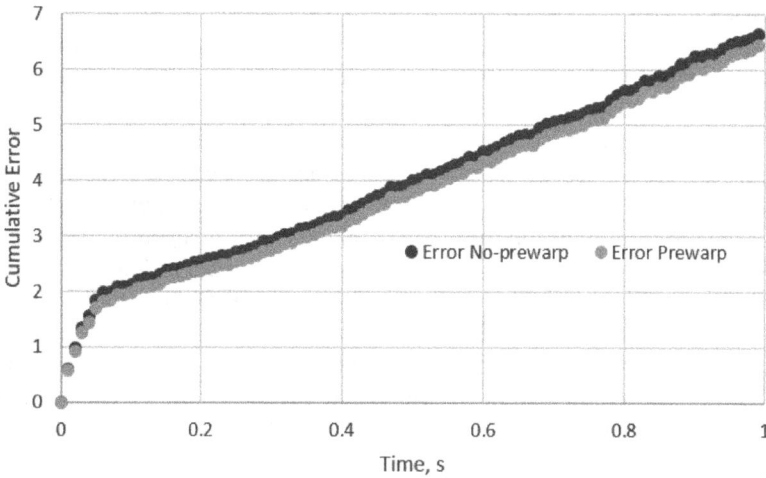

Fig. 4.13. Frequency warping comparison.

Apply inverse z-transform to obtain the difference equation of the digital filter:

$$y_n = 0.3333\, y_{n-1} + 0.3333\ (x_n + x_{n-1}) \tag{4.139}$$

Figure 4.13 shows the cumulative error of the filtered output with and without frequency warping versus the input signal. As can be seen, there is a little more precision when frequency warping is considered.

In summary, from the applied procedure, we obtain a first-order filter with a pre-warped analog cut off frequency $\omega_a = 109.26$, that, when converted to a digital filter using bilinear transform, gives a transfer function with the specified cut off frequency of $\omega_d = 100.00$, and then can be converted to a difference equation for its computational implementation in the control unit.

7. Conclusions

Discrete time linear time invariant (LTI) systems are of particular interest since the properties of linearity and time invariance allow using the difference equations and their computational implementation as an important and powerful tool in digital control and signal processing. Linear constant coefficient difference equations are discrete-time equivalents of differential equations in continuous time and reproduce the response of dynamical systems; these types of discrete-time systems are particularly useful for modeling and computer simulations.

Discrete transfer functions in the z-domain are practical since they are mathematical expressions used to model dynamical systems, system identification, controller design, and filter design. Discrete transfer functions are widely applied in digital signal processing to design digital filters. Digital controllers can be directly designed in discrete time as difference equations or as a transfer function in the complex frequency. The computer simulation of a mathematical model requires the conversion of the differential equations into difference equations. The conversion can be done directly in the time domain or indirectly in the frequency domain. For direct conversion in the time domain, the Euler backward differences can be used, and for indirect conversion, the bilinear transform or Tustin method is the most practical. This last method implies converting the transfer function in continuous frequency to a transfer function in discrete frequency and then to a difference equation.

Digital filters are required in measurement systems and for data acquisition to filter sensor signals. They are also desired to be incorporated into control systems to enhance the performance of the closed loop controllers. The sampled input signal in a microprocessor or microcontroller can be filtered with an algorithm derived from a difference equation. One of the most common approaches to filtering noise is to program an algorithm to serve as a low pass filter to allow the frequency components that are considered to convey valid information and to cut off any frequencies above a certain design frequency. Analog to digital filter conversion is subject to frequency warping; the design of digital filters with higher precision must be compensated by frequency pre-warping.

References

Arenas-Rosales, F., Martell-Chávez, F., Sánchez-Chávez, I.Y., López-Padilla, R. and Valentín-Coronado, L.M. 2022. Comparison of Euler's backward difference and bilinear transform discretization methods for modeling and simulation of a DC motor. *In*: Flores Rodríguez, K.L., Ramos Alvarado, R., Barati, M., Segovia Tagle, V. and Velázquez González, R.S. (eds.). *Recent Trends in Sustainable Engineering. ICASAT 2021. Lecture Notes in Networks and Systems, 297*. Springer, Cham. https://doi.org/10.1007/978-3-030-82064-0_5.

Bishop, R.H. 2007. *The Mechatronics Handbook*. Second Edition. CRC Press.

Hsu, H.P. 2011. *Signals and Systems*. Second Edition. McGraw-Hill.

Ogata, K. 1996. *Sistemas de Control en Tiempo Discreto*. México: Prentice Hall Hispanoamericana.

Oppenheim, A.V., Willsky, A.S. and Nawab, S.H. 1997. *Signals and Systems*. Second Edition. Prentice Hall.

Chapter 5
System Identification

1. Introduction

System identification is the process of extracting and processing information for modeling a system with measurements of input and output data. An identification model may be static or dynamic, linear or nonlinear, deterministic or stochastic, single input single output (SISO), or multivariable (Zhu, 2001). The generated model is used for system analysis, computer simulation, or controller design. A linear system modeling obtained from system identification is of particular interest because a control system design should part from the knowledge and representation of the dynamics of the process, that is, the effect of manipulation and disturbance inputs on the process variable outputs, in terms of sensibility and speed of response, is the basis for devising control schemes.

Physical, chemical, and biological principles can be applied to state the process behavior mathematically. However, important assumptions and simplifications are necessary for the resultant theoretical models to achieve a balance between accuracy and complexity, for example, uniform or constant properties, ideal behaviors, etc. On the other hand, an experimental procedure can be used to approximate a model of a predefined practical linear structure. This procedure is designated as process identification, and the use of experimental data provides a real basis for characterizing the dynamics. Linear mathematical models can be restricted to operation conditions, but combining these models can be a strategy to approach the nonlinear nature of the processes.

This chapter presents identification methods in continuous and discrete time for producing linear models. In Section 2, the system identification problem is briefly explained in terms of the convenience of specifying a model from experimental measurements of the system inputs and outputs. Section 3 on continuous-time identification reviews the step response, the model selection, and the process identification. Section 4 on discrete-time identification explains the basic approach of an experimental method for parameter estimation of ARX models for first- and second-order systems. Conclusions are discussed in Section 5.

2. System Identification Problem

The system identification problem is to estimate a model of a system based on input-output data. Identification methods require testing the system by applying input signals and measuring outputs with data acquisition systems. In control engineering, the identification process can be done by approaching continuous-time models to measure data and using discrete-time data to obtain discrete-time models computationally. System identification methods are divided into two groups: parametric and nonparametric. Nonparametric methods model a system directly assuming some particular system dynamics whose parameters specifically describe a characteristic or dynamic effect. Parametric methods identify the system model with an underlying general mathematical structure associated with a set of non-descriptive coefficients or parameters that numerically fit the system response.

Figure 5.1 shows the identification problem and the related signals. If we assume the system is linear, in continuous time, the following function can be stated:

$$Y(s) = G(s)U(s) + V(s) \tag{5.1}$$

In discrete time, there is an input sampled signal: $u(k) = \{u(0), u(1), ..., u(k), ..., u(N)\}$ and an output sampled signal: $y(k) = \{y(0), y(1), ..., y(k), ..., y(N)\}$. The corresponding discrete function is:

$$Y(z) = G(z)U(z) + V(z) \tag{5.2}$$

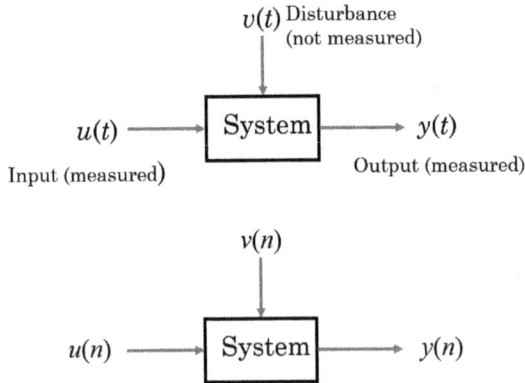

Fig. 5.1. Identification in continuous time (upper) and discrete time (lower).

3. Continuous-Time Identification

Identification of dynamical systems for engineering applications using linear systems theory in continuous time is referred to the specification of transfer functions in the s-domain or differential equation in the time domain (Dorsey, 2005). A step input is the simplest way to excite a process to observe its dynamic response. The process output may exhibit a non-oscillatory behavior, which can be approximated by a first-order continuous transfer function. In the case of an oscillatory response, an underdamped second-order system would be the simplest model to fit the data. The obtained continuous models can be discretized for computer simulation or digital control design.

Applying the step input requires an initial steady state and the absence of disturbances throughout the experiment; that is, only one input change is analyzed at a time. The approximation of transfer functions requires the use of deviation values, which means that initial nonzero values should be subtracted from the corresponding data to perform calculations. Time measurements should be made from the step application time; that is, "time zero" is the time instant when the input has acquired a new value, changing from its initial steady state value. Collected data should clearly show the initial steady state for a precise determination of initial values and initial time and the final steady state for the observation of the final value of the response.

3.1 Identification of First-Order Systems

Three different methods are presented for the identification of non-oscillatory processes. All of them are based on the step-time response of a first-order system:

$$y(t) = y(0) + k\Delta x\left(1 - e^{-(t-t_0)/\tau}\right), \ t \geq t_0 \tag{5.3}$$

where $y(t)$ is the process output or response as a function of time t, $y(0)$ is the initial steady state response value, k is the process gain, Δx is the step amplitude in the input signal x, t_0 is the dead time, and τ is the time constant of the process.

The process gain, k, is calculated in the three methods as the ratio of the total process variable change, Δy, over the total manipulation variable change, Δx. The total or final change in the input or output variable is the difference between the final and the initial values, indicated by subscripts i and f:

$$k = \frac{\Delta y}{\Delta x} = \frac{y_f - y_i}{x_f - x_i} \tag{5.4}$$

Ziegler and Nichols Method

The dead time and time constant parameters of the first order transfer function are determined graphically in the method proposed by Ziegler and Nichols (Fig. 5.2), according to the following steps:

1. The step experiment data are plotted, and the reading references are marked: horizontal lines for the initial and the final steady-state response values (base and saturation lines, respectively) and the vertical line for the initial time or time 'zero' of the test.

2. A line is traced tangent to the response curve at the point where the magnitude of the slope is maximum. This maximum response rate magnitude usually corresponds to an inflection point, in the case of a change in the curve's concavity, or to the point where the process variable starts reacting to the applied step, in case the response always has the same concavity direction. The tangent line has to be projected at both ends to intersect the baseline and the saturation line (or horizontal response reference lines).

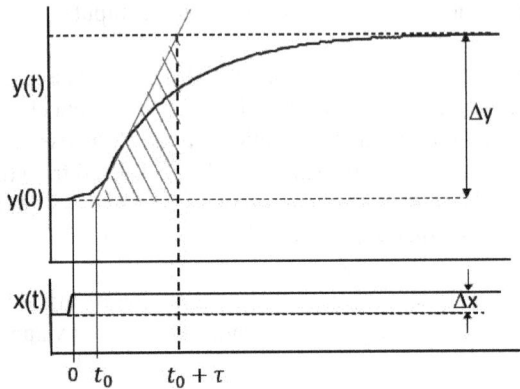

Fig. 5.2. Ziegler and Nichols Method: Determination of a first-order model's dead time and time constant.

3. The dead time of the process is measured from the initial time or time 'zero' to the time of the intersection of the tangent line with the response initial steady state line. This is because the maximum slope magnitude of the first-order system response curve occurs at a time equal to the dead time:

$$\frac{dy(t)}{dt} = \frac{k\Delta x}{\tau} e^{-(t-t_0)/\tau}, \ t \geq t_0 \tag{5.5}$$

$$\left.\frac{dy(t)}{dt}\right|_{max} = \left.\frac{dy(t)}{dt}\right|_{t-t_0} = \frac{k\Delta x}{\tau} \tag{5.6}$$

4. The time constant of the process is the time difference between the two intersection points of the tangent line with the response steady-state lines. A right triangle can be observed in the graph of Fig. 5.2, with opposite cathetus length given by Δy, and a segment of the tangent line as hypotenuse; if the slope of the hypotenuse is $\frac{k\Delta x}{\tau}$, from (5.5), and the total response change is $\Delta y = k \Delta x$, from the process gain definition (5.4), then the adjacent cathetus measures τ.

Miller Method

The Miller method is illustrated in Fig. 5.3. The calculation of the dead time is the same as in the Ziegler and Nichols method, using the tangent line to the response curve with maximum slope. The calculation of the time constant is the following:

1. Calculate the response at 63.2% of the total change of the process variable:

$$y_{63.3\%} = y(0) + 0.632\Delta y \tag{5.7}$$

2. Read the time value, $t_{63.2\%}$, corresponding to the response at 63.2% of the total change from the beginning of the step experiment or time 'zero'. This time includes the dead time and the time constant of the response, as it can be verified with the time equation of the step response:

$$y(t_{63.2\%}) = y(t_0 + \tau) = y(0) + k \Delta x(1 - e^{-1}) = y(0) + 0.632 \Delta y \tag{5.8}$$

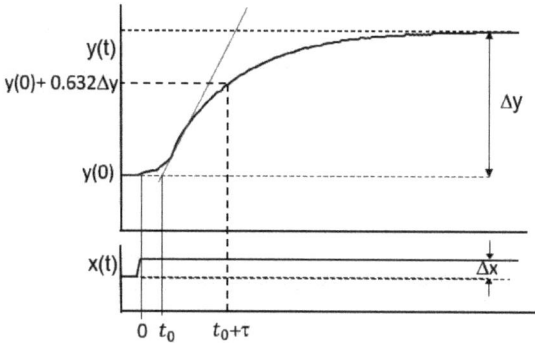

Fig. 5.3. Miller Method: Determination of a first-order model's dead time and time constant.

$$t_{63.2\%} = t_0 + \tau \qquad (5.9)$$

3. Subtract the dead time to obtain the estimation of the time constant:

$$\tau = t_{63.2\%} - t_0 \qquad (5.10)$$

Analytical Model

The analytical model is based on the reading of two representative points of the transitory response, which are assigned to the time values:

$$t_1 = t_0 + \frac{\tau}{3} \qquad (5.11)$$

and

$$t_2 = t_0 + \tau \qquad (5.12)$$

The procedure steps are based on Fig. 5.4 and are the following:

1. Calculate and locate the response value y_1 corresponding to t_1. The evaluation of the equation of the step time response indicates that this point is obtained when the 28.4% of the total response change has been achieved:

$$y(t_1) = y\left(t_0 + \frac{\tau}{3}\right) = y(0) + k\Delta x\left(1 - e^{-\frac{1}{3}}\right) = y(0) + 0.284\Delta y \qquad (5.13)$$

2. Read time t_1 from time 'zero'.
3. Calculate and locate the response y_2 corresponding to t_2. Equation (5.8) indicates that at this point, 63.2% of the total response change has just happened.
4. Read time t_2 from time 'zero' (notice that $t_2 = t_{63.2\%}$).
5. Solve equations (5.11) and (5.12) simultaneously to obtain the parameters τ and θ:

$$\tau = \frac{3}{2}(t_2 - t_1) \qquad (5.14)$$

$$t_0 = t_2 - \tau \qquad (5.15)$$

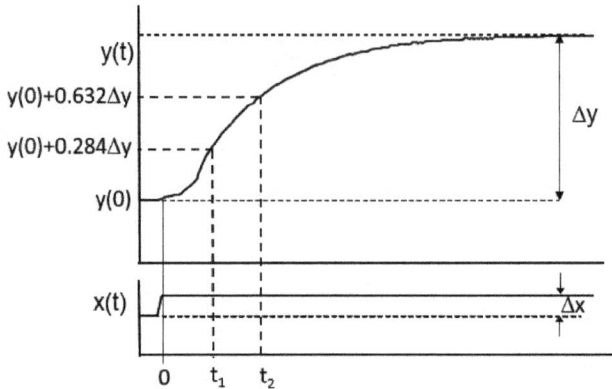

Fig. 5.4. Analytical Method: Response points for calculating dead time and time constant.

3.2 Comparison of First-Order Identification Methods

Consider a thermal process manipulated with a fuel valve and monitored with a temperature sensor. A step input is applied by changing the valve position from 10 to 40% at 105 s. The exponential response is obtained from the initial steady state value of 1400°C until the final steady temperature of 1440°C, as shown in Fig. 5.5. The process can be identified as a first-order system using the above-mentioned methods.

The curve of the process variable shows the effect of electrical noise in the measurement. A smooth non-oscillatory response is considered and represented by a filtered curve in Fig. 5.5. The reading references are marked: time 'zero' or the start time is 105 s, and the initial and final steady-state temperature lines are marked as well in the graph.

The process gain is calculated as the ratio of the total output and input changes:

$$k = \frac{(1440 - 1400)^\circ C}{(40 - 10)\%} = 1.33 \frac{^\circ C}{\%} \tag{5.16}$$

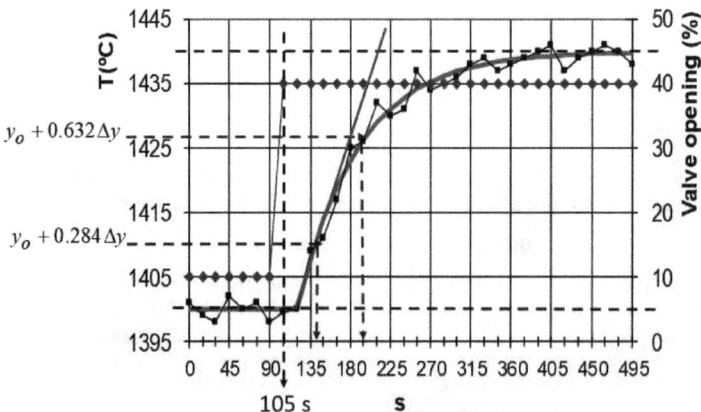

Fig. 5.5. Step Experiment on a Thermal Process: The input is the percentage of fuel valve opening (%), and the output is the temperature in Celsius degrees (T(°C)).

The Ziegler and Nichols method traces a tangent line with the maximum slope on the response curve. The intersection of this line with the initial steady state horizontal occurs at time 120 s; therefore, the dead time is 120–105 = 15 s. The intersection of the tangent line with the final steady state horizontal happens at time 210 s, and the time constant is 210–120 = 90 s.

With the Miller method, the estimation of the dead time is 15 s, and the time constant is determined based on the response value at 63.2% of the total change. This response value is:

$$y_{63.3\%} = 1400 + 0.632(1440 - 1400) = 1425.28°C \tag{5.17}$$

which is obtained at time 195 s, but this time has to be corrected by subtracting the initial time of the experiment of 105 s. Therefore, 195–105 = 90 s is the sum of the dead time plus the time constant. Since the dead time was determined as 15 s, the time constant is 90–15 = 75 s.

The analytical method is based on the response values at 28.4% and 63.2% of the total response change. The point at 63.2% is known from the procedure of the Miller method: $y(t_2) = 1425.28°C$ at $t_2 = 90$ s. The response at 28.4% of the total output change is:

$$y(t_1) = 1400 + 0.284 (1440 - 1400) = 1411.36°C \tag{5.18}$$

Moreover, the time value for this temperature is read as 142 s on the horizontal axis, but the initial time has to be subtracted; therefore, $t_1 = 142 - 105 = 37$ s. The following estimations are obtained from equations (5.14) and (5.15): $\tau = 79.5$ s and $t_0 = 10.5$ s.

These identifications show that the dead time estimation tends to be smaller for the analytical method, and the time constant method is typically bigger for the Ziegler and Nichols method. From the procedure point of view, the analytical method can be more precise because it does not depend on tracing a tangent line which can cause significant variations. An objective appreciation of the identification accuracy can be obtained by evaluating the model of the step response of the first-order system with equation (5.3) using the results of each method (gain, time constant, and dead time) and the conditions of the experiment (initial output and step input size). The model evaluations are then compared against the experiment data. The best fit to the real process behavior indicates the best model.

3.3 Identification of Second-Order Underdamped Systems

The most practical approach to model an oscillatory process is its identification as a second-order system. An underdamped response to a step input is expressed as a function of time by:

$$y(t) = y(0) + k\Delta x \left(1 - \frac{e^{-\frac{\zeta}{\tau}(t-t_0)}}{\sqrt{1-\zeta^2}} sin\left(\frac{\sqrt{1-\zeta^2}}{\tau}(t-t_0) + tan^{-1}\left(\frac{\sqrt{1-\zeta^2}}{\zeta} \right) \right) \right), t \geq t_0 \tag{5.19}$$

The oscillatory response is characterized by the following indicators, which can be measured graphically (see Fig. 5.6) as well as calculated from the dynamic parameters:

- *Overshoot* (*mp*): It characterizes the first peak point during the transitory response. It is measured as the ratio between the magnitude of the first crest or valley of the response (depending on the direction of the response curve) over the total change of the response, that is, A/B in Fig. 5.6:

$$mp = e^{-\pi\zeta/\sqrt{1-\zeta^2}} \tag{5.20}$$

- *Decay ratio* (*dr*): Ratio of the amplitude of the first two peaks in the same direction, C/A in Fig. 5.6:

$$dr = e^{-2\pi\zeta/\sqrt{1-\zeta^2}} \tag{5.21}$$

- *Rise time* (t_r): Time to cross the final response horizontal line for the first time; it is indicated as D time units in Fig. 5.6:

$$t_r = t_0 + \frac{\tau}{\sqrt{1-\zeta^2}}\left[\pi - tan^{-1}\left(\frac{\sqrt{1-\zeta^2}}{\zeta}\right)\right] \tag{5.22}$$

- *Peak time* (t_p): Time to reach the first peak of the response; E time units in Fig. 5.6:

$$t_p = t_0 + \frac{\pi\tau}{\sqrt{1-\zeta^2}} \tag{5.23}$$

- *Period time* (*T*): Time of a complete response oscillation, indicated as F time units in Fig. 5.6:

$$T = \frac{2\pi\tau}{\sqrt{1-\zeta^2}} \tag{5.24}$$

- *Settling time*: Time that the response takes to settle or to enter a band of ±2% of the final response change; approximately G time units marked in Fig. 5.6:

$$t_s = t_0 + \frac{4\tau}{\zeta} \tag{5.25}$$

The identification procedure is based on the second-order performance indicators and consists of the following steps:

1. Calculate the gain of the process as the ratio of the final change of the response over the input step size.
2. Calculate the damping factor from the overshoot value:

$$\zeta = \sqrt{\frac{ln^2(mp)}{\pi^2 + ln^2(mp)}} \tag{5.26}$$

3. Calculate the time constant from the period (or half period) measurement:

$$\tau = \frac{T}{2} \frac{\sqrt{1-\zeta^2}}{\pi} \qquad (5.27)$$

4. Calculate the dead time from the peak time and the half-period values:

$$t_0 = t_p - \frac{T}{2} \qquad (5.28)$$

For the process response shown in Fig. 5.6, the calculations for the parameter identification are the following:

- The gain from the steady state input and output values is given in units of response over units of input (it can be dimensionless, output and input have the same units):

$$k = \frac{47-37}{47-37} = 1 \qquad (5.29)$$

- The overshoot and the damping factor are:

$$mp = \frac{52-47}{47-37} = 0.5 \qquad (5.30)$$

$$\zeta = \sqrt{\frac{ln^2(0.5)}{\pi^2 + ln^2(0.5)}} = 0.2155 \qquad (5.31)$$

- Graphically, it is clearer and easier to measure the half period as the time between the first two intersections of the response curve with the final steady state line:

$$\frac{T}{2} = 14 - 11 = 3 \qquad (5.32)$$

- The time constant is calculated as a function of the damping factor and the half period:

$$\tau = 3 \frac{\sqrt{1-0.2155^2}}{\pi} = 0.9325 \qquad (5.33)$$

- The peak time is measured from the start of the step experiment until the time when the overshoot or maximum response value occurs:

$$t_p = 12 - 5 = 7 \qquad (5.34)$$

- The dead time is then calculated by subtracting half period from the peak time:

$$t_0 = 7 - 3 = 4 \qquad (5.35)$$

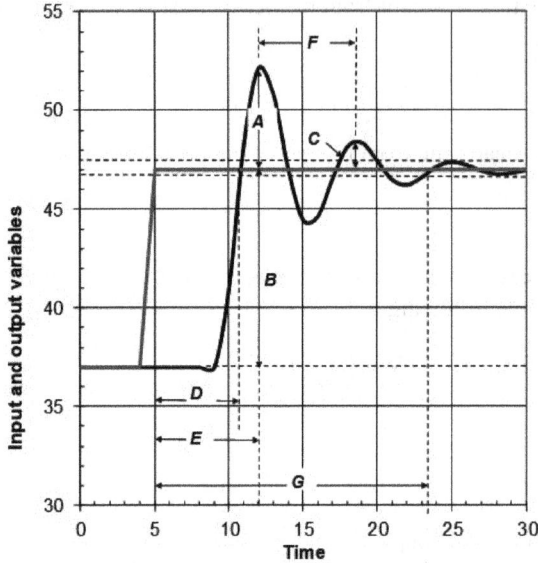

Fig. 5.6. Oscillatory step response: the step input and output are read on the same vertical axis.

Therefore, the identified second-order model is:

$$\frac{Y(s)}{X(s)} = \frac{e^{-4s}}{0.9325^2 s^2 + 2(0.9325)(0.2155)s + 1} \qquad (5.36)$$

This model can be validated by its evaluation in time with equation (5.19) and comparison with the experimental data.

4. Discrete-Time Identification

Parametric model structures in the z-domain are used to model an unknown system and are called black-box models. Identification in discrete time describes systems in terms of transfer functions in the z-domain or difference equations in discrete time. These models provide the system's dynamical response. It is often required to test different model structures to determine the most suitable.

The identification procedure is generally iterative in the sense that, starting from *a priori* information about the system to be identified, different choices, such as the order of the system, the number of coefficients, and the input delay, are successively addressed and revised until the model is validated. The complete identification procedure is composed of six main steps (Ljung, 1999): (1) experiment design; (2) input-output measurements; (3) model structure choice; (4) structure and parameter determination; (5) model parameter estimation; and (6) model validation.

A discrete transfer function or difference equation can be obtained by discretizing a continuous model from the abovementioned methods. However, the process can be identified in discrete time; that is, input and output variables are sampled with a specific sample time, T, and the resultant model can be directly a discrete-time model. The discrete model is calculated by minimizing the sum of the square errors of the model. Stochastic identification refers to the application of statistical measures to define the experiment, condition data, and validate the model, as explained in the following sections.

4.1 PRBS Experiment

The input signal for discrete identification has many variations compared to a simple step input. The input can be a pseudo-random binary sequence (PRBS) as illustrated in Fig. 5.7. From the steady state, and in the absence of disturbances, the input or manipulated variable, u, acquires one out of two possible values, calculated by incrementing or decrementing A units from the initial value, u_{ss}, with:

$$u = u_{ss} + (PRBS_i)A \tag{5.37}$$

where $PRBS_i$ can be $+1$ or -1, according to the result of the shifting operation on a register with N bits performed by the following steps:

1. The register is assigned a random value, except 0, for example, 101.

2. Then, two bits, preferably one at the middle of the register and the other at the right end of the register, are used to calculate the XOR function. This logic function has equal chances to produce a 0 or 1 value when bits are equal or different, respectively.

3. The XOR function value is fed at the opposite end (left side) of the register by shifting the bits from one position to the right. The bit at the right end leaves the register and constitutes the basis for the direction of the manipulation change with respect to its initial steady-state value, u_{ss}, in this way: $PRBS_i$ is made equal to -1, when the dropped bit is 0, and is made equal to $+1$ when the dropped bit is 1.

4. The manipulation is calculated with equation (5.37) and sent to the process.

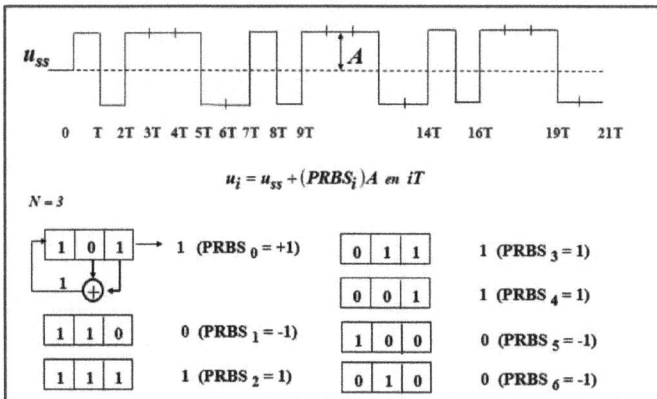

Fig. 5.7. Calculation and graph of a PRBS signal using a register of 3 bits, a Sample Time T, a Steady State Manipulation u_{ss}, and Manipulation Amplitude A.

5. With the new register value, steps 2 through 4 are repeated at subsequent sample times to obtain a sequence of $2^N - 1$ manipulation values.

6. More than one sequence can be applied to the process to produce a large number of data.

A PRBS input maintains an average value equal to the initial steady state value, which has the advantage of keeping the process response oscillating around its steady state value. This characteristic allows the application of a PRBS test with minimum interference on the normal conditions of the process, that is, practically without affecting its productivity. The same effect can be achieved by using the computer random number generation to define a random binary signal:

$$u = u_{ss} + (RBS_i)A \tag{5.38}$$

where,

$$RBS_i = \begin{cases} 1, & r \le 0.5 \\ -1, & r > 0.5 \end{cases} \tag{5.39}$$

and r is a random number in the range between 0 and 1. In this case, a sequence length of manipulation values is not repeated, but the calculations can be performed on each sample time to acquire a predefined amount of data.

Although the manipulation amplitude should be small to avoid large response variations, the amplitude should be big enough to have a more significant effect on the measured response than the electrical noise. The test amplitude should not imply a change in the operation zone.

As for the sample time, it is important to consider the natural velocity of the response of the process. It is recommended to choose a sample time as a fraction of the expected time constant.

During the application of a PRBS or RBS input, the output or response value is also monitored at each sample time. As a result of the test, input-output data is collected at a discrete time. These data should be numerically conditioned for the model calculation.

The data conditioning may consist of the normalization of the input u and output y values:

$$Y_i = \frac{y_i - \overline{y}}{\sigma_y}, U_i = \frac{u_i - \overline{u}}{\sigma_u} \tag{5.40}$$

where $\overline{y}$ and $\overline{u}$ are the mean values for the output and input, respectively, and σ_y and σ_u are the corresponding standard deviations.

However, for simulation and control purposes, the use of deviation values is more practical for a direct comparison with real values of process variable measurements and application of manipulation signals:

$$Y_i = y_i - \overline{y}, \quad U_i = u_i - \overline{u} \tag{5.41}$$

The experiment results consist of the conditioned data, which is the basis for the identification of the process dynamics. If the data show no bias with time, then the experiment is correct, and the analysis or modeling can be performed.

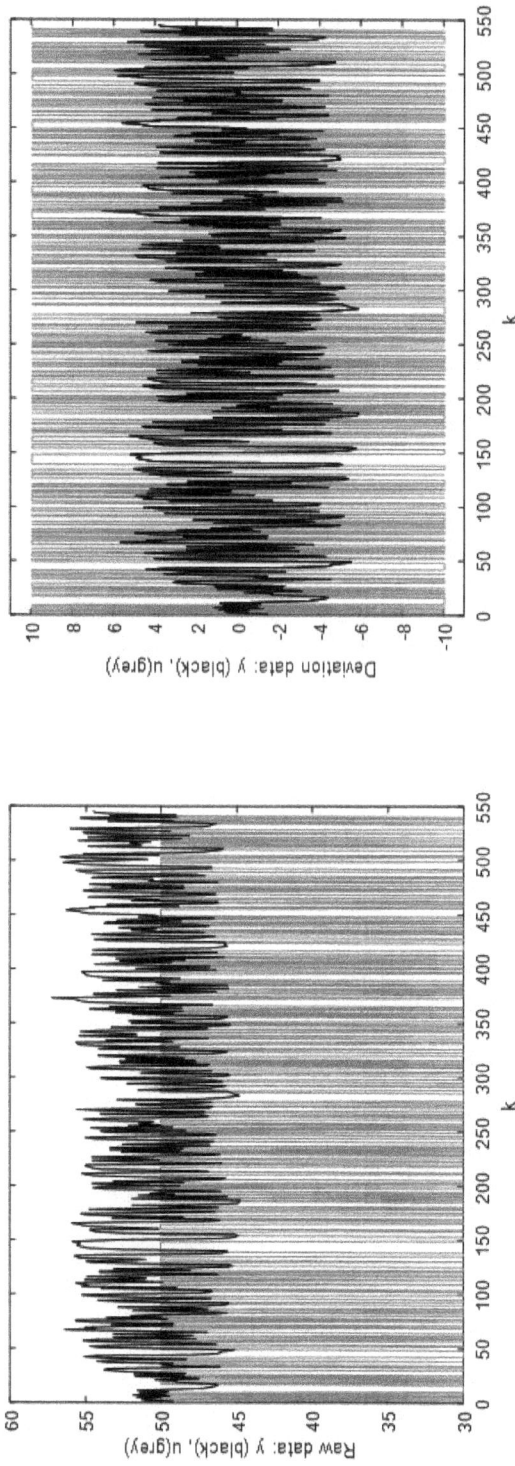

Fig. 5.8. Identification experiment based on a PRBS sequence. Left: Raw Data. Right: Deviation Data.

Figure 5.8 shows a correct experiment since the data do not tend upwards or downwards but oscillate around a constant mean value. The mean value is zero for conditioned data as deviation values.

4.2 Model Approximation by Minimum Square Error Method

The set of deviation values is processed to fit a process model. The model structure is designated as an autoregressive model with an exogenous variable or ARX. The exogenous variable is the manipulation or input variable. The autoregressive characteristic consists of the dependence of the output variable estimation on previous values of the same output variable. The ARX model may be nonlinear (Zhu, 2002), but the linear version is enough for restricted process operation conditions.

The linear ARX model is the simplest model incorporating the input signal (see Fig. 5.9). The ARX model is the most efficient of the polynomial estimation methods because it is the result of solving linear regression equations in analytic form. The problem is to find the coefficients of the polynomials, $A(z)$ and B; the solution always satisfies the minimum error of the model.

The ARX structure is defined with three parameters: na or the number of a coefficients or autoregressive terms, nb or the number of b coefficients or input terms, and d or the number of sample times included in the dead time of the process (that is, dT is approximated to the dead time t_0 of the process). The $ARX(na, nb, d)$ model is expressed as the difference equation:

$$y(n) = a_1 y(n-1) + \cdots + a_{na} y(n - na) + b_1 u(n-1-d) + \cdots + b_{nb} u(n - nb - d) \quad (5.42)$$

where y is the output deviation variable, and u is the input deviation variable. Working with deviation values is consistent with the data conditioning procedure given by equation (5.40). The discrete transfer function in the z domain is:

$$\frac{Y(z)}{U(z)} = z^{-d} \frac{b_1 z^{-1} + b_2 z^{-2} + \cdots + b_{nb} z^{-nb}}{1 - az^{-1} - a_2 z^{-2} - \cdots - a_{na} z^{-na}} \quad (5.43)$$

For a discrete system to have a continuous counterpart, $nb = na + 1$ or $nb = na$. This is an additional characteristic of the model, not a necessary one. Notice that na is the order of the system.

Equation (5.42) can be rewritten to acknowledge the error of the model inherent to the estimation $\hat{y}(n)$:

$$\hat{y}(n) = a_1 y(n-1) + \cdots + b_1 u(n-1-d) + \cdots \quad (5.44)$$

The minimum square error method has the following objective function F:

$$F = \sum_{i_{min}}^{numdata} e_k^2 = \sum_{i_{min}}^{numdata} (y_k - \hat{y}_k)^2 = \sum_{i_{min}}^{numdata} (y_k - a_1 y_{k-1} - \cdots - b_1 u_{k-1-d} - \cdots)^2 \quad (5.45)$$

where i_{min} is the greatest between na and $nb+d$, and $numdata$ is the total number of data points or input-output pairs.

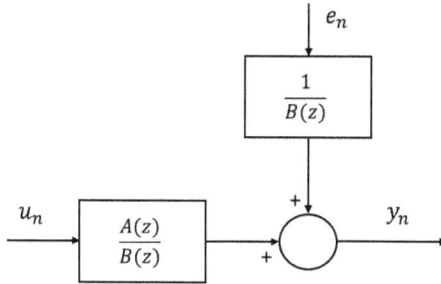

Fig. 5.9. ARX inputs and outputs.

The matrix representation of the model and the model error is:

$$
\begin{bmatrix} y_{i_{min}} \\ \vdots \\ y_{numdata} \end{bmatrix} =
$$

$$
\begin{bmatrix} y_{i_{min}-1} & \cdots & y_{i_{min}-na} & u_{i_{min}-1-d} & \cdots & u_{i_{min}-nb-d} \\ \vdots & \cdots & \vdots & \vdots & \cdots & \vdots \\ y_{numdata-1} & \cdots & y_{numdata-na} & u_{numdata-1-d} & \cdots & u_{numdata-nb-d} \end{bmatrix} \begin{bmatrix} a_1 \\ \vdots \\ a_{na} \\ b_1 \\ \vdots \\ b_{nb} \end{bmatrix} \tag{5.46}
$$

$$
+ \begin{bmatrix} e_{i_{min}} \\ \vdots \\ e_{numdata} \end{bmatrix}
$$

$$
Y = A\theta + E \tag{5.47}
$$

where the matching of equations (5.46) and (5.47) defines the output vector Y, the data matrix A, the parameter vector θ, and the error vector E.

The solution to the minimization problem, by the first-order derivative theorem, is:

$$
\theta = (A^T A)^{-1} A^T Y \tag{5.48}
$$

However, the evaluation of equation (5.48) is not practical because of the size of the data arrays. Alternatively, a recursive solution to the minimum square error problem can be used as an efficient computational algorithm for online and offline identification. As a new process data pair is considered in each iteration n, the model parameters are updated.

The recursive implementation of the minimum square error method is given by equations (5.4) through (5.11), which allow the calculation of the model coefficient vector $\theta(n) = [a_1, a_2, ..., a_{na}, b_1, b_2, ..., b_{nb}]'$, with adjustment fixed positive parameters and λ, model error e, and correction gain k_{LS} (Haykin, 2014). For initialization or step 0:

$$\theta(0) \tag{5.49}$$

$$P(0) \tag{5.50}$$

Subsequent calculations in terms of the iteration index n (from the greatest between na and $nb+d$ to the number of data) are:

$$v(n) = [y(n-1),\dots, y(n-na), u(n-1-d)\dots u(n-nb-d)]' \tag{5.51}$$

$$Q(n) = P(n-1)v(n) \tag{5.52}$$

$$k_{LS}(n) = \frac{Q(n)}{\lambda + v'(n)Q(n)} \tag{5.53}$$

$$e(n) = y(n) - \theta'^{(n-1)} v(n) \tag{5.54}$$

$$\theta(n) = \theta(n-1) + k_{LS}(n)e(n) \tag{5.55}$$

$$P(n) = \lambda^{-1} P(n-1) - \lambda^{-1} k_{LS}(n) v'(n)P(n-1) \tag{5.56}$$

Once the ARX model coefficients have been obtained, the model has to be validated. If the model is not adequate, the ARX structure should be modified.

4.3 Model Validation by Regression Analysis

If the proposed model is adequate, it should produce random errors; that is, the error at any instant should be independent of previous errors as well as of previous input values. Lag values between 0 and 20 sample times are typically considered for the inspection of the dependence or correlation of the process model error with respect to past output estimations and input levels.

The autocorrelation of the error R_e is calculated as:

$$R_{e_j} = \sum_{k=i_{min}+j}^{numdata} e_k e_{k-j} \tag{5.57}$$

where j is the lag or number of sample times the error values are apart. For $j = 0$, R_{e_0} is positive. For $j \neq 0$, R_{e_j} can be positive or negative if the errors are random. The normalized autocorrelation R_{en} is typically used to standardize its range and get $R_{en_0} = 1$, and R_{en_j} closed to zero. A tolerance of ± 0.2 for R_{en_j} for $j \neq 0$ is generally accepted to conclude the absence of autocorrelation of the model error (Fig. 5.10):

$$R_{en_j} = \frac{R_{e_j}}{R_{e_0}} \tag{5.58}$$

The cross-correlation between the error of the model and the input applied to the process during the identification experiment should also be examined to ensure the model does not produce a bias or a systematic error. The cross-correlation is calculated with:

$$R_{eu_j} = \sum_{k=i_{min}+j}^{numdata} e_k u_{k-j}$$ (5.59)

and normalized by:

$$R_{eun_j} = \frac{R_{eu_j}}{\sqrt{R_{e_0} R_{u_0}}}$$ (5.60)

where,

$$R_{u_0} = \sum_{k=i_{min}}^{numdata} u_k^2$$ (5.61)

Since the manipulation input changed randomly during the experiment, positive and deviation values of the same magnitude are expected with the same probability. A good model would have small errors with positive and negative signs also equally probable. Therefore, ideal cross-correlation values are very close to zero, with a tolerance ±0.2 in the normalized scale (Fig. 5.10).

Besides indicating whether the model is correct, the correlation analysis also guides the search for a better structure. If the error autocorrelation is out of range at a particular j or lag value, the term containing y_{k-j} should be included in the model; that is, *na* should be modified. If the error-input cross-correlation is not within the tolerance limits, then the term containing u_{k-j} should be considered in the model by modifying *nb*.

Figures 5.11 and 5.12 show the correlation analysis for the models ARX(1,1,0) and ARX(1,2,1), respectively, from the example experiment of Fig. 5.10. The error-input cross-correlation of the first set of graphs of Fig. 5.11 suggests that the term u_{k-2} and u_{k-3} should be included. Since the autocorrelation is within range, the process model is correctly proposed as a first-order system. The second set of correlation graphs in Fig. 5.12 validates the corrected model ARX(1,2,1).

Finally, another validation can be made by evaluating the model and comparing its output against the experimental conditioned data. If the model evaluation practically matches the output data from the same input, the model is considered helpful in reproducing the behavior of the process. Figure 5.13 shows this end-use validation type.

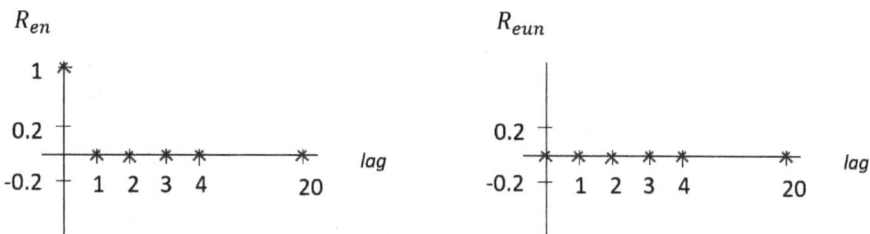

Fig. 5.10. Ideal error autocorrelation (left) and error-input cross-correlation (right) graphs.

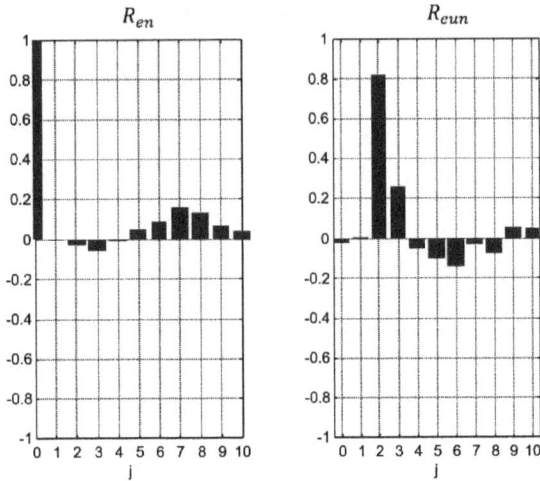

Fig. 5.11. Example correlation graphs for ARX(1, 1, 0).

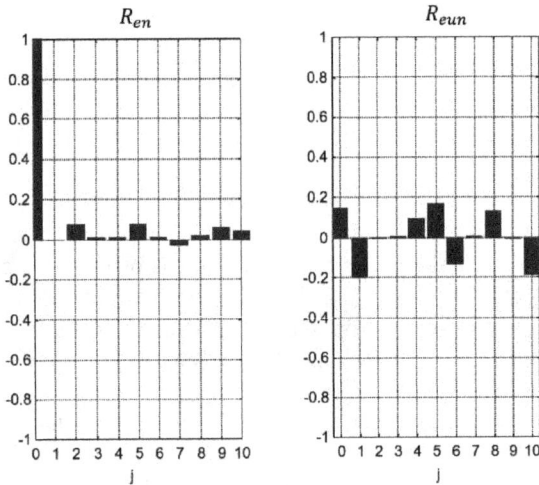

Fig. 5.12. Example correlation graphs for ARX(1, 2, 1).

5. Conclusions

System identification methods are divided into two groups: parametric and nonparametric. Parametric methods identify system models with a mathematical structure associated with a coefficient or parameter set. Nonparametric methods model a system directly with its dynamical response. System identification can also extract information other than a system model. The generated information can be used to synthesize feedback or feedforward controller gains directly from a set of input-output data without obtaining an intermediate model of the system first.

Models are important for different purposes. Some control strategies take advantage of models to make predictions that can be involved in the calculation of the

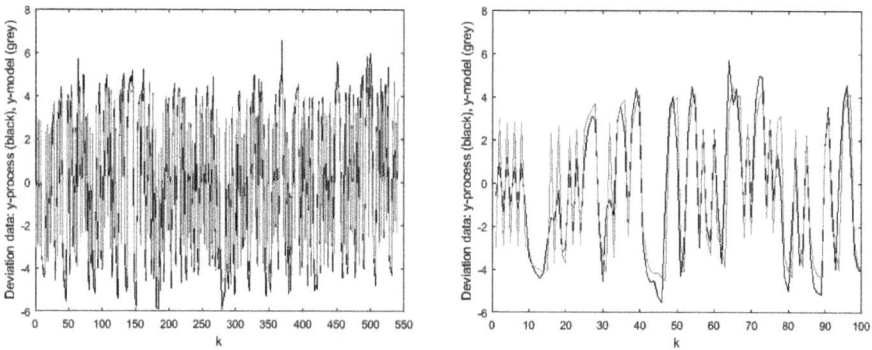

Fig. 5.13. Graphical comparison of model output and process output for model validation: for the complete data set (left) and for the first 100 data (right).

controller output (Qina and Badgwell, 2003). Virtual machines require realistic 3D graphical representations and dynamic models to animate them and show the proper behavior (Herbus and Ociepka, 2015). The development and control of complex systems, such as robots (Kashiwagi, EOLSS), are benefitted from the availability of dynamic models, which are often decoupled or separated for their components like in the case of robot articulations.

Although experimental modeling is considered a black-box process analysis, it effectively reproduces the process output. The experimental modeling approach is not subjected to any ideal or simplified behavior assumptions. Discrete modeling requires a large amount of data and computational resources for its analysis, while continuous graphical methods are suitable for manual approximations of low-order models. The relation between the continuous and the discrete domains allows the exportation of models making the process dynamics identification available for the design of continuous and discrete control systems.

The identification method for the ARX model is the least squares method, which has the advantage of providing a unique solution. The least squares method is one of the most used polynomial estimation method because this method is efficient in solving linear regression equations. The main disadvantage of the ARX model is that it allows disturbances to be part of the dynamical response of the model; this is because the denominator polynomial of the transfer function is the same for both the deterministic part and for the stochastic part, which rarely occurs in physical systems. For this reason, when the disturbance of the system is random noise, it is recommended to use a model structure of higher order. However, the drawback is that higher-order models can compromise a stable dynamical response.

References

Dorsey, J. 2005. *Continuous and Discrete Control Systems*. McGraw Hill.

Haykin, S. 2014. *Adaptive Filter Theory*. Fifth Edition. Prentice Hall.

Herbus, K. and Ociepka, P. 2015. Integration of the virtual 3D model of a control system with the virtual controller. *IOP Conf. Series: Materials Science and Engineering* 95: 012084.

Kashiwagi, H. *Control Systems, Robotics, and Automation*. Vol. VI –Nonparametric System Identification ©Encyclopedia of Life Support Systems (EOLSS).

Ljung, L. 1999. *System Identification: Theory for the User.* Second Edition. Englewood Cliffs, N.J. Prentice-Hall.

Qina, S.J. and Badgwell, T.A. 2003. A survey of industrial model predictive control technology. *Control Engineering Practice* 11: 733–764.

Zhu, Y.C. 2001. *Multivariable System Identification for Process Control.* Oxford. Elsevier Science Ltd.

Zhu Y.C. 2002. Estimation of nonlinear ARX models. *Proceedings of the 41st IEEE Conference on Decision and Control* 2: 2214–2219.

Chapter 6
State Space Modeling

1. Introduction

State space modeling and control methods are mainly applied to multiple input and multiple output systems (MIMO). The state space methods, like the methods based on transfer function in s and z domains, are tools to analyze, model, and design feedback control systems. State space refers to the space of dimensions whose coordinate axes are formed by state variables and uses linear algebra for mathematical analysis. Unlike the frequency domain modeling methods, the state space representation is not limited to linear systems; it is possible to apply state space techniques to a non-linear system, time-variant linear systems, or other classes of systems suitable to be analyzed and modeled with state space methods.

The state space representation is a mathematical tool for modeling physical systems described by a set of inputs, state variables, and outputs, related by a set of first-order differential equations that are combined into a first-order matrix differential equation. Multiple inputs, outputs, and state variables are expressed as vectors, and the algebraic equations are written in matrix form (Ogata, 2010). Controllability is the property that indicates whether the behavior of a system can be controlled using its inputs. Simultaneously, observability is the property that indicates whether the internal behavior of the system can be detected by its outputs.

This chapter presents a review of mathematical tools for state space modeling. Section 2 reviews definitions like state variables, state vector, and state space equations. Section 3 reviews procedures to solve the space state equation and methods to find the state transition matrix. Section 4 reviews the concepts of controllability and observability and the similarity transformation to find the controllable and observable canonical forms of state space systems. Section 5 reviews basic discretization procedures for converting continuous time-space systems to other discrete-time equivalents. Final comments are presented in Section 6.

2. State Space Model Formulation

State space modeling applies to multiple input multiple output systems, which may be linear or nonlinear, time invariant or time varying (Williams and Lawrence,

2007). However, the linear time-invariant systems ease the introduction of state space methods and the relation to other analysis tools.

2.1 Definitions

State of a System

The state of a system at the initial time, t_0, is the information required at time t_0, which, together with the input signal, $u(t)$, determines the system's output for all $t > t_0$.

State Variable

The state variables of a dynamical system are the minimum set of variables necessary to determine the state and response of the dynamical system.

State Vector

It is the set of n variables that can be considered as n components of a vector, x, needed to describe the behavior of the dynamic system completely.

State Space

The state space is an n-dimensional space in which the components of the state vector represent its coordinate axes. State space is a Cartesian coordinate system with an axis for each state variable. Each state vector element corresponds to a state variable of the system. For a two-order state space system, x_1 and x_2 are the state variables that completely define the system's state (Fig. 6.1).

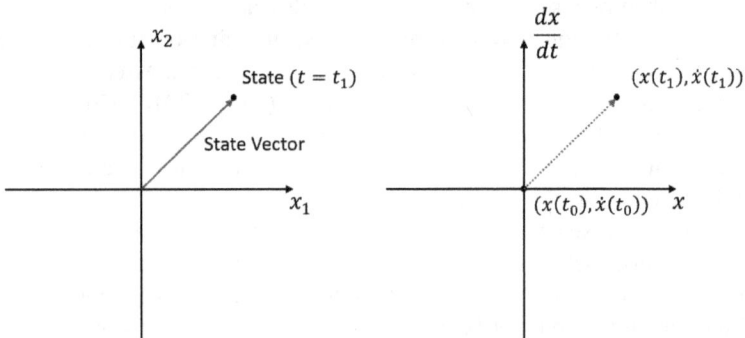

Fig. 6.1. State space for two state variables.

2.2 State Space Equations

In state space analysis for modeling dynamical systems, there are three types of variables: input variables, state variables, and output variables. The dynamical system must involve elements that memorize the values of the input for $t > t_0$. Since integrators in continuous-time control systems are memory elements, the outputs of integration functions can be used as the variables that define the state of the dynamical system.

Assume that a MIMO system involves n integrators. Assume that there are r inputs $u_1(t), u_2(t), \cdots, u_r(t)$ and m outputs $y_1(t), y_2(t), \cdots, y_m(t)$. Define n outputs of the integrators as state variables: $x_1(t), x_2(t), \cdots, x_n(t)$; the number of state variables is equal to the number of integrators, and the system is expressed as:

$$\dot{x}_1(t) = f_1(x_1, x_2, \ldots x_n; u_1, u_2, \ldots u_r; t) \tag{6.1}$$

$$\dot{x}_2(t) = f_2(x_1, x_2, \ldots x_n; u_1, u_2, \ldots u_r; t) \tag{6.2}$$

$$\cdots$$

$$\dot{x}_n(t) = f_n(x_1, x_2, \ldots x_n; u_1, u_2, \ldots u_r; t) \tag{6.3}$$

The outputs $y_1(t), y_2(t), \cdots, y_m(t)$ of the system are given by:

$$y_1(t) = g_1(x_1, x_2, \ldots x_n; u_1, u_2, \ldots u_r; t) \tag{6.4}$$

$$y_2(t) = g_2(x_1, x_2, \ldots x_n; u_1, u_2, \ldots u_r; t) \tag{6.5}$$

$$\cdots$$

$$y_m(t) = g_m(x_1, x_2, \ldots x_n; u_1, u_2, \ldots u_r; t) \tag{6.6}$$

If previous equations are linearized about the operating state, then the set of equations of both states and outputs can be expressed in matrix form:

$$\dot{x}(t) = A(t)x(t) + B(t)u(t) \tag{6.7}$$

$$y(t) = C(t)x(t) + D(t)u(t) \tag{6.8}$$

The vector-matrix functions $A_{n \times n}(t)$, $B_{n \times r}(t)$, $C_{m \times n}(t)$, $D_{m \times r}(t)$ have the size that allows the multiplication with the following vectors: a state vector, x, of dimension n; an input vector, u, of dimension r: and an output vector, y, of dimension m:

$$x(t) = \begin{bmatrix} x_1 \\ x_2 \\ \dot{x}_n \end{bmatrix}; \quad u(t) = \begin{bmatrix} u_1 \\ u_2 \\ \dot{u}_r \end{bmatrix}; \quad y(t) = \begin{bmatrix} y_1 \\ y_2 \\ \dot{y}_m \end{bmatrix} \tag{6.9}$$

If vector-matrix functions $A(t)$, $B(t)$, $C(t)$, $D(t)$, do not depend on time, t, then the system is a time-invariant system, and the state and output equations can be simplified to:

$$\dot{x}(t) = Ax(t) + Bu(t) \tag{6.10}$$

$$y(t) = Cx(t) + Du(t) \tag{6.11}$$

Figure 6.2 shows the block diagram of the state space system. As seen, matrix A implements a feedback signal and is directly related to the system dynamics.

Example. Consider the mechanical system shown in Fig. 6.3. The external force, $u(t)$, is the input to the system, and the mass displacement, $y(t)$, is the output. The

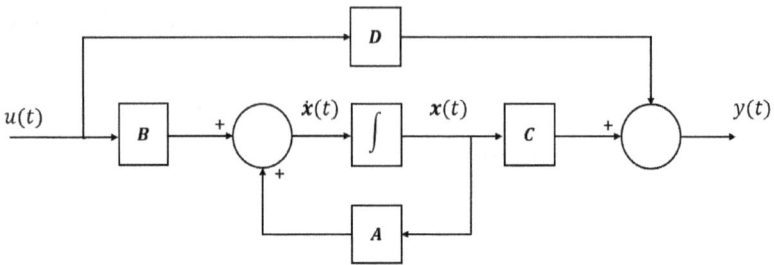

Fig. 6.2. Block diagram of a state space system.

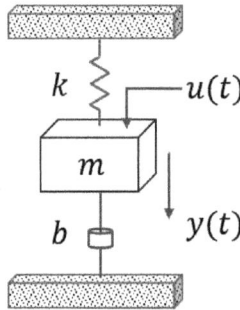

Fig. 6.3. Spring-damper mechanical system.

displacement is measured from the equilibrium position in the absence of the external force. This system is a single-input single-output system with parameters k for the spring stiffness, b for the damper coefficient, and m for the body's mass.

From the diagram of Fig. 6.3, the second-order differential equation can be expressed as conventionally deduced for mass-spring-dampers systems:

$$m\ddot{y}(t) + b\dot{y}(t) + ky(t) = u(t) \tag{6.12}$$

The second-order differential equation involves two integrators. Let us define two state variables $x_1(t)$ and $x_2(t)$ as:

$$x_1(t) = y(t) \tag{6.13}$$

$$x_2(t) = \dot{y}(t) \tag{6.14}$$

Then we identify the relationships among the states and express the differential equation in terms of the state variables:

$$\dot{x}_1(t) = x_2(t) \tag{6.15}$$

$$\dot{x}_2(t) = -\frac{b}{m}\dot{y}(t) - \frac{k}{m}y(t) + \frac{1}{m}u(t) \tag{6.16}$$

$$\dot{x}_2(t) = -\frac{b}{m}x_2(t) - \frac{k}{m}x_1(t) + \frac{1}{m}u(t) \tag{6.17}$$

with the output equation:

$$y(t) = x_1(t) \tag{6.18}$$

The system can be expressed in a vector-matrix form:

$$\begin{bmatrix} \dot{x}_1(t) \\ \dot{x}_2(t) \end{bmatrix} = \begin{bmatrix} 0 & 1 \\ -\dfrac{k}{m} & -\dfrac{b}{m} \end{bmatrix} \begin{bmatrix} x_1(t) \\ x_2(t) \end{bmatrix} + \begin{bmatrix} 0 \\ \dfrac{1}{m} \end{bmatrix} u(t) \tag{6.19}$$

$$y(t) = \begin{bmatrix} 1 & 0 \end{bmatrix} \begin{bmatrix} x_1(t) \\ x_2(t) \end{bmatrix} \tag{6.20}$$

Notice that the system has one input and one output. And finally, for a more compact form using vector-matrix parameters:

$$\dot{x}(t) = Ax(t) + Bu(t) \tag{6.21}$$

$$y(t) = Cx(t) \tag{6.22}$$

with the matrix and vectors:

$$A = \begin{bmatrix} 0 & 1 \\ -\dfrac{k}{m} & -\dfrac{b}{m} \end{bmatrix}; \ B = \begin{bmatrix} 0 \\ \dfrac{1}{m} \end{bmatrix} \text{ and } C = \begin{bmatrix} 1 & 0 \end{bmatrix} \tag{6.23}$$

Example. Find the state space model of the second-order system given by the following differential equation:

$$\ddot{y}(t) + 4\dot{y}(t) + 3y(t) = u(t) \tag{6.24}$$

Define the two state variables $x_1(t)$ and $x_2(t)$ as integrator outputs:

$$x_1(t) = y(t) \tag{6.25}$$

$$x_2(t) = \dot{y}(t) \tag{6.26}$$

Then express the set of first-order differential equations in terms of the state variables:

$$\dot{x}_1(t) = x_2(t) \tag{6.27}$$

$$\dot{x}_2(t) = -4\dot{y}(t) - 3y(t) + u(t) = -3x_1(t) - 4x_2(t) + u(t) \tag{6.28}$$

with the output equation:

$$y(t) = x_1(t) \tag{6.29}$$

The system can be expressed in a vector-matrix form:

$$\begin{bmatrix} \dot{x}_1(t) \\ \dot{x}_2(t) \end{bmatrix} = \begin{bmatrix} 0 & 1 \\ -3 & -4 \end{bmatrix} \begin{bmatrix} x_1(t) \\ x_2(t) \end{bmatrix} + \begin{bmatrix} 0 \\ 1 \end{bmatrix} u(t) \tag{6.30}$$

$$y(t) = \begin{bmatrix} 1 & 0 \end{bmatrix} \begin{bmatrix} x_1(t) \\ x_2(t) \end{bmatrix} \tag{6.31}$$

Example. A mechanical coupler can model the coupling of train wagons in railways systems, a mechanism placed at each end of the wagons to form a train. The mechanical coupling can be modeled by the free body diagram of Fig. 6.4:

The mechanical coupling requires two differential equations since the forces are applied to each mass, and both masses share the damper and the spring:

$$m_1 \ddot{y}_1(t) + b(\dot{y}_1(t) - \dot{y}_2(t)) + k(y_1(t) - y_2(t)) = F_1(t) \tag{6.32}$$

$$m_2 \ddot{y}_2(t) + b(\dot{y}_2(t) - \dot{y}_1(t)) + k(y_2(t) - y_1(t)) = F_2(t) \tag{6.33}$$

For each second-order differential equation, two state variables need to be defined: $x_1(t)$ and $x_2(t)$ for the first ODE, and $x_3(t)$ and $x_4(t)$ for the second ODE:

$$x_1(t) = y_1(t) \tag{6.34}$$

$$x_2(t) = \dot{y}_1(t) \tag{6.35}$$

$$x_3(t) = y_2(t) \tag{6.36}$$

$$x_4(t) = \dot{y}_2(t) \tag{6.37}$$

Then relate the state variables by pairs and express the system dynamics in terms of the four state variables:

$$\dot{x}_1(t) = x_2(t) \tag{6.38}$$

$$\dot{x}_2(t) = -\frac{k}{m_1}(x_1(t) - x_3(t)) - \frac{b}{m_1}(x_2(t) - x_4(t)) + \frac{1}{m_1}F_1(t) \tag{6.39}$$

$$\dot{x}_3(t) = x_4(t) \tag{6.40}$$

$$\dot{x}_2(t) = -\frac{k}{m_2}(x_3(t) - x_1(t)) - \frac{b}{m_2}(x_4(t) - x_2(t)) + \frac{1}{m_2}F_2(t) \tag{6.41}$$

The outputs are selected as:

$$y_1(t) = x_1(t) \tag{6.42}$$

$$y_2(t) = x_3(t) \tag{6.43}$$

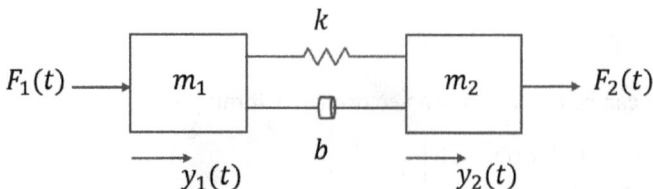

Fig. 6.4. Free body diagram of mechanical coupling of two train wagons.

The fourth order state space system in matrix form is:

$$
\begin{bmatrix} \dot{x}_1(t) \\ \dot{x}_2(t) \\ \dot{x}_3(t) \\ \dot{x}_4(t) \end{bmatrix} =
\begin{bmatrix} 0 & 1 & 0 & 0 \\ -k/m_1 & -b/m_1 & k/m_1 & b/m_1 \\ 0 & 0 & 0 & 1 \\ k/m_2 & b/m_2 & -k/m_2 & -b/m_2 \end{bmatrix}
\begin{bmatrix} x_1(t) \\ x_2(t) \\ x_3(t) \\ x_4(t) \end{bmatrix} +
\begin{bmatrix} 0 & 0 \\ 1/m_1 & 0 \\ 0 & 0 \\ 0 & 1/m_2 \end{bmatrix}
\begin{bmatrix} F_1(t) \\ F_2(t) \end{bmatrix}
\tag{6.44}
$$

$$
\begin{bmatrix} y_1(t) \\ y_2(t) \end{bmatrix} =
\begin{bmatrix} 1 & 0 & 0 & 0 \\ 0 & 0 & 1 & 0 \end{bmatrix}
\begin{bmatrix} x_1(t) \\ x_2(t) \\ x_3(t) \\ x_4(t) \end{bmatrix}
\tag{6.45}
$$

Now, to review the system response, assume some parameters that may correspond to small wagons, i.e., a scaled physical model of a train or a toy train: $k = 0.2$, $b = 0.1$, $m_1 = 1$, $m_2 = 2$, $F_1 = 1$, $F_2 = -1$:

$$
\begin{bmatrix} \dot{x}_1(t) \\ \dot{x}_2(t) \\ \dot{x}_3(t) \\ \dot{x}_4(t) \end{bmatrix} =
\begin{bmatrix} 0 & 1 & 0 & 0 \\ -0.2 & -0.1 & 0.2 & 0.1 \\ 0 & 0 & 0 & 1 \\ 0.1 & 0.05 & -0.1 & -0.05 \end{bmatrix}
\begin{bmatrix} x_1(t) \\ x_2(t) \\ x_3(t) \\ x_4(t) \end{bmatrix} +
\begin{bmatrix} 0 & 0 \\ 1 & 0 \\ 0 & 0 \\ 0 & 0.5 \end{bmatrix}
\begin{bmatrix} F_1(t) \\ F_2(t) \end{bmatrix}
\tag{6.46}
$$

$$
\begin{bmatrix} y_1(t) \\ y_2(t) \end{bmatrix} =
\begin{bmatrix} 1 & 0 & 0 & 0 \\ 0 & 0 & 1 & 0 \end{bmatrix}
\begin{bmatrix} x_1(t) \\ x_2(t) \\ x_3(t) \\ x_4(t) \end{bmatrix}
\tag{6.47}
$$

Figure 6.5 shows the response of the system to the applied forces. The state space model obtains some oscillation before reaching stable final displacements for the defined combination of input forces.

2.3 State Space to Transfer Function

Transfer functions can be obtained from the state space system equations by computing the following matrix algebra operations in the Laplace Domain, where I is the identity matrix of the required size corresponding to the same order of the state space system:

$$
G(s) = C\,(sI - A)^{-1}\,B + D
\tag{6.48}
$$

To illustrate the use of the above equation, find the transfer function of the following state space system (obtained previously in a modeling example):

$$
\dot{x}(t) = A\,x(t) + B\,u(t)
\tag{6.49}
$$

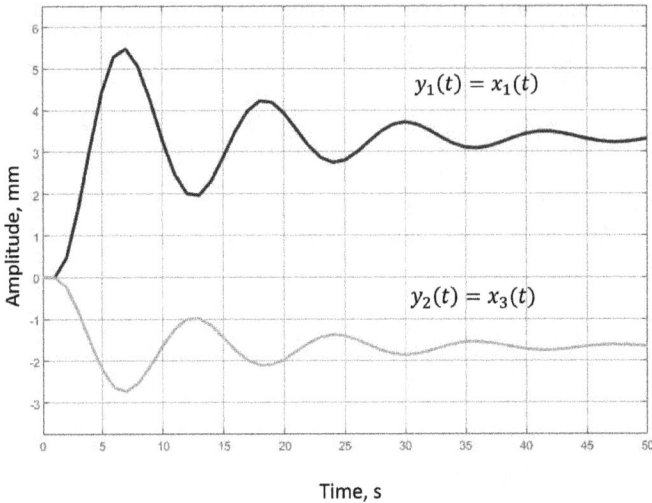

Fig. 6.5. Movement response of the coupled train wagon system.

$$y(t) = C\,x(t) \tag{6.50}$$

with the matrix and vectors:

$$A = \begin{bmatrix} 0 & 1 \\ -3 & -4 \end{bmatrix}; B = \begin{bmatrix} 0 \\ 1 \end{bmatrix} \text{ and } C = \begin{bmatrix} 1 & 0 \end{bmatrix} \tag{6.51}$$

First, construct the matrix $(sI - A)$:

$$(sI - A) = s \begin{bmatrix} 1 & 0 \\ 0 & 1 \end{bmatrix} - \begin{bmatrix} 0 & 1 \\ -3 & -4 \end{bmatrix} = \begin{bmatrix} s & -1 \\ 3 & s+4 \end{bmatrix} \tag{6.52}$$

Find the cofactor matrix of $(sI - A)$:

$$cof(sI - A) = \begin{bmatrix} (-1)^2(s+4) & (-1)^3(3) \\ (-1)^3(-1) & (-1)^4(s) \end{bmatrix} = \begin{bmatrix} s+4 & -3 \\ 1 & s \end{bmatrix} \tag{6.53}$$

Find the determinant of $(sI - A)$:

$$det(sI - A) = Det \begin{bmatrix} s & -1 \\ 3 & s+4 \end{bmatrix} = s(s+4) - 3(-1) \tag{6.54}$$

$$det(sI - A) = s^2 + 4s + 3 \tag{6.55}$$

Notice that the calculated determinant is already the characteristic polynomial of the transfer function. Now find the inverse of matrix $(sI - A)$:

$$(sI - A)^{-1} = \frac{(Cof(sI - A))^T}{Det(sI - A)} = \frac{\begin{bmatrix} s+4 & -3 \\ 1 & s \end{bmatrix}^T}{s^2 + 4s + 3} = \begin{bmatrix} \dfrac{s+4}{s^2+4s+3} & \dfrac{1}{s^2+4s+3} \\ \dfrac{-3}{s^2+4s+3} & \dfrac{s}{s^2+4s+3} \end{bmatrix} \tag{6.56}$$

The next step is to post-multiply by **B**:

$$
(s\mathbf{I} - \mathbf{A})^{-1}\mathbf{B} =
\begin{bmatrix}
\dfrac{s+4}{s^2+4s+3} & \dfrac{1}{s^2+4s+3} \\[2ex]
\dfrac{-3}{s^2+4s+3} & \dfrac{s}{s^2+4s+3}
\end{bmatrix}
\begin{bmatrix} 0 \\ 1 \end{bmatrix}
=
\begin{bmatrix}
\dfrac{1}{s^2+4s+3} \\[2ex]
\dfrac{s}{s^2+4s+3}
\end{bmatrix}
\tag{6.57}
$$

The final step is to pre-multiply by **C**:

$$
G(s) = \mathbf{C}(s\mathbf{I} - \mathbf{A})^{-1}\mathbf{B} =
\begin{bmatrix} 1 & 0 \end{bmatrix}
\begin{bmatrix}
\dfrac{1}{s^2+4s+3} \\[2ex]
\dfrac{s}{s^2+4s+3}
\end{bmatrix}
= \dfrac{1}{s^2+4s+3}
\tag{6.58}
$$

The procedure to obtain the transfer function from the state space equations is naturally more elaborated when dealing with higher-order systems. The advantage is that the transfer functions relating each output with any input can be obtained for MIMO systems.

3. State Space Equation Solution

The solution of the state space vector from the state space equation can be obtained by applying methods to find the solution of ordinary differential equations. Conventional methods find first the solution to the homogenous differential equation and then the solution to the non-homogenous differential equation.

3.1 Solutions to the State Space Equation

The solution to the state space equations can be found in two parts, obtaining the homogenous solution and then the non-homogenous solution.

State Space Homogeneous Solution

The state space equation is a set of simultaneous first-order differential equations:

$$
\frac{d}{dt}x(t) - \mathbf{A}\,x(t) = \mathbf{B}\,u(t)
\tag{6.59}
$$

and the output equation is stated algebraically as:

$$
y(t) = \mathbf{C}\,x(t) + \mathbf{D}\,u(t)
\tag{6.60}
$$

The homogenous differential equation can be expressed as follows:

$$
\frac{d}{dt}x(t) - \mathbf{A}\,x(t) = 0
\tag{6.61}
$$

Then apply the Laplace Transform considering the initial state, $x(0)$:

$$
sX(s) - x(0) - \mathbf{A}\,X(s) = 0
\tag{6.62}
$$

$$
sX(s) - \mathbf{A}\,X(s) = x(0)
\tag{6.63}
$$

$$X(s)\,(s\boldsymbol{I} - \boldsymbol{A}) = \boldsymbol{x}(0) \tag{6.64}$$

and solve for $X(s)$:

$$X(s) = (s\boldsymbol{I} - \boldsymbol{A})^{-1}\,\boldsymbol{x}(0) \tag{6.65}$$

The state vector can be expressed by taking the inverse Laplace Transform of $X(s)$, and since we have a set of first-order LTI equations, the solution of the inverse Laplace Transform implies real or complex exponentials. However, at this point, matrix A can be treated as any constant, and the inverse transform leads to the exponential of a matrix, which is also a matrix:

$$x(t) = \boldsymbol{L}^{-1}\left[(s\boldsymbol{I} - \boldsymbol{A})^{-1}\,\boldsymbol{x}(0)\right] \tag{6.66}$$

$$x(t) = e^{\boldsymbol{A}t}\,\boldsymbol{x}(0) \tag{6.67}$$

The exponential function $e^{\boldsymbol{A}t}$ is called the transition matrix, $\boldsymbol{\Phi}(t)$, and represents the characteristic response of a particular state space system.

State Space Non-Homogeneous Solution

The non-homogenous differential equation can be expressed as follows:

$$\frac{d}{dt}x(t) - \boldsymbol{A}\,x(t) = \boldsymbol{B}\,\boldsymbol{u}(t) \tag{6.68}$$

Then apply Laplace Transform considering the initial state $x(0)$:

$$sX(s) - x(0) - \boldsymbol{A}\,X(s) = \boldsymbol{B}\,U(s) \tag{6.69}$$

$$sX(s) - \boldsymbol{A}\,X(s) = x(0) + \boldsymbol{B}\,U(s) \tag{6.70}$$

$$X(s)\,(s\boldsymbol{I} - \boldsymbol{A}) = x(0) + \boldsymbol{B}\,U(s) \tag{6.71}$$

and solve for $X(s)$:

$$X(s) = (s\boldsymbol{I} - \boldsymbol{A})^{-1}\,x(0) + (s\boldsymbol{I} - \boldsymbol{A})^{-1}\,\boldsymbol{B}\,U(s) \tag{6.72}$$

The state vector, $x(t)$, can be expressed by taking the inverse Laplace Transform of $X(t)$:

$$x(t) = \boldsymbol{L}^{-1}\left[(s\boldsymbol{I} - \boldsymbol{A})^{-1}\,x(0) + (s\boldsymbol{I} - \boldsymbol{A})^{-1}\,\boldsymbol{B}\,U(s)\right] \tag{6.73}$$

The solution has two components: one is the previously obtained homogeneous solution, and the second part is the convolution in the time domain of the transition matrix with the input vector:

$$x(t) = e^{\boldsymbol{A}t}\,x(0) + \int_{o}^{t} e^{\boldsymbol{A}(t-\tau)}\boldsymbol{B}\,\boldsymbol{u}(\tau)d\tau \tag{6.74}$$

$$x(t) = \boldsymbol{\Phi}(t)\,x(0) + \int_{o}^{t} \boldsymbol{\Phi}(t-\tau)\boldsymbol{B}\,\boldsymbol{u}(\tau)d\tau \tag{6.75}$$

To find the system response, that is, the output vector, the C vector or matrix needs to be considered by pre-multiplying the previous state space vector solution:

$$y(t) = C\,\Phi(t-t_0)\,x(0) + \int_{t_0}^{t} C\,\Phi(t-\tau)B\,u(\tau)d\tau \tag{6.76}$$

The first component is also called the "zero input response" and the part with the convolution integral is called the "zero state response".

3.2 State Transition Matrix

The solution to the state space vector and the system response requires finding an analytical expression for the state transition matrix. There are two conventional methods: the first is to apply Laplace Transform, as previously explained, to obtain the solution to the state vector, and the second is to find the eigenvalues and eigenvectors for a state space transformation to find a diagonal matrix.

Method 1: Laplace Transform

The first method to obtain the transition matrix requires applying matrix algebra in the Laplace domain and using the inverse Laplace Transform:

$$\Phi(t) = e^{At}, = \mathcal{L}^{-1}\left[(sI-A)^{-1}\right] \tag{6.77}$$

This method requires computing the inverse of the matrix $[sI-A]$, which implies finding the cofactor matrix and the determinant, and finally applying the inverse Laplace Transform. A detailed procedure consists of the following steps:

1) Construct the matrix: $M = [sI - A]$
2) Find the cofactor matrix: $Cof(M)$
3) Find the determinant: $det(M)$
4) Find the inverse matrix: $M^{-1} = (Cof(M))^T/det(M)$
5) Obtain the inverse Laplace Transform: $\mathcal{L}^{-1}[M^{-1}]$

Example. Find the state transition matrix of the following state space system:

$$\dot{x}(t) = A\,x(t) + B\,u(t) \tag{6.78}$$

$$y(t) = C\,x(t) \tag{6.79}$$

with the matrix and vectors:

$$A = \begin{bmatrix} 0 & 1 \\ -2 & -3 \end{bmatrix}, B = \begin{bmatrix} 0 \\ 1 \end{bmatrix} \text{ and } C = \begin{bmatrix} 1 & 0 \end{bmatrix} \tag{6.80}$$

Construct the matrix $(sI - A)$:

$$(sI - A) = s\begin{bmatrix} 1 & 0 \\ 0 & 1 \end{bmatrix} - \begin{bmatrix} 0 & 1 \\ -2 & -3 \end{bmatrix} = \begin{bmatrix} s & -1 \\ 2 & s+3 \end{bmatrix} \tag{6.81}$$

Find the cofactor matrix of $(s\boldsymbol{I} - \boldsymbol{A})$:

$$Cof(s\boldsymbol{I} - \boldsymbol{A}) = \begin{bmatrix} (-1)^2(s+3) & (-1)^3(2) \\ (-1)^3(-1) & (-1)^4(s) \end{bmatrix} = \begin{bmatrix} s+3 & -2 \\ 1 & s \end{bmatrix} \tag{6.82}$$

Find the determinant of $(s\boldsymbol{I} - \boldsymbol{A})$:

$$det(s\boldsymbol{I} - \boldsymbol{A}) = Det \begin{bmatrix} s & -1 \\ 2 & s+3 \end{bmatrix} = s(s+3) - 2(-1) \tag{6.83}$$

$$det(s\boldsymbol{I} - \boldsymbol{A}) = s^2 + 3s + 2 \tag{6.84}$$

Now, find the inverse of the constructed matrix:

$$(s\boldsymbol{I} - \boldsymbol{A})^{-1} = \frac{(Cof(s\boldsymbol{I} - \boldsymbol{A}))^T}{Det(s\boldsymbol{I} - \boldsymbol{A})} = \frac{\begin{bmatrix} s+3 & -2 \\ 1 & s \end{bmatrix}^T}{s^2 + 3s + 2} = \begin{bmatrix} \dfrac{s+3}{s^2+3s+2} & \dfrac{1}{s^2+3s+2} \\ \dfrac{-2}{s^2+3s+2} & \dfrac{s}{s^2+3s+2} \end{bmatrix} \tag{6.85}$$

The transition matrix is the inverse Laplace of the obtained inverse matrix.

$$\Phi(t) = \mathcal{L}^{-1}[(s\boldsymbol{I} - \boldsymbol{A})^{-1}] = \mathcal{L}^{-1} \begin{bmatrix} \dfrac{s+3}{s^2+3s+2} & \dfrac{1}{s^2+3s+2} \\ \dfrac{-2}{s^2+3s+2} & \dfrac{s}{s^2+3s+2} \end{bmatrix} \tag{6.86}$$

To obtain the inverse Laplace Transform, we can expand each matrix element by the partial fraction method, beginning with the first row and first column:

$$\Phi_{1,1}(t) = \mathcal{L}^{-1} \left[\frac{s+3}{s^2+3s+2} \right] \tag{6.87}$$

$$\Phi_{1,1}(s) = \frac{s+3}{s^2+3s+2} = \frac{s+3}{(s+1)(s+2)} = \frac{A_1}{s+1} + \frac{A_2}{s+2} \tag{6.88}$$

$$A_1 = (s+1)\frac{s+3}{(s+1)(s+2)} \bigg|_{s=-1} = \frac{s+3}{(s+2)} \bigg|_{s=-1} = \frac{2}{(-1+2)} = 2 \tag{6.89}$$

$$A_2 = (s+2)\frac{s+3}{(s+1)(s+2)} \bigg|_{s=-2} = \frac{s+3}{(s+1)} \bigg|_{s=-2} = \frac{1}{(-2+1)} = -1 \tag{6.90}$$

$$\Phi_{1,1}(s) = \frac{2}{s+1} + \frac{-1}{s+2} \tag{6.91}$$

Then, apply the inverse Laplace Transform:

$$\Phi_{1,1}(t) = A_1 e^{s_1 t} + A_2 e^{s_2 t} = (2e^{-t} - e^{-2t})u(t) \tag{6.92}$$

Now for the element of the first row and second column:

$$\Phi_{1,2}(t) = \mathcal{L}^{-1}\left[\frac{1}{s^2 + 3s + 2}\right] \tag{6.93}$$

$$\Phi_{1,2}(s) = \frac{1}{s^2 + 3s + 2} = \frac{1}{(s+1)(s+2)} = \frac{A_3}{s+1} + \frac{A_4}{s+2} \tag{6.94}$$

$$A_3 = (s+1)\frac{1}{(s+1)(s+2)}\bigg|_{s=-1} = \frac{1}{(s+2)}\bigg|_{s=-1} = \frac{1}{(-1+2)} = 1 \tag{6.95}$$

$$A_4 = (s+2)\frac{1}{(s+1)(s+2)}\bigg|_{s=-2} = \frac{1}{(s+1)}\bigg|_{s=-2} = \frac{1}{(-2+1)} = -1 \tag{6.96}$$

$$\Phi_{1,2}(s) = \frac{1}{s+1} + \frac{-1}{s+2} \tag{6.97}$$

Then apply the inverse Laplace Transform:

$$\Phi_{1,2}(t) = (e^{-t} - e^{-2t})u(t) \tag{6.98}$$

For the matrix element at the second row and first column:

$$\Phi_{2,1}(t) = \mathcal{L}^{-1}\left[\frac{-2}{s^2 + 3s + 2}\right] \tag{6.99}$$

$$\Phi_{2,1}(s) = \frac{-2}{s^2 + 3s + 2} = \frac{-2}{(s+1)(s+2)} = \frac{A_5}{s+1} + \frac{A_6}{s+2} \tag{6.100}$$

$$A_5 = (s+1)\frac{-2}{(s+1)(s+2)}\bigg|_{s=-1} = \frac{-2}{(s+2)}\bigg|_{s=-1} = \frac{-2}{(-1+2)} = -2 \tag{6.101}$$

$$A_6 = (s+2)\frac{-2}{(s+1)(s+2)}\bigg|_{s=-2} = \frac{-2}{(s+1)}\bigg|_{s=-2} = \frac{-2}{(-2+1)} = 2 \tag{6.102}$$

$$\Phi_{2,1}(s) = \frac{-2}{s+1} + \frac{2}{s+2} \tag{6.103}$$

Then, apply inverse Laplace Transform:

$$\Phi_{2,1}(t) = 2(-e^{-t} + e^{-2t})u(t) \tag{6.104}$$

Work with the element at position (2, 2):

$$\Phi_{2,2}(t) = \mathcal{L}^{-1}\left[\frac{s}{s^2 + 3s + 2}\right] \tag{6.105}$$

$$\Phi_{2,2}(s) = \frac{s}{s^2 + 3s + 2} = \frac{s}{(s+1)(s+2)} = \frac{A_7}{s+1} + \frac{A_8}{s+2} \tag{6.106}$$

$$A_7 = (s+1)\frac{s}{(s+1)(s+2)}\bigg|_{s=-1} = \frac{s}{(s+2)}\bigg|_{s=-1} = \frac{-1}{(-1+2)} = -1 \tag{6.107}$$

$$A_8 = (s+2)\frac{s}{(s+1)(s+2)}\bigg|_{s=-2} = \frac{s}{(s+1)}\bigg|_{s=-2} = \frac{-2}{(-2+1)} = 2 \tag{6.108}$$

$$\Phi_{2,2}(s) = \frac{-1}{s+1} + \frac{2}{s+2} \tag{6.109}$$

Then, apply inverse Laplace Transform:

$$\Phi_{2,2}(t) = (-e^{-t} + 2e^{-2t})u(t) \tag{6.110}$$

Finally, express the 2 × 2 matrix with the four obtained elements:

$$\Phi(t) = \begin{bmatrix} 2e^{-t} - e^{-2t} & e^{-t} - e^{-2t} \\ -2e^{-t} + 2e^{-2t} & -e^{-t} + 2e^{-2t} \end{bmatrix} u(t) \tag{6.111}$$

Eigenvalues and Eigenvectors

An alternative method to find the transition matrix requires finding explicitly the eigenvalues and the eigenvectors of the system (Kuo, 1992). Before giving an example, we recall some basic concepts of eigenvalues and eigenvectors.

The eigenvectors of a *n*-order square matrix are *n* orthogonal vectors, which, after being multiplied by the matrix, remain proportional to the original vectors (i.e., change only in magnitude, not in direction). For each eigenvector, the corresponding eigenvalue, λ, is the factor by which the eigenvector changes when multiplied by the matrix (Fig. 6.6).

$$A\vec{u} = \lambda\vec{u} \tag{6.112}$$

For example, consider a two-dimensional system, the eigenvalues can be found by solving the simultaneous equation homogeneous system:

$$\begin{bmatrix} a_{11} & a_{12} \\ a_{21} & a_{22} \end{bmatrix}\begin{bmatrix} u_1 \\ u_2 \end{bmatrix} = \lambda\begin{bmatrix} u_1 \\ u_2 \end{bmatrix} \tag{6.113}$$

This represents a system of two linear equations:

$$a_{11} u_1 + a_{12} u_2 = \lambda u_1 \tag{6.114}$$

$$a_{21} u_1 + a_{22} u_2 = \lambda u_2 \tag{6.115}$$

After a convenient rearrangement:

$$a_{11} u_1 - \lambda u_1 + a_{12} u_2 = 0 \tag{6.116}$$

$$a_{21} u_1 + a_{22} u_2 - \lambda u_2 = 0 \tag{6.117}$$

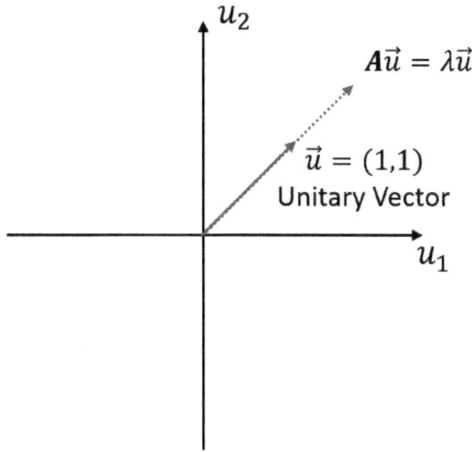

Fig. 6.6. Eigenvalues and eigenvectors.

The matrix equation is:

$$\begin{bmatrix} a_{11} - \lambda & a_{12} \\ a_{21} & a_{22} - \lambda \end{bmatrix} \begin{bmatrix} u_1 \\ u_2 \end{bmatrix} = 0 \tag{6.118}$$

$$(A - \lambda I)\vec{u} = 0 \tag{6.119}$$

The above system has a solution if the $(A - \lambda I)$ matrix is singular; that is, if the determinant equals zero. Therefore, to find the eigenvalues, compute the determinant of the matrix and find the roots of the characteristic polynomial:

$$det(A - \lambda I) = 0 \tag{6.120}$$

To find the eigenvectors, solve the simultaneous homogeneous equation system for each eigenvalue, λ_i. Since it is a homogenous system, there may be an infinite number of solutions. Therefore, arbitrary values for the eigenvector elements must be chosen.

$$\begin{bmatrix} a_{11} - \lambda_i & a_{12} \\ a_{21} & a_{22} - \lambda_i \end{bmatrix} \begin{bmatrix} u_1 \\ u_2 \end{bmatrix} = 0 \tag{6.121}$$

Example. Find the eigenvalues and eigenvectors of the *A* matrix:

$$A = \begin{bmatrix} 0 & 1 \\ -3 & -4 \end{bmatrix} \tag{6.122}$$

To find the eigenvalues, compute the determinant of the constructed matrix $A - \lambda I$:

$$(A - \lambda I) = \begin{bmatrix} 0 & 1 \\ -3 & -4 \end{bmatrix} - \begin{bmatrix} \lambda & 0 \\ 0 & \lambda \end{bmatrix} = \begin{bmatrix} -\lambda & 1 \\ -3 & -4 - \lambda \end{bmatrix} \tag{6.123}$$

$$det(A - \lambda I) = \begin{bmatrix} -\lambda & 1 \\ -3 & -4 - \lambda \end{bmatrix} = (-\lambda)(-4 - \lambda) - (-3)(1) = 0 \qquad (6.124)$$

$$\lambda^2 + 4\lambda + 3 = 0 \qquad (6.125)$$

Find the roots of the characteristic polynomial:

$$\lambda_i = \frac{-b \pm \sqrt{b^2 - 4ac}}{2a} = \frac{-4 \pm \sqrt{16 - 12}}{2} \qquad (6.126)$$

$$\lambda_1 = \frac{-4 + 2}{2} = -1 \qquad (6.127)$$

$$\lambda_2 = \frac{-4 - 2}{2} = -3 \qquad (6.128)$$

There are two eigenvalues: $\lambda_1 = -1$ and $\lambda_2 = -3$. Now find the eigenvector corresponding to each eigenvalue.

Substitute the first eigenvalue, $\lambda_1 = -1$, in the equation $(A - \lambda I)u = 0$. To find the first eigenvector, solve the system:

$$\begin{bmatrix} -\lambda_1 & 1 \\ -3 & -4 - \lambda_1 \end{bmatrix} \begin{bmatrix} u_1 \\ u_2 \end{bmatrix} = \begin{bmatrix} 1 & 1 \\ -3 & -3 \end{bmatrix} \begin{bmatrix} u_1 \\ u_2 \end{bmatrix} = 0 \qquad (6.129)$$

Express the two linear scalar equations:

$$u_1 + u_2 = 0 \qquad (6.130)$$

$$-3u_1 - 3u_2 = 0 \qquad (6.131)$$

The direction vector $\vec{u}_1 = [u_1 \quad u_2]^T = [1 \quad -1]^T$ satisfies the two equations and can be defined as the first eigenvector.

Substitute the second eigenvalue, $\lambda_2 = -3$, in the simultaneous homogeneous equation system. To find the second eigenvector, solve the system:

$$\begin{bmatrix} -\lambda_2 & 1 \\ -3 & -4 - \lambda_2 \end{bmatrix} \begin{bmatrix} u_1 \\ u_2 \end{bmatrix} = \begin{bmatrix} 3 & 1 \\ -3 & -1 \end{bmatrix} \begin{bmatrix} u_1 \\ u_2 \end{bmatrix} = 0 \qquad (6.132)$$

The linear scalar equations are:

$$3u_1 + u_2 = 0 \qquad (6.133)$$

$$-3u_1 - u_2 = 0 \qquad (6.134)$$

The direction vector $\vec{u}_2 = [u_1 \quad u_2]^T = [1 \quad -3]^T$ satisfies the two equations and can be defined as the second eigenvector.

One way to validate the correct selection of the eigenvectors is to find a diagonalized system, $D = P^{-1} A P$, that must contain the different eigenvalues in the main diagonal. For this second-order system, the P matrix is constructed with the previously found eigenvectors.

$$P = \begin{bmatrix} 1 & 1 \\ -1 & -3 \end{bmatrix} \tag{6.135}$$

Find the inverse:

$$P^{-1} = \frac{(cof(P))^T}{det(P)} = \frac{\begin{bmatrix} -3 & 1 \\ -1 & 1 \end{bmatrix}^T}{-3-(-1)} = \frac{\begin{bmatrix} -3 & -1 \\ 1 & 1 \end{bmatrix}}{-2} = \begin{bmatrix} \dfrac{3}{2} & \dfrac{1}{2} \\ -\dfrac{1}{2} & -\dfrac{1}{2} \end{bmatrix} \tag{6.136}$$

Find the diagonal matrix, $D = P^{-1} AP$:

$$AP = \begin{bmatrix} 0 & 1 \\ -3 & -4 \end{bmatrix}\begin{bmatrix} 1 & 1 \\ -1 & -3 \end{bmatrix} = \begin{bmatrix} -1 & -3 \\ 1 & 9 \end{bmatrix} \tag{6.137}$$

$$D = P^{-1}AP = \begin{bmatrix} \dfrac{3}{2} & \dfrac{1}{2} \\ -\dfrac{1}{2} & -\dfrac{1}{2} \end{bmatrix}\begin{bmatrix} -1 & -3 \\ 1 & 9 \end{bmatrix} = \begin{bmatrix} -1 & 0 \\ 0 & -3 \end{bmatrix} = \begin{bmatrix} \lambda_1 & 0 \\ 0 & \lambda_2 \end{bmatrix} \tag{6.138}$$

Method 2: Diagonal Matrix

The second method applies tools from linear matrix algebra and requires explicitly finding the eigenvalues and eigenvectors. This alternative method requires the calculation of a 'diagonalized' state space equivalent to the original state space:

$$\Phi(t) = e^{At} = P^{-1} e^{Dt} P \tag{6.139}$$

The method requires the following steps:

1. Find eigenvalues of A
2. Find eigenvectors to construct a P matrix
3. Find P^{-1} matrix
4. Find the diagonal matrix: $D = P^{-1} AP$
5. Express the transition matrix of the diagonalized state space: e^{Dt}
6. Obtain the expression for the transition matrix: $P\, e^{Dt}\, P^{-1}$

Example. Find the eigenvalues and eigenvectors of matrix A:

$$A = \begin{bmatrix} 0 & 1 \\ -2 & -3 \end{bmatrix} \tag{6.140}$$

The eigenvalues of A are calculated from the determinant of matrix $A - \lambda I$:

$$(A - \lambda I) = \begin{bmatrix} 0 & 1 \\ -2 & -3 \end{bmatrix} - \begin{bmatrix} \lambda & 0 \\ 0 & \lambda \end{bmatrix} = \begin{bmatrix} -\lambda & 1 \\ -2 & -3-\lambda \end{bmatrix} \tag{6.141}$$

$$det(A - \lambda I) = \begin{bmatrix} -\lambda & 1 \\ -2 & -3-\lambda \end{bmatrix} = (-\lambda)(-3-\lambda) - (-2)(1) = 0 \qquad (6.142)$$

$$\lambda^2 + 3\lambda + 2 = 0 \qquad (6.143)$$

Find the roots of the characteristic polynomial:

$$\lambda_i = \frac{-b \pm \sqrt{b^2 - 4ac}}{2a} = \frac{-3 \pm \sqrt{9-8}}{2} \qquad (6.144)$$

$$\lambda_1 = \frac{-3+1}{2} = -1 \qquad (6.145)$$

$$\lambda_2 = \frac{-3-1}{2} = -2 \qquad (6.146)$$

Substitute the first eigenvalue, $\lambda_1 = -1$, in $(A - \lambda I)u = 0$. Solve the equation system to find the first eigenvector:

$$\begin{bmatrix} -\lambda_1 & 1 \\ -2 & -3-\lambda_1 \end{bmatrix}\begin{bmatrix} u_1 \\ u_2 \end{bmatrix} = \begin{bmatrix} 1 & 1 \\ -2 & -2 \end{bmatrix}\begin{bmatrix} u_1 \\ u_2 \end{bmatrix} = 0 \qquad (6.147)$$

The two simultaneous linear equations are:

$$u_1 + u_2 = 0 \qquad (6.148)$$

$$-2u_1 - 2u_2 = 0 \qquad (6.149)$$

The direction vector $\vec{u}_1 = [u_1 \quad u_2]^T = [1 \quad -1]^T$ satisfies the two equations and is the first eigenvector.

Now, substitute the second eigenvalue $\lambda_2 = -2$ in $(A - \lambda I)u = 0$, and solve the system:

$$\begin{bmatrix} -\lambda_2 & 1 \\ -2 & -3-\lambda_2 \end{bmatrix}\begin{bmatrix} u_1 \\ u_2 \end{bmatrix} = \begin{bmatrix} 2 & 1 \\ -2 & -1 \end{bmatrix}\begin{bmatrix} u_1 \\ u_2 \end{bmatrix} = 0 \qquad (6.150)$$

The two linear scalar equations are:

$$2u_1 + u_2 = 0 \qquad (6.151)$$

$$-2u_1 - u_2 = 0 \qquad (6.152)$$

The direction vector $\vec{u}_1 = [u_1 \quad u_2]^T = [1 \quad -2]^T$ satisfies the two equations and is the second eigenvector.

To validate the correct selection of the eigenvectors, the diagonalized system $D = P^{-1}AP$ must contain the different eigenvalues in the main diagonal. For this second-order system, the P matrix is constructed with the previously found eigenvectors.

$$P = \begin{bmatrix} 1 & 1 \\ -1 & -2 \end{bmatrix} \tag{6.153}$$

The inverse of P is calculated as:

$$P^{-1} = \frac{(cof(P))^T}{det(P)} = \frac{\begin{bmatrix} -2 & 1 \\ -1 & 1 \end{bmatrix}^T}{-2-(-1)} = \frac{\begin{bmatrix} -2 & -1 \\ 1 & 1 \end{bmatrix}}{-1} = \begin{bmatrix} 2 & 1 \\ -1 & -1 \end{bmatrix} \tag{6.154}$$

The diagonal matrix is $D = P^{-1} AP$:

$$AP = \begin{bmatrix} 0 & 1 \\ -2 & -3 \end{bmatrix}\begin{bmatrix} 1 & 1 \\ -1 & -2 \end{bmatrix} = \begin{bmatrix} -1 & -2 \\ 1 & 4 \end{bmatrix} \tag{6.155}$$

$$D = P^{-1} AP = \begin{bmatrix} 2 & 1 \\ -1 & -1 \end{bmatrix}\begin{bmatrix} -1 & -2 \\ 1 & 4 \end{bmatrix} = \begin{bmatrix} -1 & 0 \\ 0 & -2 \end{bmatrix} \tag{6.156}$$

Find the exponential diagonal matrix e^{Dt}:

$$e^{Dt} = e^{\begin{bmatrix} -1 & 0 \\ 0 & -2 \end{bmatrix}t} = \begin{bmatrix} e^{-t} & 0 \\ 0 & e^{-2t} \end{bmatrix} \tag{6.157}$$

Finally, find the transition matrix e^{At}:

$$e^{At} = Pe^{Dt}P^{-1} = \begin{bmatrix} 1 & 1 \\ -1 & -2 \end{bmatrix}\begin{bmatrix} e^{-t} & 0 \\ 0 & e^{-2t} \end{bmatrix}\begin{bmatrix} 2 & 1 \\ -1 & -1 \end{bmatrix} \tag{6.158}$$

Multiply P, and e^{Dt} then multiply by P^{-1}:

$$e^{At} = \begin{bmatrix} e^{-t} & e^{-2t} \\ -e^{-t} & -2e^{-2t} \end{bmatrix} P^{-1} = \begin{bmatrix} e^{-t} & e^{-2t} \\ -e^{-t} & -2e^{-2t} \end{bmatrix}\begin{bmatrix} 2 & 1 \\ -1 & -1 \end{bmatrix} \tag{6.159}$$

$$\Phi(t) = e^{At} = \begin{bmatrix} 2e^{-t} - e^{-2t} & e^{-t} - e^{-2t} \\ -2e^{-t} + 2e^{-2t} & -e^{-t} + 2e^{-2t} \end{bmatrix} u(t) \tag{6.160}$$

The obtained transition matrix based on eigenvalues and eigenvectors (method 2) is equal to the obtained transition matrix in the Laplace domain and by inverse Laplace Transform (method 1), as shown in (6.111).

4. Controllability and Observability

The concepts of controllability and observability are fundamental in the design of state space control systems. Although most physical systems are controllable and observable, the corresponding mathematical models may not possess the property of controllability and observability. It is necessary to know the conditions under which a system is controllable and observable since the controllability and observability conditions determine the existence of a complete solution for a control system design problem. In other words, there is no control solution to the considered system if it is uncontrollable. Observability, therefore, refers to the ability to estimate the state variables.

4.1 Controllability

A system is completely controllable if there is an unconstrained control law, $u(t)$, that can transfer any initial state $x(t_0)$ to any other desired state $x(t)$ in a finite time interval nT (n sample times and T is the sample time). Consider a state space system in continuous time with matrices A and B. The system is controllable, given any initial condition, if and only if the rank of the $n \times n$ controllability matrix, P_c, is n, which implies that the determinant is different from zero.

$$P_c = [B \quad AB \quad .. \quad A^{n-1} B] \tag{6.161}$$

$$rank(P_c) = n \tag{6.162}$$

Example. Find the controllability matrix and determine the controllability of the following state space system:

$$\dot{x}(t) = A x(t) + B u(t) \tag{6.163}$$

$$y(t) = C x(t) \tag{6.164}$$

with the matrix and vectors:

$$A = \begin{bmatrix} 0 & 1 \\ -3 & -4 \end{bmatrix}, \quad B = \begin{bmatrix} 0 \\ 1 \end{bmatrix} \text{ and } C = \begin{bmatrix} 1 & 0 \end{bmatrix} \tag{6.165}$$

The controllability matrix is calculated in the following way:

$$P_c = [B \quad AB] = \begin{bmatrix} \begin{bmatrix} 0 \\ 1 \end{bmatrix} & \begin{bmatrix} 0 & 1 \\ -3 & -4 \end{bmatrix}\begin{bmatrix} 0 \\ 1 \end{bmatrix} \end{bmatrix} \tag{6.166}$$

$$P_c = \begin{bmatrix} 0 & 1 \\ 1 & -4 \end{bmatrix} \tag{6.167}$$

The determinant of P_c must be different from zero:

$$det(P_c) = det\begin{bmatrix} 0 & 1 \\ 1 & -4 \end{bmatrix} = -1 \tag{6.168}$$

Since $det(P_c) = -1$, the system is controllable.

State controllability implies that it is possible by the available system inputs to steer the states from any initial value to any final value within some finite time window. Because of the integral or derivate relation among the state variables, there can be more state variables than input variables. However, the input variables need to be able to alter the state variables (Williams and Lawrence, 2007).

4.2 Observability

A system is completely observable if and only if there is a finite time interval, nT, such that the initial state $x(t_0)$ can be determined from the observation history of $y(t)$, given the control law $u(t)$. Consider the state space system in continuous time with matrices A and C. The system is said to be completely state observable, given any initial condition, if and only if the rank of the $n \times n$ matrix, commonly called observability matrix, P_o, is n, which implies that the determinant is different from zero:

$$P_o = [C \quad CA \quad .. \quad CA^{n-1}]^T \tag{6.169}$$

$$rank(P_o) = n \tag{6.170}$$

Example. Find the observability matrix and determine the observability of the state space system with the following matrix and vectors:

$$A = \begin{bmatrix} 0 & 1 \\ -3 & -4 \end{bmatrix}, \ B = \begin{bmatrix} 0 \\ 1 \end{bmatrix} \text{ and } C = \begin{bmatrix} 1 & 0 \end{bmatrix} \tag{6.171}$$

The controllability matrix is:

$$P_o = [C \quad CA]^T = \left[\begin{bmatrix} 1 & 0 \end{bmatrix} \begin{bmatrix} 1 & 0 \end{bmatrix} \begin{bmatrix} 0 & 1 \\ -3 & -4 \end{bmatrix} \right]^T \tag{6.172}$$

$$P_o = [C \quad CA]^T = \left[\begin{bmatrix} 1 & 0 \end{bmatrix} \begin{bmatrix} 0 & 1 \end{bmatrix} \right]^T = \begin{bmatrix} 1 & 0 \\ 0 & 1 \end{bmatrix} \tag{6.173}$$

The determinant of P_o must be different from zero:

$$det(P_o) = det \begin{bmatrix} 1 & 0 \\ 0 & 1 \end{bmatrix} = 1 \tag{6.174}$$

Since $det(P_o) = 1$, the system is observable.

Observability measures how well a system's internal states can be estimated by measuring the outputs of the system. The controllability and observability of a system are mathematical counterparts. Controllability implies that an input signal can drive any initial state to any desired final state, and observability implies that measuring the output signal conveys enough information to estimate the initial state of the system (Veness, 2017).

4.3 Similarity Transformation and Canonical Forms

State space modeling takes advantage of vector spaces and linear algebra. A state space system can be transformed into another state space representation conserving the system's dynamical characteristics. This transformation procedure was already employed in Section 4 in the method to obtain the transition matrix with the eigenvalues and eigenvectors and with the calculation of a diagonal matrix. Similarity transformation is a procedure to transform state space systems into other equivalent systems with certain properties. In state space, certain canonical forms with specific advantages are convenient for use in particular analysis or design techniques. These canonical forms of state space models are:

- Phase variable canonical form
- Controllable canonical form
- Observable canonical form
- Diagonal canonical form

Phase Variable Canonical Form

The method of phase variables possesses mathematical advantages over other representations. This type of representation can be obtained directly from differential equations and the direct decomposition of a transfer function. State models are not unique since they depend on the selected state variables. The state variables, referred to as phase variables, are conventionally selected as the output of the integrators and appear multiplied directly by the coefficients of the differential equation.

Consider the second-order ODE:

$$\frac{d^n y(t)}{dt^n} + a_1 \frac{d^{n-1} y(t)}{dt^{n-1}} + \cdots + a_{n-1} \frac{dy(t)}{dt} + a_n y(t) = u(t) \tag{6.175}$$

Define the state variables as outputs of integrators:

$$x_1(t) = y(t) \tag{6.176}$$

$$x_2(t) = \frac{dy(t)}{dt} \tag{6.177}$$

$$\cdots$$

$$x_{n-1}(t) = \frac{d^{n-2} y(t)}{dt^{n-2}} \tag{6.178}$$

$$x_n(t) = \frac{d^{n-1} y(t)}{dt^{n-1}} \tag{6.179}$$

The state space matrix and vectors for a single input and single output system are:

$$\begin{bmatrix} \dot{x}_1(t) \\ \dot{x}_2(t) \\ \vdots \\ \dot{x}_{n-1}(t) \\ \dot{x}_n(t) \end{bmatrix} = \begin{bmatrix} 0 & 1 & 0 & \cdots & 0 \\ 0 & 0 & 1 & \cdots & 0 \\ \vdots & \vdots & \vdots & \ddots & \vdots \\ 0 & 0 & 0 & 0 & 1 \\ -a_n & -a_{n-1} & \cdots & -a_2 & -a_1 \end{bmatrix} \begin{bmatrix} x_1(t) \\ x_2(t) \\ \vdots \\ x_{n-1}(t) \\ x_n(t) \end{bmatrix} + \begin{bmatrix} 0 \\ 0 \\ 0 \\ 0 \\ 1 \end{bmatrix} u(t) \tag{6.180}$$

$$y(t) = \begin{bmatrix} 1 & 0 & \cdots & 0 & 0 \end{bmatrix} \begin{bmatrix} x_1(t) \\ x_2(t) \\ \vdots \\ x_{n-1}(t) \\ x_n(t) \end{bmatrix} \qquad (6.181)$$

Controllable Canonical Form

The controllable canonical form is a minimal realization in which all model states are controllable. Consider the transfer function:

$$H(s) = \frac{b_0 s^n + b_1 s^{n-1} + \cdots + b_{n-1} s + b_n}{s^n + a_1 s^{n-1} + \cdots + a_{n-1} s + a_n} \qquad (6.182)$$

The controllable canonical form of the corresponding LTI system is:

$$\begin{bmatrix} \dot{x}_1(t) \\ \dot{x}_2(t) \\ \vdots \\ \dot{x}_{n-1}(t) \\ \dot{x}_n(t) \end{bmatrix} = \begin{bmatrix} 0 & 1 & 0 & \cdots & 0 \\ 0 & 0 & 1 & \cdots & 0 \\ \vdots & \vdots & \vdots & \ddots & \vdots \\ 0 & 0 & 0 & 0 & 1 \\ -a_n & -a_{n-1} & \cdots & -a_2 & -a_1 \end{bmatrix} \begin{bmatrix} x_1(t) \\ x_2(t) \\ \vdots \\ x_{n-1}(t) \\ x_n(t) \end{bmatrix} + \begin{bmatrix} 0 \\ 0 \\ 0 \\ 0 \\ 1 \end{bmatrix} u(t) \qquad (6.183)$$

$$y(t) = \begin{bmatrix} b_n - a_n b_0 & b_{n-1} - a_{n-1} b_0 & \cdots & b_2 - a_2 b_0 & b_1 - a_1 b_0 \end{bmatrix} \begin{bmatrix} x_1(t) \\ x_2(t) \\ \vdots \\ x_{n-1}(t) \\ x_n(t) \end{bmatrix} + b_0\, u(t) \quad (6.184)$$

The transformation of a system to the controllable canonical form using the transformation matrix Q, which can be obtained with the controllability matrix P_c:

$$Q = P_c\, W \qquad (6.185)$$

where the W matrix is obtained with the coefficient of the characteristic polynomial of the related transfer function:

$$det(sI - A) = s^n + a_1 s^{n-1} + \cdots + a_{n-1}\, s + a_n \qquad (6.186)$$

$$W = \begin{bmatrix} a_{n-1} & a_{n-2} & \cdots & a_1 & 1 \\ a_{n-2} & a_{n-3} & \cdots & 1 & 0 \\ \vdots & \vdots & \ddots & \vdots & \vdots \\ a_1 & 1 & \cdots & 0 & 0 \\ 1 & 0 & \cdots & 0 & 0 \end{bmatrix} \qquad (6.187)$$

The transformed system can be obtained by:

$$A_c = Q^{-1} AQ, \quad B_c = Q^{-1} B, \quad C_c = CQ \quad \text{and} \quad D_c = D \tag{6.188}$$

Example. Consider the state space system with:

$$A = \begin{bmatrix} 0 & 1 & 0 \\ 0 & 0 & 1 \\ -6 & -11 & -6 \end{bmatrix}, \quad B = \begin{bmatrix} 0 \\ 0 \\ 1 \end{bmatrix} \text{ and } C = \begin{bmatrix} 1 & 0 & 0 \end{bmatrix} \tag{6.189}$$

The controllability matrix is:

$$P_c = \begin{bmatrix} B & AB & A^2B \end{bmatrix} = \begin{bmatrix} 0 & 0 & 1 \\ 0 & 1 & -6 \\ 1 & -6 & 25 \end{bmatrix} \tag{6.190}$$

The characteristic polynomial is:

$$det(sI - A) = det \begin{bmatrix} s & -1 & 0 \\ 0 & s & -1 \\ 6 & 11 & s+6 \end{bmatrix} = s^2(s+6) + 6(-1)(-1) - (-11s) \tag{6.191}$$

$$det(sI - A) = s^3 + 6s + 11s + 6 \tag{6.192}$$

The coefficients matrix is:

$$W = \begin{bmatrix} 11 & 6 & 1 \\ 6 & 1 & 0 \\ 1 & 0 & 0 \end{bmatrix} \tag{6.193}$$

The transformation matrix is:

$$Q = P_c W = \begin{bmatrix} 0 & 0 & 1 \\ 0 & 1 & -6 \\ 1 & -6 & 25 \end{bmatrix} \begin{bmatrix} 11 & 6 & 1 \\ 6 & 1 & 0 \\ 1 & 0 & 0 \end{bmatrix} = \begin{bmatrix} 1 & 0 & 0 \\ 0 & 1 & 0 \\ 0 & 0 & 1 \end{bmatrix} \tag{6.194}$$

Suppose the Q matrix is an identity matrix. In that case, the inverse is also an identity matrix; the meaning is that A_c, the matrix of the converted system's matrix, has the same elements as the original A matrix. B_c and C_c matrices also have the same values as B and C matrices.

Observable Canonical Form

The observable canonical form of a system is the transpose of its controllable canonical form. The coefficients of the characteristic polynomial of the system appear in the last column of the A matrix.

The observable canonical form of the LTI system corresponding to the transfer function of (6.182) is:

$$
\begin{bmatrix} \dot{x}_1(t) \\ \dot{x}_2(t) \\ \vdots \\ \dot{x}_{n-1}(t) \\ \dot{x}_n(t) \end{bmatrix} = \begin{bmatrix} 0 & 0 & \cdots & 0 & -a_n \\ 1 & 0 & \cdots & 0 & -a_{n-1} \\ \vdots & \vdots & \ddots & \vdots & \vdots \\ 0 & 0 & \cdots & 0 & -a_2 \\ 0 & 0 & \cdots & 1 & -a_1 \end{bmatrix} \begin{bmatrix} x_1(t) \\ x_2(t) \\ \vdots \\ x_{n-1}(t) \\ x_n(t) \end{bmatrix} + \begin{bmatrix} b_n - a_n b_0 \\ b_{n-1} - a_{n-1} b_0 \\ 0 \\ b_2 - a_2 b_0 \\ b_1 - a_1 b_0 \end{bmatrix} u(t) \tag{6.195}
$$

$$
y(t) = \begin{bmatrix} 0 & 0 & \cdots & 0 & 1 \end{bmatrix} \begin{bmatrix} x_1(t) \\ x_2(t) \\ \vdots \\ x_{n-1}(t) \\ x_n(t) \end{bmatrix} + b_0 u(t) \tag{6.196}
$$

The transformation of a system to the controllable canonical form uses the transformation matrix R, which can be obtained from the observability matrix P_o, and the W matrix obtained with the coefficients of the characteristic polynomial of the related transfer function:

$$
R = (WP_o)^{-1} \tag{6.197}
$$

The transformed system can be obtained by:

$$
A_o = R^{-1} AR, \quad B_o = R^{-1} B, \quad C_o = CR \quad \text{and} \quad D_o = D \tag{6.198}
$$

Example. Consider the system with:

$$
A = \begin{bmatrix} 0 & 1 & 0 \\ 0 & 0 & 1 \\ -6 & -11 & -6 \end{bmatrix}, \quad B = \begin{bmatrix} 0 \\ 0 \\ 1 \end{bmatrix} \quad \text{and} \quad C = \begin{bmatrix} 1 & 0 & 0 \end{bmatrix} \tag{6.199}
$$

The observability matrix is:

$$
P_o = \begin{bmatrix} C & CA & CA^2 \end{bmatrix}^T = \begin{bmatrix} 1 & 0 & 0 \\ 0 & 1 & 0 \\ 0 & 0 & 1 \end{bmatrix} \tag{6.200}
$$

The coefficient matrix W is:

$$
W = \begin{bmatrix} 11 & 6 & 1 \\ 6 & 1 & 0 \\ 1 & 0 & 0 \end{bmatrix} \tag{6.201}
$$

The transformation matrix is:

$$R = (WP_o)^{-1} = \left(\begin{bmatrix} 11 & 6 & 1 \\ 6 & 1 & 0 \\ 1 & 0 & 0 \end{bmatrix} \begin{bmatrix} 1 & 0 & 0 \\ 0 & 1 & 0 \\ 0 & 0 & 1 \end{bmatrix} \right)^{-1} = \begin{bmatrix} 0 & 0 & 1 \\ 0 & 1 & -6 \\ 1 & -6 & 25 \end{bmatrix} \qquad (6.202)$$

The transformed system is:

$$A_o = R^{-1}AR = \begin{bmatrix} 0 & 0 & 1 \\ 0 & 1 & -6 \\ 1 & -6 & 25 \end{bmatrix}^{-1} \begin{bmatrix} 0 & 1 & 0 \\ 0 & 0 & 1 \\ -6 & -11 & -6 \end{bmatrix} \begin{bmatrix} 0 & 0 & 1 \\ 0 & 1 & -6 \\ 1 & -6 & 25 \end{bmatrix} \qquad (6.203)$$

$$A_o = R^{-1}AR = \begin{bmatrix} 0 & 0 & -6 \\ 1 & 0 & -11 \\ 0 & 1 & -6 \end{bmatrix} \qquad (6.204)$$

$$B_o = R^{-1}B = \begin{bmatrix} 0 & 0 & 1 \\ 0 & 1 & -6 \\ 1 & -6 & 25 \end{bmatrix}^{-1} \begin{bmatrix} 0 \\ 0 \\ 1 \end{bmatrix} = \begin{bmatrix} 1 \\ 0 \\ 0 \end{bmatrix} \qquad (6.205)$$

$$C_o = CR = \begin{bmatrix} 1 & 0 & 0 \end{bmatrix} \begin{bmatrix} 0 & 0 & 1 \\ 0 & 1 & -6 \\ 1 & -6 & 25 \end{bmatrix} = \begin{bmatrix} 0 & 0 & 1 \end{bmatrix} \qquad (6.206)$$

5. State Space Models in Discrete Time

State space methods are suitable for computer implementation, particularly in platforms with programming languages that can perform calculations using matrix and vector arrays. These calculations usually are part of high level languages. The state space models can be implemented in discrete time in other more computationally limited programming languages, such as those found in electronic control units (Dorf and Bishop, 2001).

5.1 Discretization of a State Space System

Now consider a state space system:

$$\dot{x}(t) = A\,x(t) + B\,u(t) \qquad (6.207)$$

$$y(t) = C\,x(t) \qquad (6.208)$$

The discrete-time equivalent of the first derivative with a sampling time T is applied to the state vector:

$$\dot{x}(t) \approx \frac{x(t+T) - x(t)}{T} \tag{6.209}$$

$$\frac{x(t+T) - x(t)}{T} = A x(t) + B u(t) \tag{6.210}$$

$$x(t+T) - x(t) = TA x(t) + TB u(t) \tag{6.211}$$

$$x(t+T) = TA x(t) + x(t) + TB u(t) \tag{6.212}$$

Expressing the previous equation as a function of discrete time k (n is reserved to indicate the order of the system or number of state variables):

$$x(kT+T) = (TA + I) x(kT) + TB u(kT) \tag{6.213}$$

$$x(k+1) = (TA + I) x(k) + TB u(k) \tag{6.214}$$

In simplified notation:

$$x_{k+1} = (TA + I) x_k + TB u_k \tag{6.215}$$

The output equation is:

$$y_k = C x_k + D u_k \tag{6.216}$$

For convenience, the discretized system now can be expressed as:

$$x_{k+1} = G x_k + H u_k \tag{6.217}$$

where, $G = TA + I$, and, $H = TB$.

Example. Consider the discretization of a two order single input and single output system, with the matrix and vectors:

$$A = \begin{bmatrix} 0 & 1 \\ -3 & -4 \end{bmatrix}, \quad B = \begin{bmatrix} 0 \\ 1 \end{bmatrix} \quad \text{and} \quad C = \begin{bmatrix} 1 & 0 \end{bmatrix} \tag{6.218}$$

Consider a sample time $= 0.1$,

$$G = TA + I = \begin{bmatrix} (0.1)0 + 1 & (0.1)1 \\ (0.1)(-3) & (0.1)(-4) + 1 \end{bmatrix} = \begin{bmatrix} 1 & 0.1 \\ -0.3 & 0.6 \end{bmatrix} \tag{6.219}$$

$$H = TB = \begin{bmatrix} (0.1)0 \\ (0.1)1 \end{bmatrix} = \begin{bmatrix} 0 \\ 0.1 \end{bmatrix} \tag{6.220}$$

The system can be expressed in a vector-matrix form:

$$\begin{bmatrix} x_{1,k+1} \\ x_{2,k+1} \end{bmatrix} = \begin{bmatrix} 1 & 0.1 \\ -0.3 & 0.6 \end{bmatrix} \begin{bmatrix} x_{1,k} \\ x_{2,k} \end{bmatrix} + \begin{bmatrix} 0 \\ 0.1 \end{bmatrix} u_k \tag{6.221}$$

$$y_k = \begin{bmatrix} 1 & 0 \end{bmatrix} \begin{bmatrix} x_{1,k} \\ x_{2,k} \end{bmatrix} \tag{6.222}$$

The first order difference equations and the output equation are:

$$x_{1,k+1} = x_{1,k} + 0.1\, x_{2,k} \tag{6.223}$$

$$x_{2,k+1} = -0.3\, x_{1,k} + 0.6\, x_{2,k} + 0.1\, u_k \tag{6.224}$$

$$y_k = x_{1,k} \tag{6.225}$$

When the set of equations is implemented computationally, it is preferable to compute first the state equations since the input affects the state variables, then the output equation.

The discretization using backward Euler method (BEM) to get the difference equations of a second order state space system is simpler than the discretization by the bilinear transform explained next.

5.2 Discretization with Bilinear Transform

The discretization procedure of a state space model using bilinear transform is detailed below.

The state equations in the s-domain are given by:

$$sX(s) = A\,X(s) + B\,U(s) \tag{6.226}$$

$$Y(s) = C\,X(s) + D\,U(s) \tag{6.227}$$

Substitute the derivative term of the left side of the state equation by the bilinear transform equivalence: $s = \dfrac{2}{T}\dfrac{(z-1)}{(z+1)}$, and express the state space system in z-domain:

$$\left[\frac{2}{T}\frac{(z-1)}{(z+1)} \right] X(z) = A\ X(z) + B\ U(z) \tag{6.228}$$

$$Y(z) = C\,X(z) + D\,U(z) \tag{6.229}$$

Multiply by $(z + 1)$ and rearrange the terms:

$$\left[\frac{2}{T}(z-1) \right] X(z) = (z+1)A\ X(z) + (z+1)B\ U(z) \tag{6.230}$$

$$\left[\frac{2}{T}(z-1) \right] X(z) = A(z+1)\ X(z) + B(z+1)\ U(z) \tag{6.231}$$

Make algebraic operations to move the expression $z\,X(z)$ to the left side:

$$\frac{2}{T}zX(z) - \frac{2}{T}X(z) = Az\ X(z) + AX(z) + B(z+1)\ U(z) \tag{6.232}$$

$$\frac{2}{T}zX(z) - Az\ X(z) = \frac{2}{T}X(z) + A\,X(z) + B(z+1)\ U(z) \qquad (6.233)$$

Incorporate the identity matrix and reduce and factorize terms:

$$\left(\frac{2}{T}I - A\right)zX(z) = \left(\frac{2}{T}I + A\right)X(z) + B(z+1)\ U(z) \qquad (6.234)$$

Pre-multiply by $\left(\dfrac{2}{T}I - A\right)^{-1}$

$$zX(z) = \left(\frac{2}{T}I - A\right)^{-1}\left(\frac{2}{T}I + A\right)X(z) + \left(\frac{2}{T}I - A\right)^{-1}B(z+1)\ U(z) \qquad (6.235)$$

In the last expression in the z-domain, G can be identified:

$$G = \left(\frac{2}{T}I - A\right)^{-1}\left(\frac{2}{T}I + A\right) \qquad (6.236)$$

The state space equation is:

$$zX(z) = G\,X(z) + \left(\frac{2}{T}I - A\right)^{-1}B(z+1)\ U(z) \qquad (6.237)$$

The factor (z+1) pre-multiplying $U(z)$ in the state equation can be passed to the output equation pre-multiplying $X(z)$; the new state space system discrete equations are:

$$zX(z) = G\,X(z) + \left(\frac{2}{T}I - A\right)^{-1}B\ U(z) \qquad (6.238)$$

$$Y(z) = C\,(z+1)\,X(z) + D\ U(z) \qquad (6.239)$$

From this new system, we identify H:

$$H = \left(\frac{2}{T}I - A\right)^{-1}B \qquad (6.240)$$

From the output equation:

$$Y(z) = C\,z\,X(z) + C\,X(z) + D\ U(z) \qquad (6.241)$$

We can substitute $z\,X(z)$ from the state equation:

$$Y(z) = C\,[G\,X(z) + H\ U(z)] + C\,X(z) + D\ U(z) \qquad (6.242)$$

Reducing and factorizing terms:

$$Y(z) = C\,[G + I]\,X(z) + [C\,H + D]\ U(z) \qquad (6.243)$$

From this final output equation, E and F can be identified as:

$$E = C\,[G + I] \qquad (6.244)$$

$$F = CH + D \tag{6.245}$$

Example. Now consider the discretization of a two order single input and single output system, with the following matrix and vectors and a sample time of 0.1.

$$A = \begin{bmatrix} 0 & 1 \\ -3 & -4 \end{bmatrix}, \quad B = \begin{bmatrix} 0 \\ 1 \end{bmatrix} \text{ and } C = \begin{bmatrix} 1 & 0 \end{bmatrix} \tag{6.246}$$

With $T = 0.1$,

$$G = \left(\frac{2}{T} I - A \right)^{-1} \left(\frac{2}{T} I + A \right) = \begin{bmatrix} 20 & -1 \\ 3 & 20+4 \end{bmatrix}^{-1} \begin{bmatrix} 20 & 1 \\ -3 & 20-4 \end{bmatrix} \tag{6.247}$$

$$G = \begin{bmatrix} 20 & -1 \\ 3 & 24 \end{bmatrix}^{-1} \begin{bmatrix} 20 & 1 \\ -3 & 16 \end{bmatrix} = \begin{bmatrix} 0.0497 & 0.0021 \\ -0.0062 & 0.0414 \end{bmatrix} \begin{bmatrix} 20 & 1 \\ -3 & 16 \end{bmatrix} \tag{6.248}$$

$$G = \begin{bmatrix} 0.09876 & 0.0828 \\ -0.2484 & 0.6563 \end{bmatrix} \tag{6.249}$$

$$H = \left(\frac{2}{T} I - A \right)^{-1} B = \begin{bmatrix} 0.0497 & 0.0021 \\ -0.0062 & 0.0414 \end{bmatrix} \begin{bmatrix} 0 \\ 1 \end{bmatrix} = \begin{bmatrix} 0.0021 \\ 0.0414 \end{bmatrix} \tag{6.250}$$

$$E = C(G+I) = \begin{bmatrix} 1 & 0 \end{bmatrix} \left(\begin{bmatrix} 0.9876 & 0.0828 \\ -0.2484 & 0.6563 \end{bmatrix} + \begin{bmatrix} 1 & 0 \\ 0 & 1 \end{bmatrix} \right) \tag{6.251}$$

$$E = C(G+I) = \begin{bmatrix} 1 & 0 \end{bmatrix} \begin{bmatrix} 1.9876 & 0.0828 \\ -0.2484 & 1.6563 \end{bmatrix} = \begin{bmatrix} 1.9876 & 0.0828 \end{bmatrix} \tag{6.252}$$

$$F = CH + D = \begin{bmatrix} 1 & 0 \end{bmatrix} \begin{bmatrix} 0.0021 \\ 0.0414 \end{bmatrix} + 0 = 0.0021 \tag{6.253}$$

The discrete time system can be expressed in a vector-matrix form:

$$\begin{bmatrix} x_{1,k+1} \\ x_{2,k+1} \end{bmatrix} = \begin{bmatrix} 0.09876 & 0.0828 \\ -0.2484 & 0.6563 \end{bmatrix} \begin{bmatrix} x_{1,k} \\ x_{2,k} \end{bmatrix} + \begin{bmatrix} 0.0021 \\ 0.0414 \end{bmatrix} u_k \tag{6.254}$$

$$y_k = \begin{bmatrix} 1.9876 & 0.0828 \end{bmatrix} \begin{bmatrix} x_{1,k} \\ x_{2,k} \end{bmatrix} + 0.0021 \, u_k \tag{6.255}$$

The scalar first order difference equations and the output equation are:

$$x_{1,k+1} = 0.09876 \, x_{1,k} + 0.0828 \, x_{2,k} + 0.0021 \, u_k \tag{6.256}$$

$$x_{2,k+1} = -0.2484 \, x_{1,k} + 0.6563 \, x_{2,k} + 0.0414 \, u_k \tag{6.257}$$

$$y_k = 1.9876 \, x_{1,k} + 0.0828 \, x_{2,k} + 0.0021 \, u_k \tag{6.258}$$

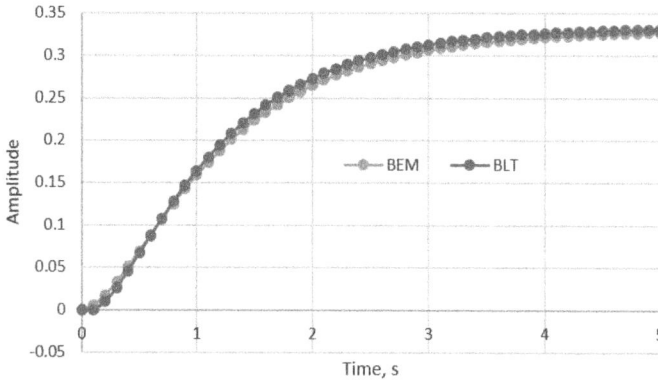

Fig. 6.7. System response from discretizations by Backward Euler Method (BEM) and Bilinear Transform (BLT).

Figure 6.7 shows the step response of both discrete-time state space systems, obtained by Euler backward differences and bilinear transform discretization methods.

6. Conclusions

The major benefit of state space analysis, in contrast, to transfer function analysis, is that it can be extended from single input single output systems to multivariable systems with many inputs and outputs, and, therefore, its applicability encompasses a wide range of systems: linear and nonlinear, time-varying and time-invariant. In control engineering, a state space representation is a mathematical tool for modeling physical systems described by a set of inputs, outputs, and state variables related by first-order differential equations. The set of first-order differential equations are combined into a first-order matrix differential equation. Linear time-invariant systems are ideal to be expressed as vectors-matrix systems. The state of the system can be represented as a vector; for this reason, the state space representation provides a compact and convenient way to model and analyze systems with multiple inputs and outputs.

Controllability and observability are dual concepts that describe the interaction between the external inputs and outputs and the internal state space variables. The system is controllable if the rank of the controllability matrix is the same as the order of the system. Controllability implies that a state space system can be taken from any initial state to a desired state by means of a control vector without restrictions, in an interval of finite time. The system is observable if the rank of the observability matrix is the same as the order of the system. Observability implies that it is possible to determine the state of the system by observing the outputs during a finite time interval. State space transformations are possible and convenient for different purposes. State space models have a more direct representation of the phase variable canonical form, but can be converted to controllable or observable canonical forms through matrix algebra procedures. These alternative representations can later be used for model analysis and state feedback and output feedback control schemes, as reviewed in Chapter 8.

References

Dorf, R. and Bishop, R. 2001. *Modern Control Systems*. Upper Saddle River, NJ: Prentice Hall.

Kuo, B.C. 1992. *Digital Control Systems*. Second Edition. Saunders College Pub., Oxford University Press.

Ogata, K. 2010. *Modern Control Engineering*. Fifth Edition, Upper Saddle River, Pearson.

Veness, T. 2017. *Practical Guide to State-space Control*. Graduate-Level Control Theory for High Schoolers.

Williams II, R.L. and Lawrence, D.A. 2007. *Linear State-Space Control Systems*. John Wiley & Sons, Inc.

Chapter 7

PID Control

1. Introduction

Controllers are used to reduce the difference between the process variable or response and its set point or desired value. In the process industry, more than 90% of applications are implemented with a controller that combines proportional, integral, and derivative control actions, which is called a PID controller. The PID control of position and speed of motors has been researched with an important impact in many plants (Hassan et al., 2017; Maung et al., 2018; Metha et al., 2013; Saranya and Pamela, 2012; Razmjooy et al., 2021). Tuning the PID parameters requires knowledge of the process or system under control, which can be gained by experimentation or theoretical analysis. Continuous time process models are commonly used to implement closed loop digital PID controllers.

Diverse practical tuning rules have been derived for a wide variety of processes that respond similarly to positive and negative changes in the manipulation variable. Some examples are the tuning rules according to the Chien, Hrones, and Reswick method, the Ziegler and Nichols method, and minimum error criteria (Daful, 2018). Although there are several PID tuning techniques, a well-known method is direct synthesis (Chen and Seeborg, 2002). This method uses the model of the plant to obtain the controller that will generate the desired closed loop response, defined by the time constant λ, which must be realizable and adjusted for good disturbance rejection.

In this chapter, the fundamentals of continuous PID controllers are introduced, as well as their discretization and programming. Basic concepts of first-and second-order systems are applied for the controller design. Section 2 explains a novel deduction and interpretation of the proportional-integral (PI) control. Section 3 derives the PI controller for first-order models using the direct synthesis method. Section 4 explains the effect of the derivative term. Section 5 deducts tuning parameters using the direct synthesis method for the second-order system. Section 6 explains how to deal with the first-order process with time delays. Section 7 discusses some considerations for PID implementation and performance evaluation, and Section 8 provides the conclusions.

2. Fundamentals of Proportional and Integral Control

The PID control equation is tuned with three parameters: controller gain k_c, integral time constant τ_i, and derivative time constant τ_d, see (7.1). The error $E(s)$ is the difference of reference $R(s)$ minus the process variable $Y(s)$. The manipulation, $U(s)$, is the superposition of the three controller effects. The controller equation in the complex frequency s-domain, using the Laplace Transform, is given by:

$$U(s) = k_c \left(1 + \frac{1}{\tau_i s} + \tau_d s \right) E(s) \tag{7.1}$$

The proportional and integral terms are commonly used. The derivative effect is often more difficult to use properly. To explain all the controller terms, reviewing the open and closed loop control schemes is necessary.

2.1 Open Loop Control

In an open loop control scheme, the controller is a system that has a reference signal as input, r_n, which is the desired value to be reached by the process output, y_n. The controller requires to compute a manipulation value, m_n, which is the input to be applied to the process (see Fig. 7.1).

Let us assume that the process is a first-order system in continuous time, with a time constant $\tau = 1$ and unit gain:

$$G_p(s) = \frac{1}{s+1} \tag{7.2}$$

If the reference is generated with a step signal, $u(t)$, and the controller is only a constant unit gain, the same unit step signal is applied to the process; in this case, the process variable will take the form of the unit step response, $s(t)$:

$$y(t) = s(t) = (1 - exp(-t))u(t) \tag{7.3}$$

A deviation or error signal between the reference and the output can be computed as follows:

$$e(t) = r(t) - s(t) = u(t) - (1 - exp(-t))u(t) = exp(-t)u(t) \tag{7.4}$$

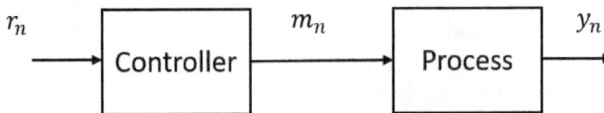

Fig. 7.1. Open loop control scheme.

Under the previously mentioned conditions, the error will take the form of an exponentially decaying function, and, interestingly, the unit step signal required to be generated by the controller can also be obtained mathematically by adding the exponentially decaying error signal to the step response:

$$m(t) = e(t) + s(t) = (exp(-t) + 1 - exp(-t))u(t) = u(t) \tag{7.5}$$

In other words, a control signal can be generated in terms of the error and the integral of the error, since $\int_0^t exp(-t/\tau)dt = \frac{1}{\tau}[1 - exp(-t/\tau)]$ and $\tau = 1$. We can express the manipulation as the addition of a component proportional to the error signal and a component with the integral of the error signal (weighted by the constant factor k_i):

$$m(t) = e(t) + k_i \int e(t)dt \tag{7.6}$$

2.2 Closed Loop Control

In a closed loop system, the error signal is calculated by substracting the process variable from the reference signal:

$$e_n = r_n - y_n \tag{7.7}$$

The calculation of the error needed for the computational implementation of the controller implies that the system is now in a feedback control scheme, as depicted in Fig. 7.2.

Digital controllers are implemented with a sampling period, and discrete equations are needed. The manipulation variable in the continuous time PI equation (7.6) can be converted and computed in discrete time with the following difference equation:

$$m_n = m_{n-1} + \left[(e_n - e_{n-1}) + \frac{T_s}{\tau_i}(e_n) \right] \tag{7.8}$$

where T_s is the sample time, and τ_i is the integral time constant ($\tau_i = 1/k_i$). This controller can have a component proportional to the error (with unit gain) added to an integral of the error ($k_i = 1$); graphically, this can be seen in Fig. 7.3 with $T_s = 0.1$. Under this scenario, the manipulation is a unit step signal in discrete time, that is obtained by adding the error and the integral of the error. Therefore, the proportional and integral components can be used to compute a control signal.

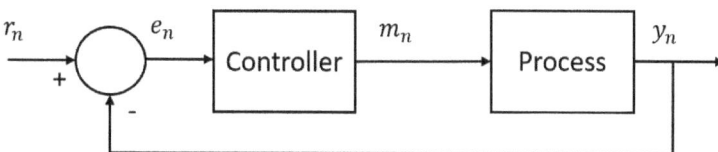

Fig. 7.2. Closed loop control scheme.

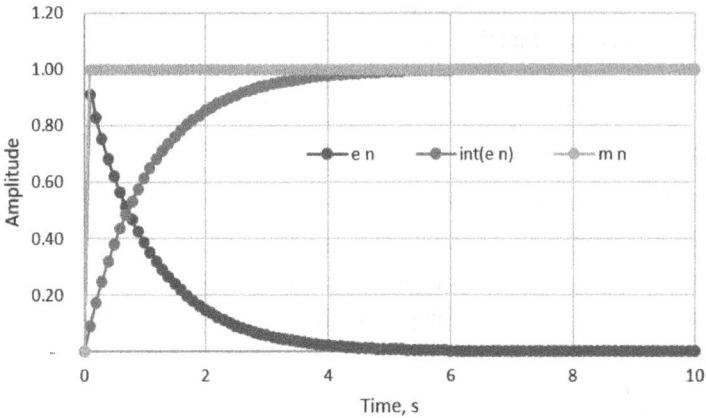

Fig. 7.3. Error and integral of the error in discrete time.

3. PI Controller for First-Order Systems

At this point, the essential proportional and integral controller components have been obtained assuming a first-order process with unit gain. The direct synthesis method uses the closed loop desired response of the plant defined by the time constant λ.

3.1 Direct Synthesis Method

A controller can be defined in the frequency domain. Let us assume that the process is a first-order system with a time constant $\tau = 1$ and unit gain:

$$G_p(s) = \frac{1}{s+1} \tag{7.9}$$

Now consider the specification of a desired closed-loop response $F(s)$:

$$F(s) = \frac{1}{\lambda s + 1} \tag{7.10}$$

The controller, $D(s)$, is required to obtain the desired closed-loop response:

$$\frac{Y(s)}{R(s)} = \frac{D(s)G_p(s)}{1 + D(s)G_p(s)} = F(s) \tag{7.11}$$

$$F(s) = \frac{D(s)G_p(s)}{1 + D(s)G_p(s)} \tag{7.12}$$

$$F(s)(1 + D(s)\,G_p(s)) = D(s)\,G_p(s) \tag{7.13}$$

$$F(s) + F(s)D(s)\,G_p(s) = D(s)\,G_p(s) \tag{7.14}$$

$$F(s) = D(s)\,G_p(s) - F(s)D(s)\,G_p(s) \tag{7.15}$$

$$F(s) = D(s)\,G_p(s)(1 - F(s)) \tag{7.16}$$

Solving for D implies inverting the plant and substituting the desired closed-loop response:

$$D(s) = G_p^{-1}(s)\frac{F(s)}{1-F(s)} = G_p^{-1}(s)\frac{\dfrac{1}{\lambda s+1}}{1-\dfrac{1}{\lambda s+1}} \tag{7.17}$$

$$D(s) = G_p^{-1}(s)\frac{1}{\lambda s+1-1} = G_p^{-1}(s)\frac{1}{\lambda s} \tag{7.18}$$

3.2 First-Order Process

When the process is a first-order system, the controller not only seeks to invert the process gain but also to produce a closed loop dynamics (with unit gain to avoid a steady state error) by canceling the pole of the plant and specifying a desired time constant λ or velocity of response.

The first-order system is expressed in terms of its gain k_p and its time constant τ_p:

$$G_p(s) = \frac{k_p}{\tau_p s+1} \tag{7.19}$$

The cancelation of the original dynamic response can be proposed using the inverse of the transfer function:

$$G_p^{-1}(s) = \frac{\tau_p s+1}{k_p} \tag{7.20}$$

A simple basic controller $D(s)$ in the frequency domain can be defined as incorporating an integrator function; this is necessary to make the order of the denominator equal the order of the numerator and specify the controller as a casual system, but, more important, to obtain the desired closed-loop response specified by the time constant λ:

$$D(s) = G_p^{-1}(s)\left(\frac{1}{\lambda s}\right) = \frac{\tau_p s+1}{k_p}\left(\frac{1}{\lambda s}\right) \tag{7.21}$$

The closed-loop system response has a unit gain and a time constant λ:

$$\frac{Y(s)}{R(s)} = \frac{\dfrac{\tau_p s+1}{k_p}\left(\dfrac{1}{\lambda s}\right)\dfrac{k_p}{\tau_p s+1}}{1+\dfrac{\tau_p s+1}{k_p}\left(\dfrac{1}{\lambda s}\right)\dfrac{k_p}{\tau_p s+1}} \tag{7.22}$$

$$\frac{Y(s)}{R(s)} = \frac{\dfrac{1}{\lambda s}}{1+\dfrac{1}{\lambda s}} = \frac{1}{\lambda s+1} \tag{7.23}$$

The controller transfer function can also be written as a proportional term plus an integral term as follows:

$$D(s) = \frac{M(s)}{E(s)} = \frac{\tau_p s + 1}{k_p \lambda s} = \frac{\tau_p}{k_p \lambda} \left(1 + \frac{1}{\tau_p s} \right)$$

(7.24)

The Proportional-Integral controller transfer function can be therefore expressed as:

$$\frac{M(s)}{E(s)} = k_c \left[1 + \frac{1}{\tau_i s} \right]$$

(7.25)

with controller gain and integral time constant parameters defined by:

$$k_c = \frac{\tau_p}{k_p \lambda}, \ \tau_i = \tau_p$$

(7.26)

The manipulation variable can be solved from (7.25) and expressed in the frequency domain as:

$$M(s) = k_c \left[1 + \frac{1}{\tau_i s} \right] E(s)$$

(7.27)

The above equation can be multiplied by a differential operator to obtain a first-order differential equation in the frequency domain to be converted to the continuous-time domain and, later, to a difference equation:

$$sM(s) = k_c \left[s + \frac{1}{\tau_i} \right] E(s)$$

(7.28)

The PI expressed as a differential equation is:

$$\frac{dm(t)}{dt} = k_c \left[\frac{de(t)}{dt} + \frac{1}{\tau_i} e(t) \right]$$

(7.29)

The discretization by the backward Euler method results in the PI control law:

$$\frac{m_n - m_{n-1}}{T_s} = k_c \left[\frac{e_n - e_{n-1}}{T_s} + \frac{1}{\tau_i} e_n \right]$$

(7.30)

$$m_n = m_{n-1} + k_c \left[(e_n - e_{n-1}) + \frac{T_s}{\tau_i} (e_n) \right]$$

(7.31)

An alternative expression for the PI control law in discrete time considers a moving average of the last two error samples in the integral term:

$$m_n = m_{n-1} + k_c \left[(e_n - e_{n-1}) + \frac{T_s}{2\tau_i} (e_n + e_{n-1}) \right]$$

(7.32)

Consider a PI controller with $k_c = 0.5$, $\tau_i = 1$, $T_s = 0.1$, and a closed loop time constant $\lambda = 1$ for a simulated first-order process with $k_p = 2$ and $\tau_p = 1$. The following program implements the PI control law in the work cycle of the controller:

```c
// Proportional-Integral Controller Code in C Programing Language
float mn=0;
float mn1=0;
float en=0;
float en1=0;
float kc=0.5;
float ti=1;
float ts=0.1;
// PI control equation
en=rn-yn;
mn=mn1+kc*((en-en1)+(ts/(2*ti)*(en+en1)))
//
If (mn>mmax) {mn=mmax;}
If (mn<mmin) {mn=mmin;}
mn1=mn;
en1=en;
```

When the closed loop time constant λ is set equal to the same value of the process time constant τ_p, and the controller gain is calculated as the inverse of the process gain ($k_c = 1/k_p$), a quasi-step manipulation signal in discrete time is produced. The control law is a difference equation based on the error that automatically generates the required step signal at the manipulation to obtain the system's step response (Fig. 7.4).

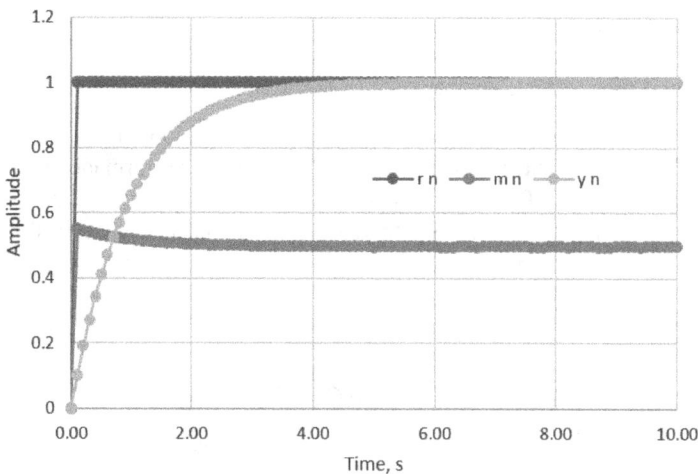

Fig. 7.4. Closed-loop response of a first-order process with a PI and $\lambda = \tau_p$.

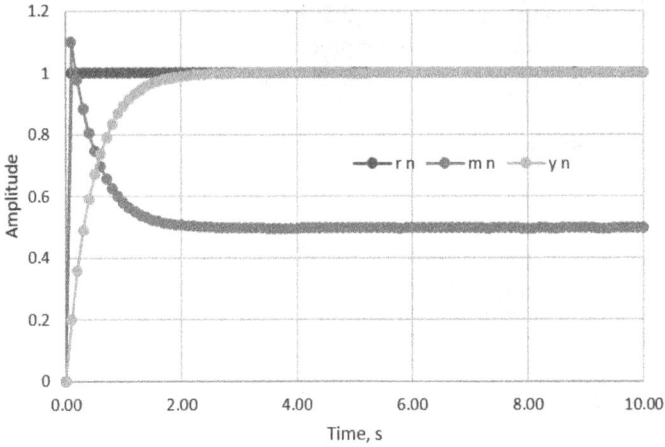

Fig. 7.5. Closed-loop response of a first-order process with a PI and $\lambda = \tau_p/2$.

Now assume a PI controller with $k_c = 1$, $\tau_i = 2$, $T_s = 0.1$, and a desired closed loop time constant $\lambda = 1$, for a simulated first-order process with $k_p = 2$ and $\tau_p = 2$. When the closed loop controlled system response is required to be faster $(\lambda = \tau_p/2)$, a manipulation variable that is initially higher than its final steady state value is computed. This can be interpreted as an over manipulation required for a faster closed-loop response (Fig. 7.5).

Example. Consider the circuit of Fig. 7.6 with a resistor and a capacitor in series, with resistance $R = 5\ k\Omega$ and capacitance $C = 100\ \mu F$, respectively. The input voltage $V_i(t)$, is and the output voltage is $V_0(t)$.

The transfer function, given the specific resistance and capacitance values, is:

$$G_p(s) = \frac{V_o(s)}{V_i(t)} = \frac{1}{RCs+1} = \frac{1}{0.5s+1} \tag{7.33}$$

with a natural process time constant of $\tau_p = 0.5\ s$.

The PI controller can be proposed to achieve a closed loop time constant $\lambda = 0.2$ s by defining $k_c = 2.5$ and $\tau_i = 0.5s$. Figure 7.7 shows the performance of the closed loop system with $T_s = 0.01\ s$ when a voltage of 10 V is required: a high initial input voltage is applied in a closed loop to reach the desired reference in 0.8 seconds instead of 2 seconds in open loop.

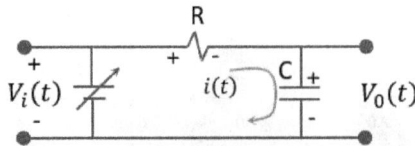

Fig. 7.6. RC circuit.

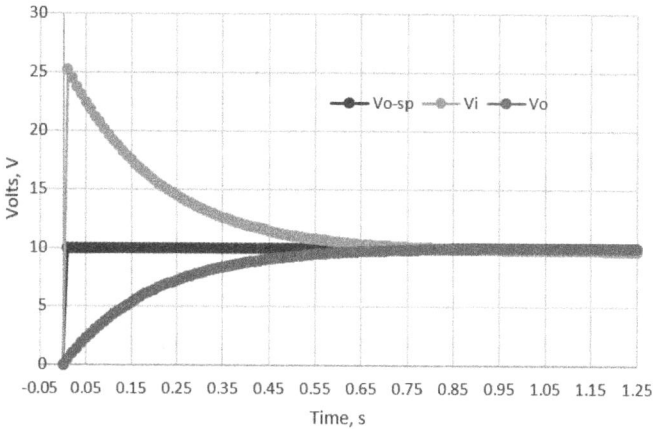

Fig. 7.7. Closed-loop response of the output voltage of an RC circuit with a PI controller.

3.3 First Order with Integrator

First-order systems consisting of a pure integrator are found in many processes, such as the speed integration into position and the volumetric flow integrated into volume. The process can be expressed in terms of the integration function of a gain k_p:

$$G_p(s) = \frac{k_p}{s} \tag{7.34}$$

The controller can be defined as a simple proportional controller since an integrator in the process already introduces a time constant in a closed loop.

$$D(s) = k_c \tag{7.35}$$

The closed-loop response is:

$$\frac{Y(s)}{R(s)} = \frac{D(s)G_p(s)}{1 + D(s)G_p(s)} = \frac{k_c \dfrac{k_p}{s}}{1 + k_c \dfrac{k_p}{s}} = \frac{k_c k_p}{s + k_c k_p} \tag{7.36}$$

The time constant depends on the controller gain if $k_c = 1/k_p$. The closed-loop response is:

$$\frac{Y(s)}{R(s)} = \frac{\dfrac{1}{s}}{1 + \dfrac{1}{s}} = \frac{1}{s + 1} \tag{7.37}$$

The proportional controller is the simplest loop controller. The proportional controller operates according to this principle: the more the actual value differs from the set point, the larger the manipulated variable will be.

4. Derivative Controller Effect

The PI behavior is derived directly from the error at any moment, which makes it fast and dynamically stable. However, certain process disturbances are not fully compensated. The derivative term can be added to a PI controller for a first-order process to improve performance under disturbances. The derivative effect is also useful when the process is modeled by a second-order system with complex conjugate poles.

4.1 Derivative Term on the Error

The Proportional-Integral-Derivative controller transfer function is:

$$\frac{M(s)}{E(s)} = k_c \left[1 + \frac{1}{\tau_i s} \tau_d s \right] = k_c \left[\frac{\tau_i \tau_d s^2 + \tau_i s + 1}{\tau_i s} \right] \qquad (7.38)$$

where τ_d is the derivative time constant, which can be seen as a prediction time horizon to estimate a future error given the current rate of change or time derivative of the error signal; a small value for this parameter is adequate for a better error forecast and an assertive manipulation change based on this prediction.

From the transfer function of the PID controller, the manipulation variable can be solved in terms of the error, and the application of a differential operator in the frequency domain produces:

$$sM(s) = k_c \left[\tau_d s^2 + s + \frac{1}{\tau_i} \right] E(s) \qquad (7.39)$$

In the time domain, the PID differential equation is given by:

$$\frac{dm(t)}{dt} = k_c \left[\tau_d \frac{d^2 e(t)}{dt^2} + \frac{de(t)}{dt} + \frac{1}{\tau_i} e(t) \right] \qquad (7.40)$$

Discretization by Euler method of backward differences leads to:

$$\frac{m_n - m_{n-1}}{T_s} = k_c \left[\tau_d \frac{e_n - 2e_{n-1} + e_{n-2}}{T_s^2} + \frac{e_n - e_{n-1}}{T_s} + \frac{1}{\tau_i} e_n \right] \qquad (7.41)$$

Solving for m_n leads to the PID control law:

$$m_n = m_{n-1} + k_c \left[(e_n - e_{n-1}) + \frac{T_s}{\tau_i}(e_n) + \frac{\tau_d}{T_s}(e_n - 2e_{n-1} + e_{n-2}) \right] \qquad (7.42)$$

Alternatively, the moving average of the last two discrete error values can be used in the integral term giving the following expression of the PID control law in discrete time:

$$m_n = m_{n-1} + k_c \left[(e_n - e_{n-1}) + \frac{T_s}{2\tau_i}(e_n + e_{n-1}) + \frac{\tau_d}{T_s}(e_n - 2e_{n-1} + e_{n-2}) \right] \qquad (7.43)$$

The PID control algorithm can be programmed as shown:

```
// Proportional-Integral-Derivative Controller Code in C Programing Language
float mn=0;
float mn1=0;
float en=0;
float en1=0;
float en2=0;
float kc=0.5;
float ti=0.5;
float td=0.1;
float ts=0.1;
// PID control equation
en=rn-yn;
mn=mn1+kc*((en-en1)+(ts/(2*ti)*(en+en1)+td*(en-2*en1+en2)/ts)
//
If (mn>mmax) { mn=mmax; }
If (mn<mmin) { mn=mmin; }
mn1=mn;
en2=en1;
en1=en;
```

Consider a closed-loop system simulated with a sample time $T_s = 0.1$ during a set point change in the presence of a variable disturbance. The process parameters are $k_p = 2$, and $\tau_p = 2$, and the controller parameters are $k_c = 0.5$, $\tau_i = 2$. The derivative term is evaluated from the second cycle or sample time after the set point change to avoid an initial abrupt manipulation. If the derivative time constant is null ($\tau_d = 0$), the presence of the additive disturbance (p_k) at the process input has a higher effect on the response (Fig. 7.8). On the other hand, if the derivative effect is used, $\tau_d \neq 0$,

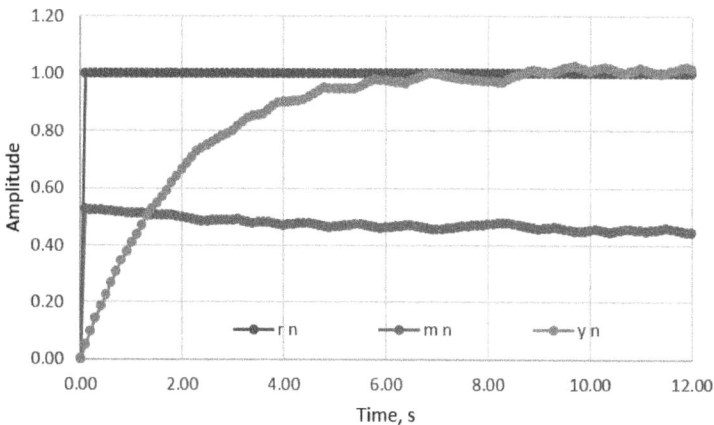

Fig. 7.8. Closed-loop response of a first-order process in the presence of disturbance with a PI.

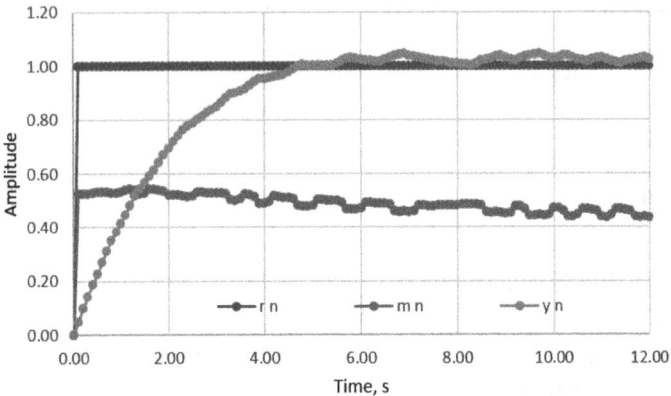

Fig. 7.9. Closed-loop response of a first-order process in the presence of disturbance with a PID.

the manipulation is adjusted (spikes in the manipulation curve) to keep a faster and tighter control of the response (Fig. 7.9).

Like the proportional and integral terms within a PID controller, the derivative term aims to correct the error. As valuable as the third term can be in maintaining adequate control, experience suggests that the appropriate use of the derivative is unclear.

As the rate of error increases or decreases, so does the size of the derivative response. This aspect of the derivative action makes it ideal for some uses. For example, the nonlinearity of the process may cause the process behavior to change in different operation zones or at different operation time stages; the derivative effect is then convenient to foresee the trend of the process variable (or its error with respect to a reference) when it is not accurately modeled and manipulate the process more effectively. Nevertheless, applying the derivative effect may not be practical or easy to implement in many cases.

To avoid big manipulations when the reference is modified, it is convenient to compute and add the derivative term two samples after the set-point change. The maximum manipulation value can be bounded to reduce the initial derivative effect. However, there are other ways to implement the derivative component.

An alternative PID controller equation includes a filter time constant τ_a. The filter time constant is conventionally assigned to be $1/10$ of the derivative time constant, or at least the execution time (sample time) in the discrete time implementation.

$$M(s) = k_c \left(1 + \frac{1}{\tau_i s} + \frac{\tau_d s}{1 + \tau_a s} \right) E(s) \tag{7.44}$$

4.2 Derivative Term on the Process Variable

When considering the use of the derivative term, it is helpful to keep the following in mind: most of the time, the reference value remains constant. Therefore, the time derivative of the error signal can be approximated to the negative of the derivative of the process response:

$$\frac{de(t)}{dt} = \frac{d(r(t) - y(t))}{dt} \approx -\frac{dy(t)}{dt} \tag{7.45}$$

With this approximation, the controller equation is:

$$M(s) = k_c \left[\left(1 + \frac{1}{\tau_i s} \right) E(s) - \tau_d s Y(s) \right] \tag{7.46}$$

Although "derivative on error" is technically correct, "derivative on measurement" is the most suitable form of the PID equation for industrial applications. From a practical point of view, the math associated with "derivative on error" can result in excessive spikes in the controller output's behavior, often referred to as derivative kick. In contrast, "derivative on measurement" reduces the sensitivity to set point changes and is more appropriate for practical applications. However, the noise in the measurement causes the derivative to have abrupt changes and may also cause control problems and wear out the actuator. For this reason, a first-order filter is often used on the derivative effect of the controller:

$$M(s) = k_c \left[\left(1 + \frac{1}{\tau_i s} \right) E(s) - \frac{\tau_d s Y(s)}{\alpha \tau_d s + 1} \right] \tag{7.47}$$

where α is a fraction, typically equal to 0.1, to make the filter time constant significantly small with respect to the derivative time constant.

The discretization attenuates the adverse measurement noise effect on the manipulation because the derivative is not instantaneous but based on a finite sample time period. For this reason, the PID controller with derivative action on the process variable can be expressed in the discrete time simply as:

$$m_n = m_{n-1} + k_c \left[(e_n - e_{n-1}) + \frac{\tau_s}{2\tau_i} (e_n + e_{n-1}) - \frac{\tau_d}{T_s} (y_n - 2y_{n-1} + y_{n-2}) \right] \tag{7.48}$$

4.3 Filtering of the Process Variable

If the noise in the process variable is high and affects the performance of the controller, an infinite impulse response (IIR) filter, with parameter β, as the filter pole, can be used to filter the noise of the process variable sensors:

$$y_n = \beta y_{n-1} + (1 - \beta) \cdot y_m \tag{7.49}$$

where y_m is the measured process variable (process output). The IIR filter can be specified considering a time constant to be at least one half the process time constant to ensure that the filter's cut-off frequency is greater than the cut-off frequency of the process. Therefore, the parameter β of the equation of the filter can be defined as:

$$\beta = \frac{\dfrac{\tau_p}{2}}{\dfrac{\tau_p}{2} + T} \tag{7.50}$$

5. PID Controller for Second-Order Systems

Consider now a damped second-order process expressed in terms of two first-order time constants:

$$G_p(s) = \frac{k_p}{(\tau_1 s + 1)(\tau_2 s + 1)} \tag{7.51}$$

where τ_1 and τ_2 are different for an overdamped system or equal for a critically damped system.

The controller is again calculated from the inverse of the process transfer function and the integrator $\dfrac{1}{\lambda s}$:

$$D(s) = \frac{(\tau_1 s + 1)(\tau_2 s + 1)}{k_p} \left(\frac{1}{\lambda s} \right) \tag{7.52}$$

The analysis of the closed-loop response transfer function leads to a simplified first-order dynamics with a desired time constant λ:

$$\frac{Y(s)}{R(s)} = \frac{\dfrac{(\tau_1 s + 1)(\tau_2 s + 1)}{k_p} \left(\dfrac{1}{\lambda s} \right) \dfrac{k_p}{(\tau_1 s + 1)(\tau_2 s + 1)}}{1 + \dfrac{(\tau_1 s + 1)(\tau_2 s + 1)}{k_p} \left(\dfrac{1}{\lambda s} \right) \dfrac{k_p}{(\tau_1 s + 1)(\tau_2 s + 1)}} = \frac{1}{\lambda s + 1} \tag{7.53}$$

The controller can be represented as a ratio of two polynomials:

$$D(s) = \frac{\tau_1 \tau_2 s^2 + (\tau_1 + \tau_2)s + 1}{k_p \lambda s} \tag{7.54}$$

This controller resembles the structure of a PID:

$$D(s) = k_c \left[\frac{\tau_i \tau_d s^2 + \tau_i s + 1}{\tau_i s} \right] \tag{7.55}$$

Algebraic manipulation allows us to determine the PID parameter values. If we consider the following:

$$\tau_i = \tau_1 + \tau_2 \tag{7.56}$$

and multiply (7.54) by $(\tau_1 + \tau_2)/(\tau_1 + \tau_2)$:

$$k_c \left[\frac{\tau_i \tau_d s^2 + \tau_i s + 1}{\tau_i s} \right] = \frac{\tau_1 + \tau_2}{k_p \lambda} \left(\frac{\tau_1 \tau_2 s^2 + (\tau_1 + \tau_2)s + 1}{(\tau_1 + \tau_2)s} \right) \tag{7.57}$$

We can further match the coefficients by assigning:

$$k_c = \frac{\tau_1 + \tau_2}{k_p \lambda} \tag{7.58}$$

$$\tau_i \tau_d = \tau_1 \tau_2 = (\tau_1 + \tau_2) \tau_d \tag{7.59}$$

From (7.59), the derivative time constant of the controller is obtained:

$$\tau_d = \frac{\tau_1 \tau_2}{\tau_1 + \tau_2} \tag{7.60}$$

Example. Consider the spring-damper-mass system shown in Fig. 7.10.

The second order differential equation for this system is given by:

$$m \frac{d^2 Z(t)}{dt^2} + b \frac{dZ(t)}{dt} + kZ(t) = F(t) + mg \tag{7.61}$$

where Z is the displacement, F is the force input, g is the gravity, m is mass, k is the spring stiffness, and b is the damping coefficient of the damper.

The evaluation at the initial steady state is:

$$kZ(t = 0) = F(t = 0) + mg \tag{7.62}$$

The subtraction of the initial steady state evaluation from the differential equation can be expressed in terms of deviation variables $z(t)$ and $f(t)$:

$$\frac{d^2 z(t)}{dt^2} + \frac{b}{m} \frac{dz(t)}{dt} + \frac{k}{m} z(t) = f(t) \tag{7.63}$$

Let us assume the physical parameters of the system determine the coefficients of the differential equation as shown next:

$$\frac{d^2 z(t)}{dt^2} + 1.5 \frac{dz(t)}{dt} + 0.5z(t) = f(t) \tag{7.64}$$

where z is the process variable (PV), and f is the control variable (CV).

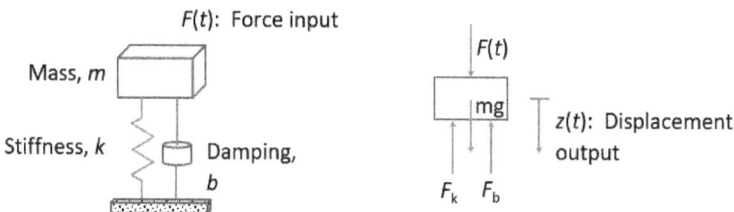

Fig. 7.10. Spring-damper-mass system and its free body diagram.

In the domain of the complex variable s, the transfer function is:

$$G_p(s) = \frac{Z(s)}{F(s)} = \frac{1}{s^2 + 1.5s + 0.5} = \frac{2}{(2s+1)(s+1)} \tag{7.65}$$

The factorization of the denominator allows us to distinguish two process time constants of 2 and 1 time units. According to (7.56) and (7.60):

$$\tau_i = 3, \quad \tau_d = 2/3 \tag{7.66}$$

The selection of the closed loop time constant affects the controller gains as in (7.58):

$$k_c = \begin{cases} 0.75, \lambda = 2 \\ 1.0, \lambda = 1.5 \\ 1.5, \lambda = 1 \end{cases} \tag{7.67}$$

The process is simulated with the following difference equation:

$$\frac{z_n - 2z_{n-1} + z_{n-2}}{T_s^2} = -1.5\frac{z_n - z_{n-1}}{T_s} - 0.5z_n + f_n \tag{7.68}$$

$$z_n = \frac{1}{1 + 0.5T_s^2 + 1.5T_s}[(1.5T_s + 2)z_{n-1} - z_{n-2} + T_s^2 f_n] \tag{7.69}$$

and the controller is given by the difference equation (7.43) with the evaluation of the derivative term two sample times after the set point change.

The closed-loop response for different values is shown in Fig. 7.10 through 7.12. First, $\lambda = 2$ is used (Fig. 7.11). Then, a higher speed response is achieved with $\lambda = 1.5$; the manipulation reaches a pick value slightly higher than its final value (Fig. 7.12). Finally, an even higher speed is attained with $\lambda = 1$, with a very high pick value in the manipulation (Fig. 7.13).

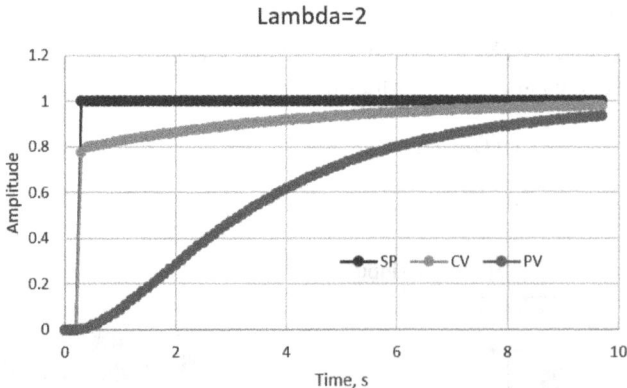

Fig. 7.11. Closed-loop response of second-order process with PID controller using $\lambda = 2$.

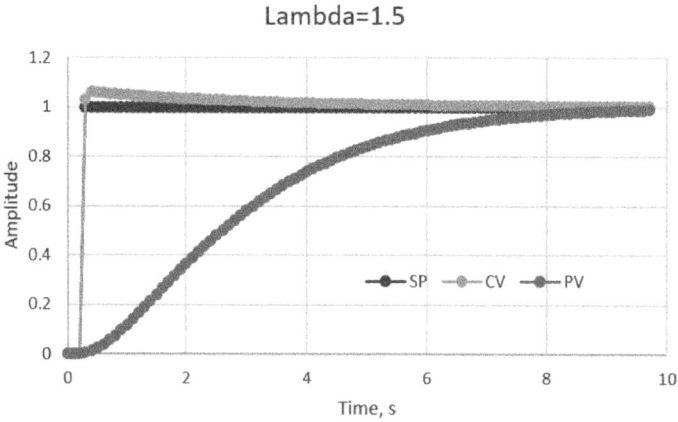

Fig. 7.12. Closed-loop response of second-order process with PID controller using $\lambda = 1.5$.

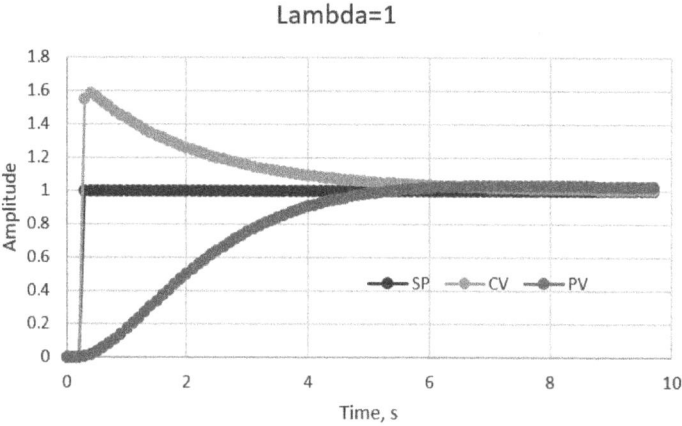

Fig. 7.13. Closed-loop response of second-order process with PID controller using $\lambda = 1$.

Example. Now consider an underdamped second-order system:

$$G_p(s) = \frac{k_p}{\tau^2 s^2 + 2\tau\zeta s + 1} \tag{7.70}$$

where τ is a second order time constant, and ζ is the damping factor less than 1 for an underdamped response (complex conjugate process poles).

The controller $D(s)$ is calculated as:

$$D(s) = \frac{\tau^2 s^2 + 2\tau\zeta s + 1}{k_p}\left(\frac{1}{\lambda s}\right) \tag{7.71}$$

Multiplying numerator and denominator by $2\tau\zeta$:

$$D(s) = \frac{2\tau\zeta}{k_p\lambda}\frac{\tau^2 s^2 + 2\tau\zeta s + 1}{2\tau\zeta s} \tag{7.72}$$

The parameters of a PID can be defined in terms of the factors and coefficients of $D(s)$:

$$D(s) = k_c \left[\frac{\tau_i \tau_d s^2 + \tau_i s + 1}{\tau_i s} \right]_i = \frac{2\tau\zeta}{k_p \lambda} \frac{\tau^2 s^2 + 2\tau\zeta s + 1}{2\tau\zeta s} \tag{7.73}$$

$$k_c = \frac{2\tau\zeta}{k_p \lambda} \tag{7.74}$$

$$\tau_i = 2\tau\zeta \tag{7.75}$$

$$\tau_i \tau_d = \tau^2 \tag{7.76}$$

$$\tau_d = \frac{\tau}{2\zeta} \tag{7.77}$$

It is worth noticing that these tuning formulas or equations (7.73) through (7.77) apply to any stable second-order system, underdamped, critically damped, and overdamped. The analysis of a second-order process in terms of its time constant and damping ratio is more general than the analysis of a second-order system with two first-order time constants.

Consider the spring-damper-mass system of Fig. 7.10 but with elements having characteristics that lead to the following process transfer function:

$$G_p(s) = \frac{1}{s^2 + s + 1} \tag{7.78}$$

with $k_p = 1$, $\tau = 1$ and $\zeta = 0.5$. The open loop settling time is equal to 8 time units. For a closed-loop stabilization in 4 time units, $\lambda = 1$ and the following controller parameters are obtained using the tuning equations (7.74) through (7.77):

$$k_c = 1, \tau_i = 1, \tau_d = 1 \tag{7.79}$$

The process difference equation needed for the simulation is obtained as follows:

$$\frac{z_n - 2z_{n-1} + z_{n-2}}{T_s^2} = -\frac{z_n - z_{n-1}}{T_s} - z_n + f_n \tag{7.80}$$

$$z_n = \frac{1}{1 + T_s^2 + T_s}[(T_s + 2)z_{n-1} - z_{n-2} + T_s^2 f_n] \tag{7.81}$$

The controller difference equation is given by (7.43). Simulation of the closed-loop system in Fig. 7.14 verifies the desired non-oscillatory response and reduced settling time of 4 time units (half with respect to the open loop stabilization time of 8 time units). In this case, the derivative term is evaluated from the beginning of the set point change and, as expected, the initial manipulation varies rapidly reaching a high peak value. The initial suppression of the derivative part of the controller would avoid the sudden manipulation change but would only approach the desired behavior with less accuracy than the one achieved in Fig. 7.14.

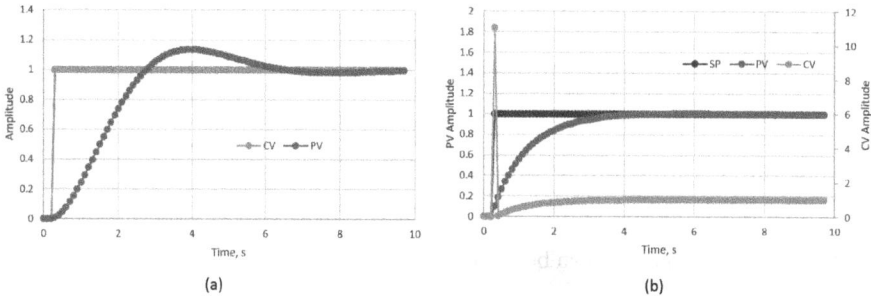

Fig. 7.14. Underdamped process simulated with sample time $T_s = 0.1$: (a) Open-loop Response, (b) Closed-loop Response with PID.

6. PID Controller for Processes with Time Delay

Generally, a dead time in the response of a process can be expected. A first-order process with a time delay can be converted to a second order with the first-order Pade approximation, and the tuning rules of a PID for second-order systems can be applied.

The first-order Pade approximation to the dead time exponential is given by:

$$e^{-\theta s} = \frac{1-0.5\theta s}{1+0.5\theta s} \tag{7.82}$$

A first-order system with dead time can be approximated by:

$$G_p(s) = \frac{k_p(1-0.5\theta s)}{(\tau s +1)(1+0.5\theta s)} \tag{7.83}$$

The controller $D(s)$ is calculated again with the inverse of $G_p(s)$ and an integrator $\dfrac{1}{\lambda s}$, but this time a first-order factor is also included to cancel the unstable pole introduced by the inverse of $G_p(s)$:

$$D(s) = \frac{(\tau s +1)(1+0.5\theta s)}{k_p(1-0.5\theta s)}\left(\frac{1-0.5\theta s}{\lambda s}\right) = \frac{(\tau s +1)(1+0.5\theta s)}{k_p \lambda s} = \frac{0.5\theta \tau s^2 + (\tau +0.5\theta)s +1}{k_p \lambda s} \tag{7.84}$$

This controller fits the PID structure with a second-order polynomial in the numerator and a first-order factor in the denominator. To define the parameters of the PID, the numerator and denominator are multiplied by the constant $\tau + 0.5\theta$ (coefficient of 's' in the numerator):

$$D(s) = k_c \frac{\tau_i \tau_d s^2 + \tau_i s + 1}{\tau_i s} = \frac{(\tau +0.5\theta)[0.5\theta \tau s^2 + (\tau +0.5\theta)s +1]}{k_p \lambda (\tau +0.5\theta)s} \tag{7.85}$$

Now the coefficients can be related to the PID parameters by defining the following tuning rules:

$$\tau_i = \tau + 0.5\theta \tag{7.86}$$

$$k_c = \frac{\tau + 0.5\theta}{k_p \lambda} \tag{7.87}$$

$$\tau_i \tau_d = 0.5\theta\tau \tag{7.88}$$

$$\tau_d = \frac{0.5\theta\tau}{\tau + 0.5\theta} \tag{7.89}$$

The closed-loop dynamics can be verified by:

$$\frac{Y(s)}{R(s)} = \frac{\frac{1}{\lambda s}(1 - 0.5\theta s)}{1 + \frac{1}{\lambda s}(1 - 0.5\theta s)} = \frac{1 - 0.5\theta s}{\lambda s + (1 - 0.5\theta s)} \tag{7.90}$$

$$\frac{Y(s)}{R(s)} = \frac{1 - 0.5\theta s}{(\lambda - 0.5\theta)s + 1} \tag{7.91}$$

This equation corresponds to first-order dynamics with time constant $\lambda - 0.5\theta$. For the closed-loop constant to be approximately equal to λ, this parameter should be defined as reasonably large, assuring stability at the same time, that is, satisfying the condition:

$$\lambda > 0.5\theta \tag{7.92}$$

The design in the case of having dead time can also be based on the use of the Taylor approximation:

$$e^{-\theta s} = 1 - \theta s \tag{7.93}$$

The process model is considered as:

$$G_p(s) = \frac{k_p(1 - \theta s)}{\tau s + 1} \tag{7.94}$$

The integral effect, absent in $G_p(s)$, is considered in the controller calculation, and the positive zero of $G_p(s)$ is canceled to avoid a positive pole in the controller transfer function:

$$D(s) = \frac{\tau s + 1}{k_p(1 - \theta s)}\left(\frac{1 - \theta s}{\lambda s}\right) = \frac{\tau s + 1}{k_p \lambda s} \tag{7.95}$$

This controller is a PI with $k_c = \dfrac{\tau}{k_p \lambda}$ and $\tau_i = \tau$. The closed loop transfer function corroborates the first-order dynamics again:

$$\frac{Y(s)}{R(s)} = \frac{\frac{1}{\lambda s}(1 - \theta s)}{1 + \frac{1}{\lambda s}(1 - \theta s)} = \frac{\lambda s(1 - \theta s)}{\lambda s + (1 - \theta s)} \tag{7.96}$$

$$\frac{Y(s)}{R(s)} = \frac{1 - \theta s}{(\lambda - \theta)s + 1} \tag{7.97}$$

The closed loop time constant, $\lambda - \theta$, is approximately equal to λ, for values considerably larger than the dead time, which meets the stability criterion:

$$\lambda > \theta \tag{7.98}$$

Using a simpler expression to approximate the dead time leads to a simpler controller structure, where the dead time is not explicitly considered for tuning the controller parameters.

Example. Assume a first-order plus dead-time process with a transfer function given by:

$$G_p(s) = \frac{3e^{-2s}}{4s + 1} \tag{7.99}$$

and represented in discrete time by the recurrence equation:

$$4\frac{y_n - y_{n-1}}{T_s} + y_n = 3x_{n-d} \tag{7.100}$$

where d is the integer number of sample times included in the dead time of the process; for $T_s = 0.1$, $d = 20$. The simplified recurrence equation is:

$$y_n = \frac{4}{4.1}y_{n-1} + \frac{3}{4.1}(0.1)x_{n-20}, \quad n \geq 20; \quad y_n = 0, \quad n < 20 \tag{7.101}$$

For the controller design, consider $\lambda = 3$. If the Pade approximation for the dead time is used, the controller parameters for the PID of (7.43), according to (7.86) to (7.89), are:

$$\tau_i = 5, k_c = \frac{5}{9}, \tau_d = 0.8 \tag{7.102}$$

If the Taylor approximation is used, the control parameters are defined as in (7.26):

$$k_c = \frac{4}{9}, \tau_i = 4 \tag{7.103}$$

for the PI implemented by equation (7.32).

Figure 7.15 shows the simulation of the closed loop system using the PID based on the model with the Pade dead-time approximation (applying the derivative effect two samples after the step reference change): the response settles in less than 12 time units; abrupt changes in the manipulation are observed due to the derivative controller action and dead-time effect.

Figure 7.16 shows the performance of the closed loop system with the PI (based on the model with the Taylor approximation for the dead time): a smoother and slower response that settles in 12 seconds after an overshoot due to the effect of the positive zero of the closed loop transfer function.

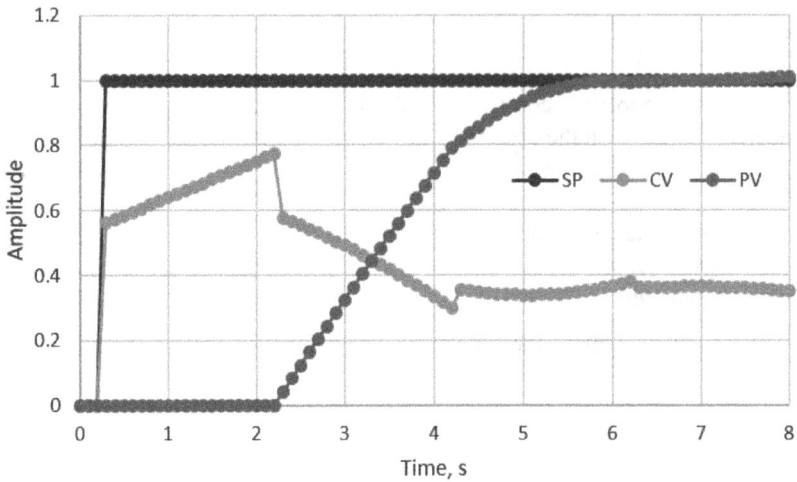

Fig. 7.15. The process with time delay and a PID controller designed using the Pade Dead-Time Approximation.

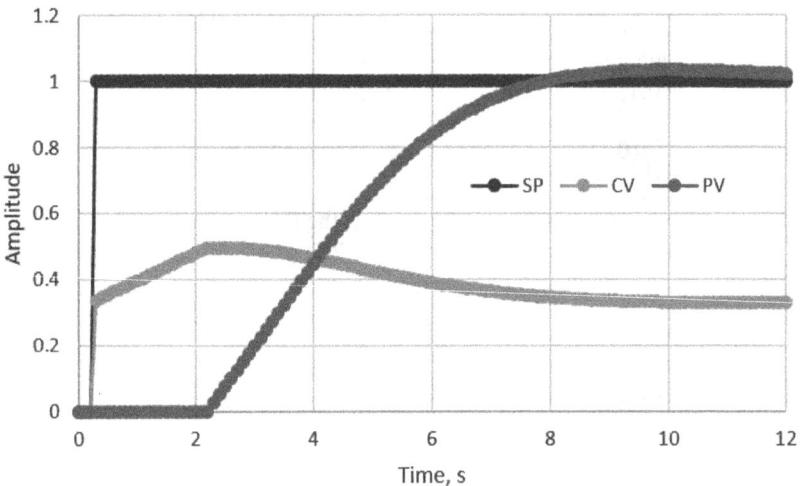

Fig. 7.16. The process with time delay and a PI controller designed using the Taylor Dead-Time Approximation.

7. Implementation Considerations and Performance Evaluation

Implementing discrete-time controllers in computational platforms, such as computers, microcontrollers, and programmable controllers, requires digital processing of the sensor and actuator variables (see Fig. 7.17). The sensors measure the process variables converted to digital values in an analog-to-digital converter (ADC). Process variables are fed back and subtracted from the reference signal to generate the error signal and passed to the controller algorithm with the difference equation that generates the controller output or manipulation that is then converted

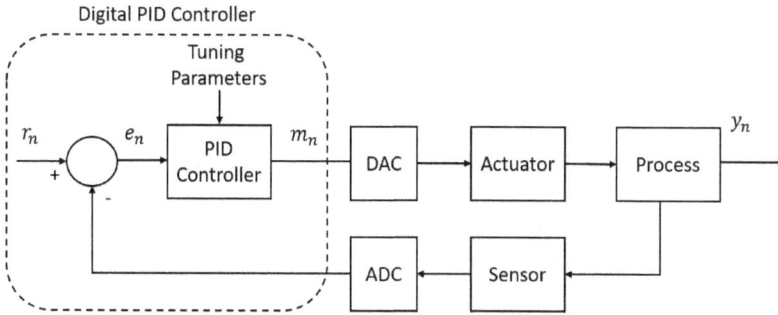

Fig. 7.17. Digital PID implementation with ADC and DAC converters.

back to an analog value in a digital-to-analog converter (DAC) or modulated by pulse width or amplitude. The range of the manipulation or control variable must be limited to the range of the actuator.

The advantages of digital controller implementation are accomplished by properly selecting computer and interface hardware (Salem, 2013). The data types of the variables related to the digital controller must be chosen preferably as floating to achieve accuracy due to the quantization and resolution of digitization. The performance of the digital controller must be high enough that its real time response is as precise as the one that an analog controller could obtain.

7.1 Hardware and Software Platform

The selection of the computer and interface hardware should observe some quality criteria related to real-time operation and accuracy.

Execution Time

The digital computer works in variable scan times or in fixed-time intervals. The execution time must be preferably fixed and high enough that a significant change of the process variable cannot occur during a time interval.

Resolution of the ADC and DAC Converters

The chosen resolution must ensure that both the process and control variables operate in the required range of the sensors and actuators.

Sampling Rate of the ADC and DAC Converters

The sampling frequency is the rate at which the analog input values are measured and digitalized at the ADC, and the manipulated output variables are applied to the actuators. This sampling rate must be high enough so the controller can respond to changes in the controlled process variable and produce a control manipulation in real time.

7.2 Bumpless Transfer and Anti-Reset Windup

Successful implementation of a PID requires additional functionalities to improve the system's stability and response. In manual operation, the system is operated in an open loop, and automatic operation, the system is operated in closed-loop control. Change from manual to automatic mode requires the implementation of a bumpless transfer to prevent an abrupt change of the manipulated variable. For a smooth transition, the set point value is equal to the value of the process variable; with this set point, the initial error is zero, and the controller manipulation remains unchanged until the user changes the set point in the automatic mode.

Set point changes can also be applied with linear or logarithmic ramps unless the process is required to respond fast, and a step change is preferred. The anti-reset windup function is used to prevent the integral term of the manipulated variable from continuing to increase by specifying a maximum value when a control deviation cannot be corrected due to the manipulated variable limitation or saturation.

7.3 Evaluating the Performance of Closed Loop Controllers

For evaluation and comparison of controllers, performance indexes are computed. Popular performance indexes are the integral of time multiplied by the squared error (7.104), the integral of the absolute magnitude of the error (7.105), and the control signal total variation (7.106). Their discrete equivalents, adapted from de Moura Oliveira et al. (2020), are the following, where N is the number of samples:

$$ITSE = \sum_{k=0}^{N} k\, e^2(k) T_s^2 \tag{7.104}$$

$$IAE = \sum_{k=0}^{N} |e(k)| T_s \tag{7.105}$$

$$TV_u = \sum_{k=1}^{N} |u_k - u_{k-1}| \tag{7.106}$$

The error indices ITSE and IAE compute the errors to quantify which control scheme or algorithm is best at set-point tracking. The TV index compares the control signals and shows which scheme applies less amplitude and variations (i.e., voltage) to the actuators.

8. Conclusions

The Proportional-Integral-Derivative controller is the workhorse controller in industry systems for both disturbances and set point changes because it is robust and relatively simple to implement and tune. PID can be programmed in devices ranging from small microcontrollers to large control systems. A PID controller is a feedback control strategy widely used in applications requiring continuously modulated control. PID uses feedback to continuously adjust the output of a process or system to match a desired set point. PID controllers are widely used in various applications, including temperature control, flow control, and motor control, due to the PID's ability to provide stable and accurate control. PID is well understood both in industry and academia. PID controllers are also used as a comparison reference for advanced controllers such as adaptive, predictive, robust, and optimal control strategies or soft computing techniques such as fuzzy logic controllers and neural networks.

The output of a PID controller is calculated using the sum of the proportional, integral, and derivative terms. Each term of the PID seeks to complement the others and adds incremental value to the manipulation or control variable toward controlling the process dynamics. The proportional term reacts proportionally to the error; the integral term is proportional to the cumulative error over time. The integral term helps to eliminate the steady state error and can improve the stability of the control system. The derivative term contributes to a fast response to error deviation since it is proportional to the rate of change of the error. The derivative term can amplify measurement noise (random fluctuations) and cause excessive output changes. Filters are important to estimate the process variable rate of change better.

A PID design procedure is presented in this chapter based on the inverse of the process model and the inclusion of an integrator in case the process does not have such an effect. The presented method takes advantage of the availability of the process model and the direct specification of the desired closed loop dynamics to define the controller. A process with an integrator requires only a proportional controller. The effect of the integral term is that a sustained control deviation is compensated. This means that manipulated variable becomes larger and larger despite the constant control deviation. If disturbances occur in a system, they can be rejected with the derivative-action component. The derivative term can be calculated with the rate of change of the error or the rate of change of the process variable. Derivative action is also required for better control of second-order systems. Practical considerations for PID control include correct selection of computer and interface hardware and some process control good practices such as bumpless transfer and the anti-reset windup. PID controllers are subject to evaluation by performance indicators based on calculating the error or the manipulation effort.

References

Chen, D. and Seborg, D.E. 2002. PI/PID controller design based on direct synthesis and disturbance rejection. *Ind. Eng. Chem. Res.* 41: 4807–4822.

Daful, A.G. 2018. Comparative Study of PID Tuning Methods for Processes with Large and Small Delay Times. *Advances in Science and Engineering Technology International Conferences (ASET)*, pp. 1–7.

de Moura Oliveira, P.B., John D. Hedengren and Solteiro Pires, E.J. 2020. Swarm-based design of proportional integral and derivative controllers using a compromise cost function: an arduino temperature laboratory case study. *Algorithms* 13(12): 315.

Hassan, A.A., Al-Shamaa, N.K. and Abdalla, K.K. 2017. Comparative study for DC motor speed control using PID controller. *IJET* 9(6): 4181–4192.

Maung, M.M., Latt, M.M. and new, C.M. 2018. DC motor angular position control using PID controller with friction compensation. *IJSRP* 8(11): 149–155.

Metha, S., Shah, P. and Vaidya, V. 2013. Design and comparative study of PID controller tuning method from IMC Tuned 2-DOF pole placement parameter structure for the DC motor speed control application. *Nirma University International Conference on Engineering (NUiCONE)*, pp. 1–4.

Razmjooy, N., Vahedi, Z., Estrela, V., Padilha, R. and Monteiro, A.C. 2021. Speed control of a DC motor using PID controller based on improved whale optimization algorithm. pp. 153–167. *In*: Razmjooy, N., Ashourian, M. and Foroozandeh, Z. (eds.). *Metaheuristics and Optimization in Computer and Electrical Engineering.*

Salem, F.A. 2013. Modeling, Simulation, controller selection, and design of electric motor for mechatronics motion applications, using different control strategies and verification using Matlab/Simulink. *European Scientific Journal* 9(27).

Saranya, M. and Pamela, D. 2012. A real time IMC tuned PID controller for DC motor. *IJRTE* 1(1): 65–69.

Chapter 8

Model-Based Control

1. Introduction

Model-based control (MBC) refers to several advanced control techniques typically implemented using a model of the process, plant, or mechanism to obtain a desired closed loop control response. MBC techniques incorporate knowledge about the process (gain, time constant, and dead time). Three well known control schemes considering a model of the process are internal model control, model reference adaptive control, and model predictive control. The internal model control (IMC) relies on the internal model principle, which expresses that a feedback controller can better compensate for disturbances using a model of the process. The IMC tuning rules have been derived for PI and PID controllers, and they have proven to be practical for designing controllers with good dynamical response and robust performance.

Model reference adaptive control asserts the control strategy of incorporating a reference model to provide additional control actions that can optionally include adaptive mechanisms. Adaptive control is a technique used for adjusting the parameters of a plant in real time to maintain a desired level of performance when the system's parameters are unknown and change with time (Shekhar and Sharma, 2018). Control methods in state space also incorporate the model of the process, that is, the state space model. There are simple approaches to controller design, one of them the full state feedback, also known as pole placement. The desired closed-loop poles that the state feedback will generate are chosen such that the system meets the performance requirements.

This chapter presents the analytical process for controller synthesis using the internal model control principle to design conventional PID controllers and other model-based controllers. Section 2 on the IMC principle explains the incorporation of the process model in the control loop. Section 3 on IMC-based PID control presents the development of the PID as a general controller and derives parameter tuning equivalences of PID direct synthesis. Section 4 presents control schemes and practical applications using the model reference adaptive control. Section 5 gives a basic introduction to state space control methods. Conclusions are in Section 6.

2. Internal Model Control Principle

The internal model principle was introduced in the early 1970s. The IMC design procedure has been developed in many forms, including single input, single output (SISO) systems, and multiple input multiple output (MIMO) systems. There are both continuous time and discrete time design procedures. Design procedures for unstable open loop systems (Tan et al., 2003; Yang et al., 2002) and combined feedback feedforward IMC design (Vilanova et al., 2009) have also been researched.

The Internal Model Control (IMC) principle is a controller design procedure that cancels open loop dynamics with the inverse of the transfer function and incorporates an integrator in the controller to achieve the specified closed-loop dynamical response. The complexity of the IMC controller depends on the order of the model and performance requirements. The IMC design procedure helps design basic feedback control schemes with PID controllers.

Consider the control scheme of Fig. 8.1, where a controller component is represented by $Q(s)$, and the process and its model are given by $G_p(s)$ and $G_m(s)$, respectively. The signals are reference $R(s)$, error $E(s)$, manipulation $M(s)$, process output $Y(s)$, and model output $\hat{Y}(s)$.

The diagram is slightly modified in Fig. 8.2 by putting the summation points together to emphasize $Y(s)$ as the process output and $\hat{Y}(s)$ or process variable estimation, as an internal signal.

A further modification in Fig. 8.3, redefining summation points with the corresponding changes of signal signs, achieves the separation of $Y(s)$ and $\hat{Y}(s)$ at different summation points.

The internal closed loop of Fig. 8.3 can be solved to obtain the controller $D(s)$, whose representation leads to the typical feedback control scheme, where the controller is concentrated in a single block as depicted in Fig. 8.4.

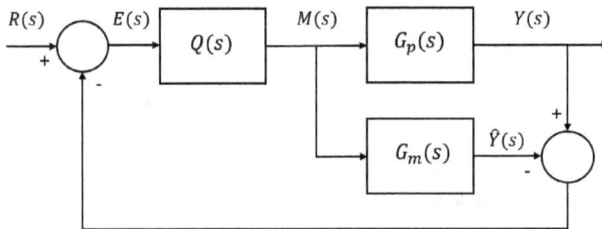

Fig. 8.1. IMC scheme.

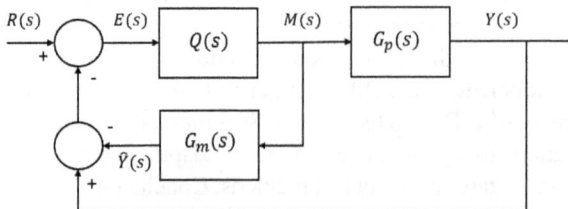

Fig. 8.2. IMC scheme rearranged.

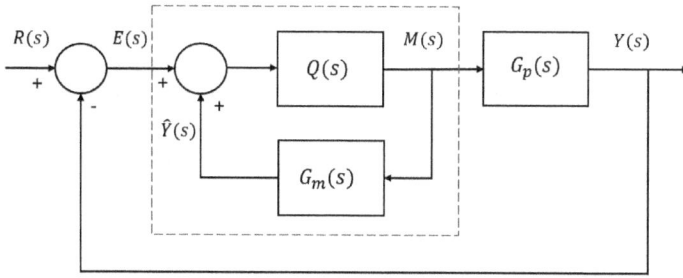

Fig. 8.3. IMC controller inside dotted square.

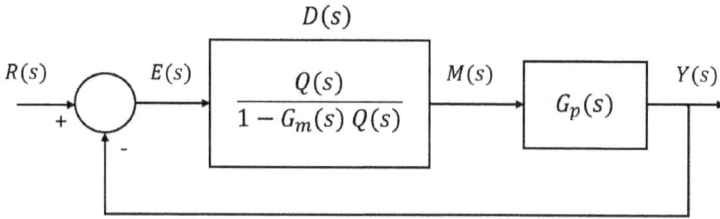

Fig. 8.4. IMC controller solved in terms of component $Q(s)$ and process model $G_m(s)$.

3. IMC-Based PID Type Controllers

Internal Model Control (IMC) for PID tuning was presented by Rivera et al. (1986), and since then it has been widely accepted for designing model-based PID controllers. The PID design methods based on the IMC principle allow controlling the closed-loop dynamics assuring the stability of the control system. IMC-based PID tuning methods play an important role in PID design due to robustness to model mismatch.

This section focuses on the design procedure for linear and stable process systems, with a particular interest in the specification of tuning criteria for PID controllers. Here we consider a first-order process with no dead time and derive the IMC controller. The standard feedback controller based on the plant model is given by:

$$D(s) = \frac{Q(s)}{1 - G_m(s)Q(s)} \tag{8.1}$$

3.1 First-Order Process

Let us assume an exact model by stating:

$$G_m(s) = G_p(s) = \frac{k_p}{\tau_p s + 1} \tag{8.2}$$

The desired closed-loop first-order dynamics with time constant λ is:

$$F(s) = \frac{1}{\lambda s + 1} \tag{8.3}$$

The component $Q(s)$ is defined as:

$$Q(s) = G_p^{-1}(s)F(s) = \frac{(\tau_p s + 1)}{k_p}\left(\frac{1}{\lambda s + 1}\right) \tag{8.4}$$

Substitution of $G_m(s)$ and $Q(s)$ in the controller $D(s)$ equation gives:

$$D(s) = \frac{\dfrac{(\tau_p s + 1)}{k_p}\left(\dfrac{1}{\lambda s + 1}\right)}{1 - \dfrac{(\tau_p s + 1)}{k_p}\left(\dfrac{1}{\lambda s + 1}\right)\dfrac{k_p}{(\tau_p s + 1)}} = \frac{\dfrac{(\tau_p s + 1)}{k_p}\left(\dfrac{1}{\lambda s + 1}\right)}{1 - \left(\dfrac{1}{\lambda s + 1}\right)} \tag{8.5}$$

Multiplying by $\lambda s + 1$ and rearranging terms and factors:

$$D(s) = \frac{(\tau_p s + 1)/k_p}{\lambda s + 1 - 1} = \frac{1}{k_p}\left(\frac{\tau_p s + 1}{\lambda s}\right) = \frac{1}{k_p \lambda}\left(\tau_p + \frac{1}{s}\right) = \frac{\tau_p}{k_p \lambda}\left(1 + \frac{1}{\tau_p s}\right) \tag{8.6}$$

It can be observed that the controller has proportional and integral components; that is, it is equivalent to a PI controller. For this reason, this controller is called IMC-based PI. The resultant PI is the same as the one previously obtained, in Chapter 7, with $k_c = \tau_p/(k_p \lambda)$ and $\tau_i = \tau_p$. The conclusion is that a PI can be considered a standard feedback controller, implying a first-order model of the process.

3.2 Second-Order Process

The IMC principle can be applied to second-order processes, obtaining the same controller structure and parameters as those obtained by the procedure of Section 5 of Chapter 7.

Consider now a damped second-order process expressed in terms of two first-order time constants:

$$G_p(s) = \frac{k_p}{(\tau_1 s + 1)(\tau_2 s + 1)} \tag{8.7}$$

where τ_1 and τ_2 are different for an overdamped system or equal for a critically damped system.

The closed-loop specification $F(s)$ is again a first-order dynamics with time constant λ, and the controller is calculated from (8.1):

$$Q(s) = G_p^{-1}(s)F(s) = \frac{(\tau_1 s + 1)(\tau_2 s + 1)}{k_p}\left(\frac{1}{\lambda s + 1}\right) \tag{8.8}$$

$$D(s) = \frac{\dfrac{(\tau_1 s + 1)(\tau_2 s + 1)}{k_p}\left(\dfrac{1}{\lambda s + 1}\right)}{1 - \dfrac{(\tau_1 s + 1)(\tau_2 s + 1)}{k_p}\left(\dfrac{1}{\lambda s + 1}\right)\dfrac{k_p}{(\tau_1 s + 1)(\tau_2 s + 1)}} = \frac{\dfrac{(\tau_1 s + 1)(\tau_2 s + 1)}{k_p}\left(\dfrac{1}{\lambda s + 1}\right)}{1 - \left(\dfrac{1}{\lambda s + 1}\right)} \tag{8.9}$$

$$D(s) = \frac{(\tau_1 s + 1)(\tau_2 s + 1)}{k_p \lambda s} \tag{8.10}$$

The controller can be represented as a ratio of two polynomials:

$$D(s) = \frac{\tau_1 \tau_2 s^2 + (\tau_1 + \tau_2)s + 1}{k_p \lambda s} \tag{8.11}$$

This controller resembles the structure of a PID:

$$D(s) = k_c \left[\frac{\tau_i \tau_d s^2 + \tau_i s + 1}{\tau_i s} \right] \tag{8.12}$$

Algebraic manipulation allows us to determine the PID parameter values. If we consider the following:

$$\tau_i = \tau_1 + \tau_2 \tag{8.13}$$

and multiply the right side of (8.11) by $(\tau_1 + \tau_2)/(\tau_1 + \tau_2)$:

$$k_c \left[\frac{\tau_i \tau_d s^2 + \tau_i s + 1}{\tau_i s} \right] = \frac{\tau_1 + \tau_2}{k_p \lambda} \left(\frac{\tau_1 \tau_2 s^2 + (\tau_1 + \tau_2)s + 1}{(\tau_1 + \tau_2)s} \right) \tag{8.14}$$

we can further match the coefficients by assigning:

$$k_c = \frac{\tau_1 + \tau_2}{k_p \lambda} \tag{8.15}$$

$$\tau_i \tau_d = \tau_1 \tau_2 = (\tau_1 + \tau_2) \tau_d \tag{8.16}$$

and solving for the derivative time constant of the controller:

$$\tau_d = \frac{\tau_1 \tau_2}{\tau_1 + \tau_2} \tag{8.17}$$

3.3 First-Order Process with Dead Time

For processes with dead time, using a dead time approximation causes the representation of a positive process zero, leading to differences between the IMC design and the tuning rules previously discussed. The IMC procedure for a process with dead time should consider keeping the process zero in the desired closed loop transfer function to avoid unstable poles in the controller. When using the Pade approximation for dead time, the closed-loop transfer function is:

$$F(s) = \frac{1 - 0.5\theta s}{\lambda s + 1} \tag{8.18}$$

In the case of using the Taylor dead-time approximation:

$$F(s) = \frac{1 - \theta s}{\lambda s + 1} \tag{8.19}$$

As exemplified above, the rest of the calculations can be applied to complete the controller design, which is left to the reader as an exercise.

3.4 IMC for High-Order Processes

Now consider a process of order three or superior. The controller is intended to compensate the dominant poles of the system. To illustrate the design, a third-order system can be used. Assume the two dominant poles, regardless of the type, real or complex, are represented by a second-order polynomial, which can be expressed in terms of its characteristic parameters τ and ζ. A third pole equal to, where τ_3 is a positive time constant, with less effect on the dynamical response (real pole further left on the negative side of the real axis of the complex plane), is also present and modeled:

$$G_m(s) = G_p(s) = \frac{k_p}{(\tau^2 s^2 + 2\tau\zeta s + 1)(\tau_3 s + 1)} \tag{8.20}$$

The closed-loop specification can still be represented by a desired approximated first-order time constant for the regulated response, with or without including the non-dominant pole. If $F(s)$ considers the first-order factor that produces the non-dominant pole as in (8.21), the controller $D(s)$ shown in (8.24) has a particular structure that does not correspond to the PID controller: third-order numerator over a second-order denominator. The calculation procedure is stated from (8.21) to (8.24):

$$F(s) = \frac{1}{(\lambda s + 1)(\tau_3 s + 1)} \tag{8.21}$$

$$Q(s) = G_p^{-1}(s)F(s) = \frac{(\tau^2 s^2 + 2\tau\zeta s + 1)(\tau_3 s + 1)}{k_p(\lambda s + 1)(\tau_3 s + 1)} = \frac{\tau^2 s^2 + 2\tau\zeta s + 1}{k_p(\lambda s + 1)} \tag{8.22}$$

$$D(s) = \frac{Q(s)}{1 - G_m(s)Q(s)} = \frac{\dfrac{\tau^2 s^2 + 2\tau\zeta s + 1}{k_p(\lambda s + 1)}}{1 - \dfrac{k_p}{(\tau^2 s^2 + 2\tau\zeta s + 1)(\tau_3 s + 1)} \dfrac{(\tau^2 s^2 + 2\tau\zeta s + 1)}{k_p(\lambda s + 1)}} \tag{8.23}$$

$$D(s) = \frac{(\tau^2 s^2 + 2\tau\zeta s + 1)(\tau_3 s + 1)}{k_p[(\tau_3 s + 1)(\lambda s + 1) - 1]} = \frac{(\tau^2 s^2 + 2\tau\zeta s + 1)(\tau_3 s + 1)}{k_p[\tau_3 \lambda s^2 + (\tau_3 + \lambda)s]} \tag{8.24}$$

Suppose $F(s)$ does not include the non-dominant pole τ_3, as in (8.25). In that case, the IMC calculation procedure leads to a controller with a third-order numerator and a first-order denominator, which is not a PID type controller. The calculation of the controller is detailed by equations (8.25) through (8.28):

$$F(s) = \frac{1}{\lambda s + 1} \tag{8.25}$$

$$Q(s) = G_p^{-1}(s)F(s) = \frac{(\tau^2 s^2 + 2\tau\zeta s + 1)(\tau_3 s + 1)}{k_p(\lambda s + 1)} \tag{8.26}$$

$$D(s) = \frac{Q(s)}{1 - G_m(s)Q(s)} = \frac{\dfrac{(\tau^2 s^2 + 2\tau\zeta s + 1)(\tau_3 s + 1)}{k_p(\lambda s + 1)}}{1 - \dfrac{k_p}{(\tau^2 s^2 + 2\tau\zeta s + 1)(\tau_3 s + 1)}\dfrac{(\tau^2 s^2 + 2\tau\zeta s + 1)(\tau_3 s + 1)}{k_p(\lambda s + 1)}} \tag{8.27}$$

$$D(s) = \frac{(\tau^2 s^2 + 2\tau\zeta s + 1)(\tau_3 s + 1)}{k_p \lambda s} \tag{8.28}$$

In any case, the controller $D(s)$ does not fit the PID structure. Instead of using the IMC principle, a simple exploration of the closed-loop transfer function with a predefined PID controller structure can be helpful in the cancellation of the dominant poles of the third-order process. The closed-loop response is given by:

$$\frac{Y(s)}{R(s)} = \frac{D(s)G_p(s)}{1 + D(s)G_p(s)} = \frac{k_c\left(\dfrac{\tau_i\tau_d s^2 + \tau_i s + 1}{\tau_i s}\right)\dfrac{k_p}{(\tau^2 s^2 + 2\tau\zeta s + 1)(\tau_3 s + 1)}}{1 + k_c\left(\dfrac{\tau_i\tau_d s^2 + \tau_i s + 1}{\tau_i s}\right)\dfrac{k_p}{(\tau^2 s^2 + 2\tau\zeta s + 1)(\tau_3 s + 1)}} \tag{8.29}$$

To compensate for the main poles of the system, the coefficients for the controller second-order polynomial are made equal to the coefficients of the process second-order polynomial to determine τ_i and τ_d:

$$\tau_i = 2\tau\zeta \tag{8.30}$$

$$\tau_i \tau_d = \tau^2 \tag{8.31}$$

$$\tau_d = \frac{\tau}{2\zeta} \tag{8.32}$$

After the cancellation of factors of the controller and the process, the closed-loop transfer function can be further simplified to a second-order system:

$$\frac{Y(s)}{R(s)} = \frac{D(s)G_p(s)}{1 + D(s)G_p(s)} = \frac{k_c k_p\left(\dfrac{1}{\tau_i s}\right)\dfrac{1}{(\tau_3 s + 1)}}{1 + k_c k_p\left(\dfrac{1}{\tau_i s}\right)\dfrac{1}{(\tau_3 s + 1)}} \tag{8.33}$$

$$\frac{Y(s)}{R(s)} = \frac{k_c k_p}{\tau_i s(\tau_3 s + 1) + k_c k_p} = \frac{1}{\dfrac{\tau_i \tau_3}{k_c k_p}s^2 + \dfrac{\tau_i}{k_c k_p}s + 1} \tag{8.34}$$

A critically damped closed-loop response can be proposed, and this condition can be used to tune the controller gain. The closed loop second-order constant τ_c, and the closed-loop damping ratio, ζ_c, are:

$$\tau_c = \sqrt{\frac{\tau_i \tau_3}{k_c k_p}} \tag{8.35}$$

$$\zeta_c = \frac{1}{2}\sqrt{\frac{\tau_i}{k_c k_p \tau_3}} \tag{8.36}$$

Then k_c can be solved for $\zeta_c = 1$ specifically:

$$k_c = \frac{1}{4}\frac{\tau_i}{k_p \tau_3} \tag{8.37}$$

with this controller gain for the critically damped condition, the settling time of the closed-loop system would be $4\tau_c$.

Example. Design a controller for the following process with output Y and input X:

$$G_p(s) = \frac{Y(s)}{X(s)} = \frac{8}{(s^2 + 1.6s + 9.64)(s+2)} = \frac{0.4149}{(0.1037s^2 + 0.1660s + 1)(0.2s + 1)} \tag{8.38}$$

The poles of the system are $s_1 = -0.8 + 3j$, $s_2 = -0.8 - 3j$, and $s_3 = -5$. The complex poles represented by the second-order factor dominate over the real pole.

According to (8.20), the parameters of this process are:

$$k_p = 0.4149,\ \tau = 0.3221,\ \zeta = 0.2577,\ \tau_3 = 0.2 \tag{8.39}$$

The expected open-loop response is oscillatory (because of $\zeta < 1$) and with a stabilization time of 5 time units from the complex dominant poles or $4\tau/\zeta$.

With (8.30), (8.32), and (8.37), the values for the controller parameters are calculated as:

$$\tau_i = 0.1660,\ \tau_d = 0.625,\ k_c = 0.5 \tag{8.40}$$

The closed-loop performance is expected to have no oscillation and a time constant given by:

$$\tau_c = 0.4 \tag{8.41}$$

and a stabilization time of 1.6 time units.

For the simulation of the closed-loop system, the discrete equation of the controller is:

$$m_n = m_{n-1} + k_c\left[(e_n - e_{n-1}) + \frac{T_s}{2\tau_i}(e_n + e_{n-1}) + \frac{\tau_d}{T_s}(e_n - 2e_{n-1} + e_{n-2})\right] \tag{8.42}$$

and the discrete equation of the third-order process has to be obtained. In terms of the parameters of (8.20), the continuous equation in the 's' domain is:

$$\tau^2 \tau_3 s^3 Y(s) = -(2\tau\zeta\tau_3 + \tau^2) s^2 Y(s) - (\tau_3 + 2\tau\zeta)sY(s) - Y(s) + k_p X(s) \qquad (8.43)$$

The Euler approximations for the first and second derivative of the output have already been used in the previous chapter. The second-order derivative changes the first-order one over one sampling period. Similarly, the third-order derivative is the change of the second derivative in a sample time interval T_s, therefore:

$$\frac{d^3Y}{dt^3} \approx \frac{\dfrac{y_k - 2y_{k-1} + y_{k-2}}{T_s^2} - \dfrac{y_{k-1} - 2y_{k-2} + y_{k-3}}{T_s^2}}{T_s} = \frac{y_k - 3y_{k-1} + 3y_{k-2} - y_{k-3}}{T_s^3} \qquad (8.44)$$

The use of the discrete approximations for the continuous derivatives leads to the following process discrete model:

$$y_k \left(\frac{\tau^2 \tau_3}{T_s^3} + \frac{2\tau\zeta\tau_3 + \tau^2}{T_s^2} + \frac{\tau_3 + 2\tau\zeta}{T_s} + 1 \right) = \left(\frac{3\tau^2 \tau_3}{T_s^3} + \frac{4\tau\zeta\tau_3 + 2\tau^2}{T_s^2} + \frac{\tau_3 + 2\tau\zeta}{T_s} \right) y_{k-1} +$$
$$\left(-\frac{3\tau^2 \tau_3}{T_s^3} - \frac{2\tau\zeta\tau_3 + \tau^2}{T_s^2} \right) y_{k-2} + \frac{\tau^2 \tau_3}{T_s^3} y_{k-3} + k_p x_k \qquad (8.45)$$

The effect of the designed model-based PID can be observed in Fig. 8.5. The closed-loop response has higher damping than the open-loop response. The closed-loop response stabilizes faster (in 1.6 time units as calculated) than the open-loop response (more than 2 time units as it can be observed; less than calculated) as expected. The manipulated variable exhibits a peak because the error value changes drastically at the time of the reference change, and the derivative term produces a high value. There are several possible ways to avoid this issue: the first is to limit this rate of change of the error to a maximum value, and the second is to apply the derivative term only on the process variable.

The disadvantage, in this case, is that the velocity of the response is not directly addressed by a closed-loop parameter. However, the characterization of the closed-loop response as a critically damped second-order system guides the

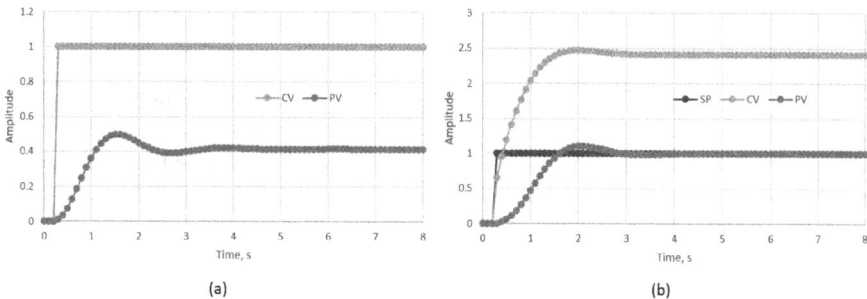

Fig. 8.5. Third-order process simulated with $T_s = 0.1$: (a) Open-Loop Response, (b) Closed-Loop response with PID controller tuned with model parameters.

determination of the value of the controller gain. A fourth or higher order process would result in an uncharacterized closed-loop response that could not guide the selection of the controller gain practically or with other specifications different from assuring a stable response.

4. Model Reference Adaptive Control

The model reference adaptive control (MRAC) is a type of model-based control (MBC) that uses a model to define the desired behavior of the control system; MRAC offers an approach for the solution of problems related to disturbance rejection or compensation for variations in the process model (Butler, 1990). By creating a closed-loop controller, MRAC tries to compare the plant's output with a standard reference response and to compensate for the change of parameters of the plant (see Fig. 8.6). MRAC includes an adaptation mechanism to compute additional control gains to improve the disturbance rejection performance.

Adaptive control is a control method for adjusting the controller when the process parameters vary or are uncertain. The foundation of adaptive control is parameter estimation, supported by system identification. There is a basic approach for the adaptive mechanism that uses and model of the plant to estimate disturbance variables. This approach is addressed with a DC motor speed control application example.

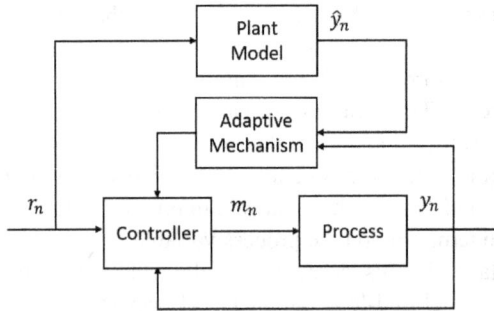

Fig. 8.6. Model reference adaptive control.

4.1 Load Disturbances in DC Motors

In some motion control applications in mechatronics and robotic systems, the load torque on the actuators may be variable and must be compensated by the controller. For example, in robotic manipulators, even though the high transmission ratios make the variable torques almost negligible when each joint is controlled independently, there are cases in which it is indispensable to compensate for them.

To show examples of compensating such disturbances, let us consider the reduced order model of a DC. In this model, the load torque, T_L, is a variable input that needs to be compensated with the electromagnetic torque generated by the armature electrical circuit. The difference equation to simulate the motor speed can

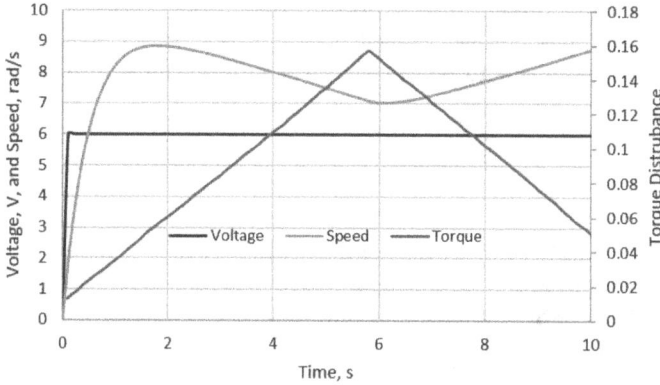

Fig. 8.7. Speed disturbance due to load variation.

be expressed using the main motor parameters and including the armature voltage, V, and the load torque, T_L, as input variables:

$$\omega_n = \alpha\, \omega_{n-1} + (1-\alpha)[(V - k_b\ \omega_{n-1})\ k_t\ k_e - T_L] \tag{8.46}$$

The effect of the load on the motor speed can be simulated by applying a constant input voltage to the motor or by generating a variable signal to be the variable load torque input. Figure 8.7 shows how the variation in the motor speed in an open loop is affected by the load variation, in this case, presented as a triangle signal. Variation in the motor load affects the developed motor speed, as expected for a real motor. The motor decelerates when the load increases and the motor accelerates when the applied load is reduced.

4.2 *Adaptive Control of DC Motor*

The discrete time model of a DC motor can be used in the control scheme as an adaptive mechanism to estimate and reject load disturbances. The load estimation can be computed in real time and be used as an additional term added to the PID controller to compensate for the disturbances caused by the real load variation (Arenas-Rosales et al., 2021).

To specify the controller, consider a PI control law for controlling the speed of a motor. The conventional control law of a discrete time proportional-integral (PI) controller is:

$$u_n = u_{n-1} + k_c\left[(e_n - e_{n-1}) + \frac{T}{\tau_i}e_n\right] \tag{8.47}$$

where k_c is the controller gain and, τ_i is the integral time constant, T is the sample time.

The proportional and integral terms can be calculated using the IMC principle. The gains of the proportional and integral terms can be set to obtain a desired closed-loop response with a λ parameter as a closed-loop time constant:

$$k_c = \frac{1}{k_m} \frac{\tau_m}{\lambda} \tag{8.48}$$

$$\tau_i = \tau_m \tag{8.49}$$

where k_m the mechanical motor gain, and τ_m is the mechanical motor time constant.

If the motor speed can be measured, equation (8.46) used to simulate the motor speed can now be used to estimate the load torque. The discrete-time calculation of the estimated load torque, T_{LE}, is:

$$T_{LE} = (V - k_b \omega_{n-1}) k_t \, k_e - \frac{\omega_n - \alpha \omega_{n-1}}{1 - \alpha} \tag{8.50}$$

For a variable torque application, to detect and compensate for variations in the load torque, it is desirable to compute the rate of change of the estimated load torque:

$$D_n = k_d \frac{T_{LE_n} - T_{LE_{n-1}}}{T} \tag{8.51}$$

where T_{LE_n} is the current estimated load torque, and $T_{LE_{n-1}}$ is the previously estimated load torque, a sample time ago. The controller is proposed as:

$$u_n = u_{n-1} + k_c \left[(e_n - e_{n-1}) + \frac{T}{\tau_i} e_n \right] + k_d \frac{T_{LE_n} - T_{LE_{n-1}}}{T} \tag{8.52}$$

The obtained controller is called PI plus D to emphasize that the derivative term is an adaptive mechanism. Figure 8.8 shows the PI plus D control scheme with the proportional and integral gains computed with the conventional IMC procedure for a first-order process and an additional term computed using the proposed equations

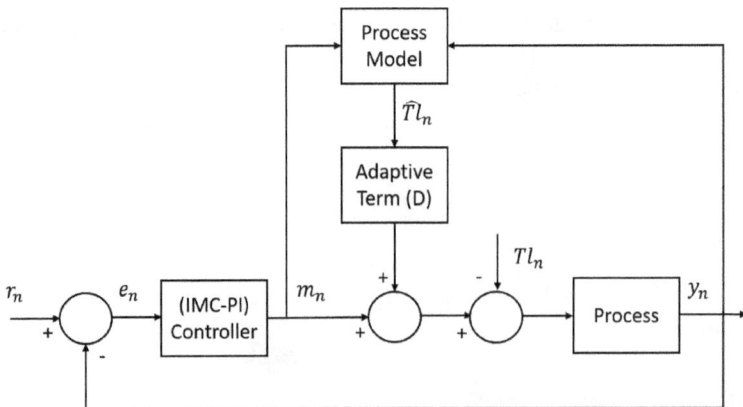

Fig. 8.8. PI controller with the load torque estimation as derivative term.

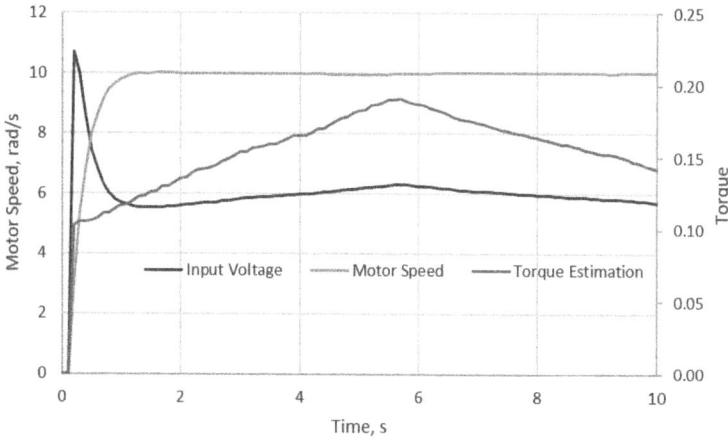

Fig. 8.9. Speed control response under load variations.

(8.50) and (8.51) to estimate the motor load torque as a derivative term and then added to the PI controller.

The response of the IMC-based PI plus D adaptive speed control is shown in Fig. 8.9. In this close loop control system response, the load disturbances are rejected mainly by the adaptive term, and the motor speed is regulated at the desired speed. When the estimated load torque changes, the adaptive term compensates for the increasing or decreasing variation to maintain the motor speed steady at the desired value.

4.3 Model Reference Control Schemes

It is possible to find MRAC schemes that do not include adaptive mechanisms. The control scheme is called Model Reference Control (MRC). The MRC is a control strategy that uses a model to impose a desired system response. To accomplish this, the error between the model reference output (the desired behavior) and the process output signal is used to generate an additional control signal added to the feedback control terms (see Fig. 8.10).

The MRC is then an advanced control strategy that, for specific applications, may lead to a better dynamical response of the process under control. The reference model often is the desired closed-loop response with unit gain and a time constant specified by the λ parameter:

$$\hat{y}_n = \left(\frac{\lambda}{\lambda + T}\right)\hat{y}_{n-1} + \left(\frac{T}{\lambda + T}\right)r_n \tag{8.53}$$

In the MRC scheme (Fig. 8.10), there are now two error calculations available for computing the control law, the error with respect to the set point and the deviation from a calculated response according to the desired dynamics:

$$e_{1,n} = r_n - y_n \tag{8.54}$$

$$e_{2,n} = \hat{y}_n - y_n \tag{8.55}$$

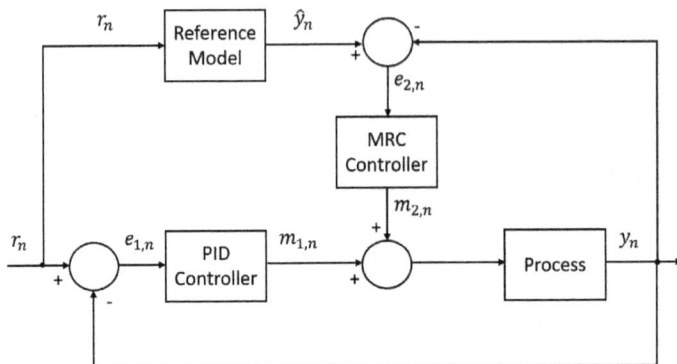

Fig. 8.10. Example of model reference control.

A conventional PI controller can be specified in the feedback loop since the MRC control action can substitute the derivate term. To keep a reduced number of control terms, the MRC control term can be only a proportional controller, but to get the intended effect of a proportional term, it has to be included as a difference of two consecutive error samples since the difference equation integrates the value:

$$m_n = m_{1,n} + m_{2,n} \tag{8.56}$$

$$m_n = m_{n-1} + k_{c1}\left[(e_{1,n} - e_{1,n-1}) + \frac{T}{\tau_i}e_{1,n}\right] + k_{c2}(e_{2,n} - e_{2,n}) \tag{8.57}$$

The PI + MRC controller can compensate for the disturbances caused by process disturbances detected by deviations from the desired closed-loop response.

4.4 Combined MRC-IMC Motor Speed Control

MRC and IMC can be combined to improve the speed control of a DC motor. This combination is convenient for servo control and axis motion control applications, especially when load variation and disturbances are present, like in many mechatronic and robotic systems. The IMC and MRC have already been combined using IMC as a nested loop in an MRAC scheme (Shekhar and Sharma, 2018). The scheme shown in Fig. 8.11 includes the MRC to make a DC motor behave as a specified reference model and the IMC to ensure the output reaches the reference value.

The manipulation variable is the sum of the PI output that can be tuned up with the IMC principle and the contribution of the PD of the MRC. The PD is chosen for two reasons, the proportional action is to compensate for internal model variations, and the derivative is to reject disturbances:

$$m_{1,n} = m_{1,n-1} + k_{c1}\left[(e_{1,n} - e_{1,n-1}) + \frac{T}{\tau_i}e_{1,n}\right] \tag{8.58}$$

$$m_{2,n} = m_{2,n-1} + k_{c2}\left[(e_{2,n} - e_{2,n-1}) + \frac{\tau_d}{T}(e_{2,n} - 2e_{2,n-1} + e_{2,n-2})\right] \tag{8.59}$$

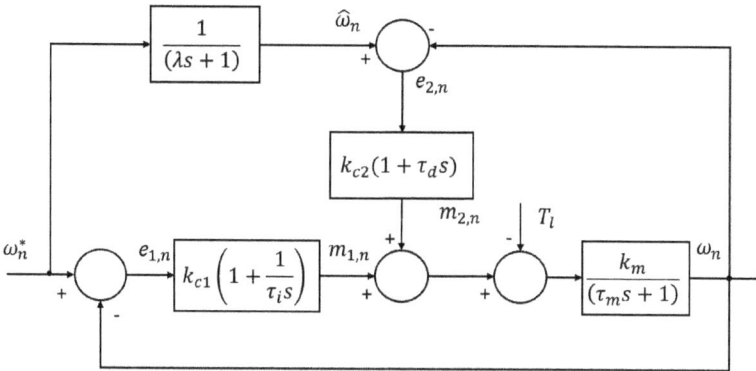

Fig. 8.11. Combined MRC-IMC block diagram.

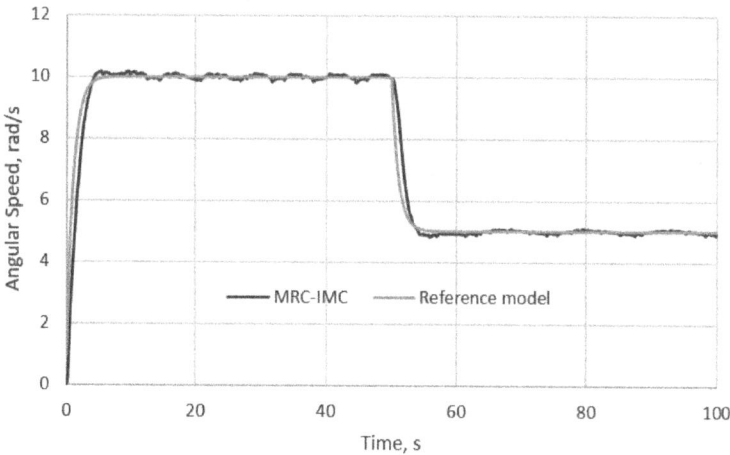

Fig. 8.12. Performance of combined MRC-IMC.

$$m_n = m_{1,n} + m_{2,n} \tag{8.60}$$

The response of the DC motor is not linear from acceleration to deceleration and is even more nonlinear under load variations, therefore a conventional PID must be carefully tuned-up with different gains. If a combined MRC-IMC is implemented, the controlled angular speed follows the reference model compensating better for internal model variations than conventional PID controllers. Figure 8.12 shows the response of a real DC motor applying the combined MRC-IMC scheme, the initial speed change is to accelerate the motor up to reach a speed of 10 rad/s and then, at time 50 seconds, a speed change is applied to decelerate and reach a speed of 5 rad/s.

5. Introduction to State Space Control

Model-based control is inherent to state space control. The effort to construct an accurate state space model benefits the design of the controller. The state space model is directly used in specifying the control laws. State space control methods aim to drive the states to zero or to a reference value. There are two basic analyses for state space control: the first considers the state feedback without a reference input, and the second considers the response to a reference input (Williams II and Lawrence, 2007).

5.1 Full State Feedback

In state feedback control, the state vector is used to compute the control manipulation for a specified system's dynamical response. Full state feedback is also known as pole placement. Figure 8.13 shows a state space time invariant linear system using a constant state feedback gain matrix, K. This methodology requires that all the state variables are measurable and available for feedback and that the system must be controllable (Ogata, 1997).

Consider the state space system with constant arrays A, B, C, and D. This is a time-invariant system with the state and output equations given by:

$$\dot{x}(t)A\,x(t) + B\,u(t) \tag{8.61}$$

$$y(t)C\,x(t) + D\,u(t) \tag{8.62}$$

Now consider the control law known as state feedback control, which implies the input reference is zero:

$$u(t) = -\,K\,x(t) \tag{8.63}$$

Now substitute the control law into state space and output equations:

$$\dot{x}(t) = A\,x(t) + B\,[-\,K\,x(t)] \tag{8.64}$$

$$y(t) = C\,x(t) + D\,[-\,K\,x(t)] \tag{8.65}$$

The closed loop state space system response to zero reference input is as follows:

$$\dot{x}(t) = [A - B\,K]\,x(t) \tag{8.66}$$

$$y(t) = [C - D\,K]\,x(t) \tag{8.67}$$

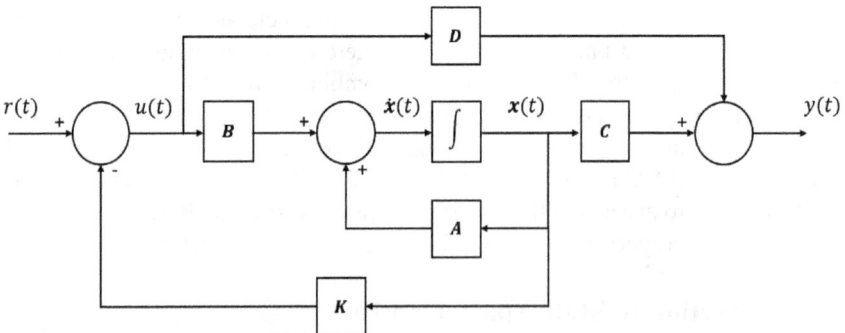

Fig. 8.13. Block diagram using state feedback.

The stability and characteristics of the transient response are determined by the characteristic values that satisfy the equation:

$$|I\lambda - (A - B\ K)| = 0 \tag{8.68}$$

The elements of K must be chosen appropriately, so the exponential matrix $e^{[A - B\ K]t}$, becomes asymptotically stable from any initial state $x(0)$.

If there is a controller that can drive the state vector to zero, it is possible to design a controller that can drive the state vector to a desired reference. We can now specify a vector of desired states r, and the system will choose values of u to make the system converge to the reference. The control law that considers a desired reference is as follows:

$$u(t) = -\ K\ [r(t) - x(t)] \tag{8.69}$$

The dynamical response of the closed-loop system to converge to a reference values is again defined by the matrix $[A - B\ K]$. The controller design requires determining the desired closed-loop poles for the dynamical response and the steady state requirements. The poles are the system's eigenvalues and determine the type of dynamical response.

5.2 Pole Placement in Discrete Time

In discrete time, the steps to specify a state feedback control law is similar to those for continuous-time systems. Consider the state equation in discrete time, in simplified notation:

$$x_{k+1} = G\ x_k + H\ u_k \tag{8.70}$$

Now consider the system in discrete time:

$$u_k = -\ K\ x_k \tag{8.71}$$

The stability and characteristics of the transient response are determined by the eigenvalues of the characteristic polynomial of the following equation:

$$|I\lambda - (G - H\ K)| = 0 \tag{8.72}$$

As for the continuous case, the controller design by pole placement consists of equating the characteristic equation of the closed-loop system to the desired characteristic equation and then finding the values of the feedback gains or array K.

Example. Consider the state space equation in discrete time:

$$x_{k+1} = \begin{bmatrix} 0 & 1 \\ -0.16 & -1 \end{bmatrix} x_k + \begin{bmatrix} 0 \\ 1 \end{bmatrix} u_k \tag{8.73}$$

Determine the gain vector for state feedback control to obtain a dynamical response defined by the eigenvalues: $0.5 \pm j0.5$.

The first step is to determine the system's controllability:

$$[H \quad GH] = \left[\begin{bmatrix} 0 \\ 1 \end{bmatrix} \begin{bmatrix} 0 & 1 \\ -0.16 & -1 \end{bmatrix} \begin{bmatrix} 0 \\ 1 \end{bmatrix} \right] = \begin{bmatrix} 0 & 1 \\ 1 & -1 \end{bmatrix} \tag{8.74}$$

the controllability matrix is rank 2. Therefore, the system is controllable.

The second step is to find the desired characteristic equation corresponding to the desired eigenvalues:

$$(\lambda - 0.5 - j0.5)(\lambda - 0.5 + j0.5) = \lambda^2 - \lambda + 0.5 \tag{8.75}$$

The third step is to determine the characteristic equation of the closed-loop system. All terms are known except the gain vector/matrix K; for this example, K is a row vector with two columns:

$$K = [k_1 \quad k_2] \tag{8.76}$$

Substitute all matrix and vector values and gain variables in equation but (8.72):

$$|I\lambda - (G - HK)| = \left| I\lambda - \left(\begin{bmatrix} 0 & 1 \\ -0.16 & -1 \end{bmatrix} - \begin{bmatrix} 0 \\ 1 \end{bmatrix} [k_1 \quad k_2] \right) \right| = 0 \tag{8.77}$$

$$|I\lambda - (G - HK)| = \left| \begin{bmatrix} \lambda & 0 \\ 0 & \lambda \end{bmatrix} - \left(\begin{bmatrix} 0 & 1 \\ -0.16 & -1 \end{bmatrix} - \begin{bmatrix} 0 & 0 \\ k_1 & k_2 \end{bmatrix} \right) \right| = 0 \tag{8.78}$$

$$|I\lambda - (G - HK)| = \left| \begin{bmatrix} \lambda & -1 \\ k_1 + 0.16 & \lambda + 1 + k_2 \end{bmatrix} \right| = 0 \tag{8.79}$$

The characteristic equation is:

$$\lambda^2 + (1 + k_2)\lambda + 0.16 + k_1 \tag{8.80}$$

The coefficients of (8.75) and (8.80) must be equal. Therefore, k_1 and k_2 can be obtained:

$$1 + k_2 = -1 \tag{8.81}$$

$$0.16 + k_1 = 0.5 \tag{8.82}$$

and the gain vector is:

$$K = [0.34 \quad -2)] \tag{8.83}$$

5.3 State Space Observers

In the state feedback or pole placement approach to control system design, all state variables are assumed to be available for feedback. However, not all state variables are measurable in practice. Since the state space equation implies an integration relation among the states, the non-measurable states can be obtained by differentiating the others. However, this practice must be avoided because of the low signal to noise ratio in measurements and the faster fluctuation of states and process variables with respect to manipulation or control variables (Tsui, 2015).

A state observer is a device that estimates or observes state variables. A state observer estimates the state variables based on measurements of the output and control variables. Here the concept of observability plays an important role, since state observers can be designed if and only if the system is observable. The observers are therefore used to estimate the state variables that are not measured instead of using a differentiation process.

Luenberger Observer

The Luenberger observer introduces an additional term in the state space equation. This term uses the difference between the estimated outputs and measured outputs to control the estimated state toward the desired state. Large values of L rely more on the measurements, while small values rely more on the model. The state and output equations with the Luenberger observer are:

$$\dot{x}(t) = A\,x(t) + B\,u(t) + L\,[y(t) - \hat{y}(t)] \tag{8.84}$$

$$y(t) = C\,x(t) + D\,u(t) \tag{8.85}$$

The Luenberger observer introduces an additional term in the state space equation. The control scheme is shown in Fig. 8.14. The observer estimation is separate from the controller. Similar to the controller gain array K, the observer design consists of evaluating the constant array L (observer gain vector/matrix) so that the transient response of the observer is faster than the response of the controlled loop to obtain an accurate estimation of the state vector.

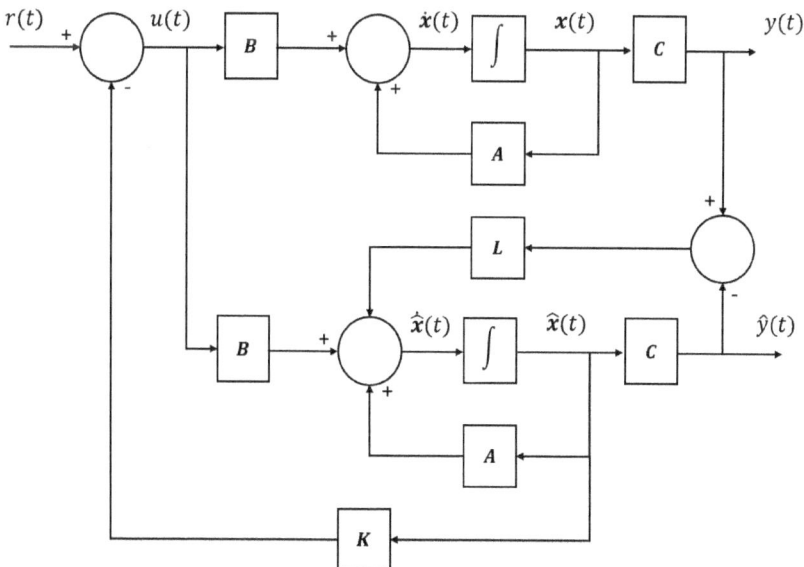

Fig. 8.14. Block diagram including the complete order observer.

A full state observer captures all system state variables, regardless of whether some are available for measurement; however, if the observer considers the output variables that are directly related to the state variables and only estimates some states, then it is called a minimum order state observer.

The observer is obtained with the model of the plant using the same control input. There is a comparing element between the actual measured output and the estimated output; this will cause the estimated states to approximate the actual states x. The dynamics of the observer error is given by the poles of $[A - LC]$. Usually, the observer poles are selected so that the observer response is much faster than the system response. A conventional rule is to choose an observer response at least 2 to 5 times faster than the system response.

In discrete time, the system can be expressed to include the Luenberger observer:

$$x_{k+1} = G\,x_k + H\,u_k + L\,[y_k - \hat{y}_k] \tag{8.86}$$

$$y_{k+1} = C\,x_k + D\,u_k \tag{8.87}$$

For discrete-time systems to be stable, the eigenvalues need to converge to a stable region in the z plane region. The eigenvalues of $[G - LC]$ must be within the unit circle. These eigenvalues represent how fast the estimator converges to the actual state of the given model. A fast estimator converges quickly, while a slow estimator avoids amplifying noise in the measurements used to estimate state variables.

5.4 State Space Control Methods

State space control methods, including state feedback control systems or output feedback systems using observers, can be designed for regulatory or set-point tracking. Regulatory systems seek to keep the output of the system constant in the presence of some external or internal disturbance, where the primary design criteria are emphasized based on the desired transient response. In a tracking or servo-controlled system, the output must follow, with a minimum of error, a prescribed path represented by a time varying reference. In this class of systems, both the transient and steady state responses must be accomplished within specific performance criteria.

State space model-based control is the base for modern control. Many advanced control techniques, including modal control, optimal control, the linear quadratic optimal regulator problem, model predictive control, etc., require state space representation and controllability and observability concepts. Advanced state space control techniques are validated with Lyapunov functions and stability theorems.

There are also advanced state space observer methods, such as the Kalman filter, that are capable of choosing optimal feedback gains when the variance of the noise affecting the system is known.

6. Conclusions

The internal model control (IMC) is a model-based scheme that can be applied to design proportional-integral-derivative (PID) controllers. Perfect control can be theoretically achieved if the controller incorporates an exact process model. IMC has been a popular design procedure in the process industry, particularly for tuning single loop PID type controllers (Rivera et al., 1986). For most common low-order models, the IMC principle leads to a controller with a PID structure. For more complicated models, the IMC design produces non-conventional controllers that can be computationally implemented. However, matching the PID structure is convenient and practical for its familiarity and easy implementation. In the case of high-order models, the controller with a PI or PID structure is applied to cancel the dominant poles of the process. The rest of the poles may give an orientation of the desired response, hopefully in terms of low order closed-loop dynamics (due to the cancellation of dominant poles) or at least in terms of stability.

Model Reference Adaptive Control (MRAC) considers a reference model with the desired closed-loop response and an adaptive mechanism to compensate for disturbances or process model variations. An example provided and referred to as IMC-PI plus a derivative term based on the real-time estimation of the load torque led to a good performance for the motor speed control under disturbances of the load torque. MRC takes advantage of the direct specification of the desired closed-loop dynamics to compute an error between both the estimated process output and the actual process output to generate an additional control signal. The controller terms can be added to conventional PID or IMC based PID to have a more robust controller. A combined MRC-IMC scheme intends to improve the speed control of DC motors and compensate for load variations, which are present in some mechatronic and robotic applications.

State space control methods are essentially model-based controllers since they incorporate the state space model. There are state space techniques to locate all the system closed-loop poles for a desired response. They can be set to any desired position by providing state feedback via a state feedback gain matrix. The pole placement approach for control system design assumes that all state variables are available for feedback; however, in practice, not all state variables are available for feedback. Therefore, state observers are incorporated. The pole placement design and the observer design are independent but can be combined to form the control system through the feedback of the observed states.

References

Arenas-Rosales, F., Martell-Chavez, F. and Sanchez-Chavez, I.Y. 2022. Discrete time DC motor model for load torque estimation for PID-IMC speed control. *Mechatron. Syst. Control.* 50(2).

Butler, H. 1990. *Model Reference Adaptive Control: Bridging the Gap Between Theory and Practice.* Doctoral Thesis. Technische Universiteit Delft.

Ogata, K. 1997. *Modern Control Engineering.* Prentice-Hall, pp. 912–915.

Rivera, D.E., Morari, M. and Skogestad, S. 1986. Internal model control: PID controller design. *Ind. Eng. Chem. Process Des. Dev.* 25: 252–265.

Shekhar, A. and Sharma, A. 2018. Review of Model Reference Adaptive Control. 2018 International Conference on Information, Communication, Engineering, and Technology (ICICET), pp. 1–5.

Tan, W., Marquez, H.J. and Chen, T. 2003. IMC design for unstable processes with time delays. *Journal of Process Control* 13: 203–221.

Tsui, CC. 2015. Observer design—A survey. *Int. J. Autom. Comput.* 12: 50–61.

Vilanova, R., Arrieta, O. and Ponsa, P. 2009. IMC based feedforward controller framework for disturbance attenuation on uncertain systems. *ISA Transactions* 48(4): 439–448.

Williams II, R.L. and Lawrence, D.A. 2007. *Linear State Space Control Systems.* John Wiley & Sons, Inc.

Yang, X., Wang, Q., Hang, C.C. and Lin, C. 2002. IMC-based control system design for unstable processes. *Ind. Eng. Chem. Res.* 41(17): 4288–4294.

Chapter 9

Multiple Loop Control Schemes

1. Introduction

Many industrial processes and mechatronic systems can be controlled with a PID in a single-loop configuration and tuned up with model based control approaches. There are certain process control applications in which alternative control schemes must be implemented due to the complexity or convenience of controlling additional process variables. In such applications, alternative strategies considering multiples loops can be designed and implemented.

Nowadays, digital technology offers enough computing capacity to accommodate multiple control algorithms in most control units. Even in small digital electronic microcontrollers, the computing time can be divided or multiplexed to allow the implementation of multiple control loops required for multivariable processes, see Fig. 9.1.

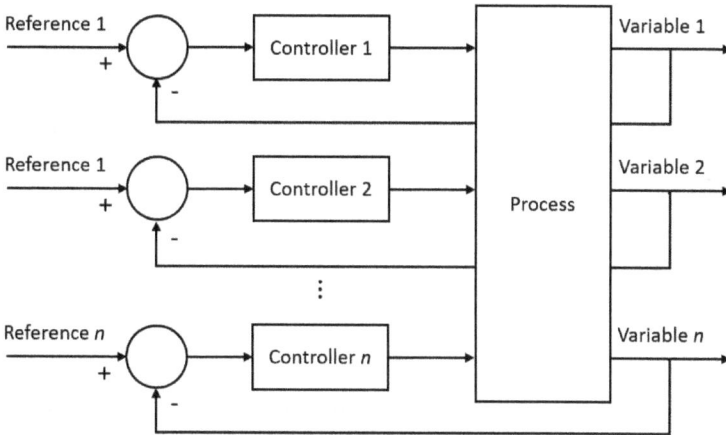

Fig. 9.1. Multi-loop control scheme for multivariable process.

The combined MRC and IMC control strategies already covered in the previous chapter can be regarded as multiple loop control strategies. Still, more conventional multiple loop control schemes are applied in industrial applications such as feedforward control, cascade control, and ratio control.

Multi-loop control schemes are suitable for applications that need to couple or decouple control loops of multiple process variables and for other process control applications that focus on disturbances that interfere directly with the manipulation variable. These disturbances are not related to external variables but to an internal variable whose variation directly affects the manipulation and therefore affects the process output.

The multi-loop control strategies considered in this chapter are feedforward control in Section 2, cascade control in Section 3, and ratio control is reviewed in Section 4. In Section 5, the Two degree of freedom PID is reviewed and configured as a feedforward or feedback controller. Section 6 presents a case study of the control of a reheat furnace, whit examples of applying multi-loop control strategies and conclusions are discussed in Section 7.

2. Feedforward Control

The disturbances over a process are generally related to external input variables, that is, other variables different from the manipulation that have an important effect on the output of the process. If the physical variable related to a disturbance can be measured, its effect can be counteracted in time to avoid a noticeable deviation of the process variable.

Since the disturbance variable measurement is assumed to be available, its dynamic effect on the process variable can be identified. That is, the open loop transfer function G_d of the output Y to the disturbance D can be calculated. The output variable can then be estimated by the superposition of the manipulation and disturbance effects, as shown in Fig. 9.2.

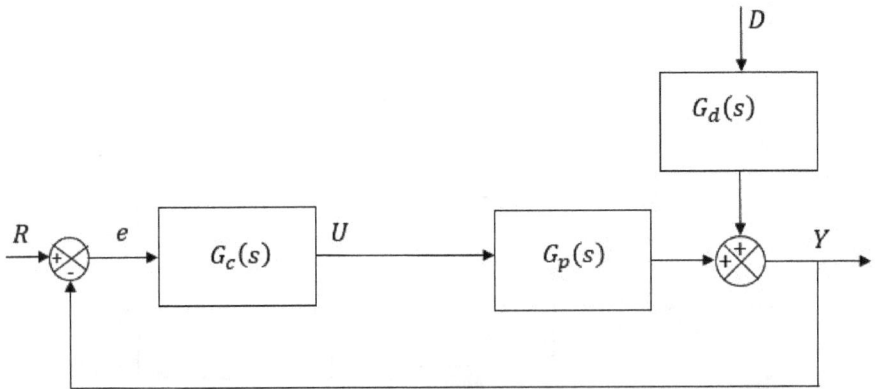

Fig. 9.2. Control system with an additive disturbance: the disturbance effect is added to the manipulation effect on the process variable. **Variables**: reference R, error e, manipulation U, disturbance D, process variable Y. Functions: controller G_c, process dynamics G_p, disturbance dynamics G_d.

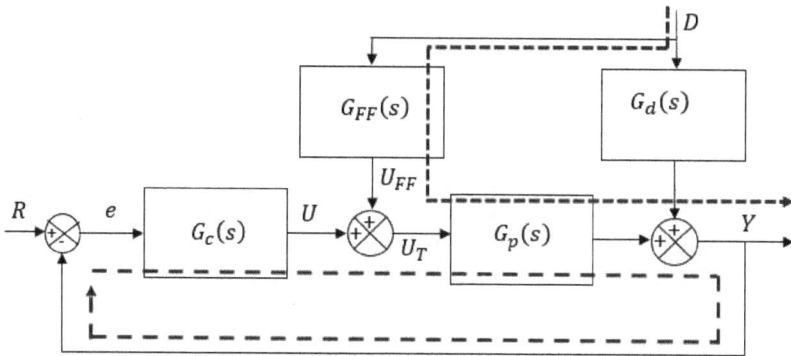

Fig. 9.3. Feedforward control strategy: feedforward trajectory and feedback loop.

The single closed-loop system would react to the disturbance to return the process variable to the desired set-point value. However, this correction implies the presence of the error; that is, the controller in the simple loop can only act when the deviation of the process variable takes place and is detected in the output measurement.

To improve the performance before the disturbance, the possible deviation of the output can be anticipated and compensated by producing an opposing effect with the manipulation (Shah and MacGregor, 2004). For this, an additional trajectory is added to the block diagram as part of the control strategy, as shown in Fig. 9.3. The feedforward function G_{ff} is the means to consider the disturbance presence for calculating the improved manipulation of the process.

The analysis of the disturbance trajectories to the output is given by:

$$D(s)\, G_{ff}(s)\, G_p(s) + D(s)\, G_d(s) = Y(s) \qquad (9.1)$$

The feedforward function is defined to cancel the disturbance effect, that is:

$$D(s)\, G_{ff}(s)\, G_p(s) + D(s)\, G_d(s) = 0 \qquad (9.2)$$

which leads to:

$$G_{ff}(s) = -\frac{G_d(s)}{G_p(s)} \qquad (9.3)$$

Therefore, the PID controller output U or U_{PID} is added to the feedforward function output U_{FF} to produce a total manipulation U_T:

$$U_{FF}(s) = D(s)\, G_{ff}(s) \qquad (9.4)$$

$$U_T(s) = U(s) + U_{FF}(s) \qquad (9.5)$$

This manipulation makes the process variable unaffected by the characterized external disturbance D:

$$U(s)\, G_p(s) + D(s)\, G_{ff}(s)\, G_p(s) + D(s)\, G_d(s) = U(s)\, G_p(s) = Y(s) \qquad (9.6)$$

This compensation for the disturbance is ideal. However, despite model errors, the function G_{ff} produces a noticeable improvement.

A feedforward function solves the problem of a particular disturbance. Other unmeasured, uncharacterized, and uncompensated disturbances may occur, and the only way to be rejected would be through the feedback loop.

Consider the heating process in an oven where the temperature is to be controlled by manipulating the combustible flow (neglecting the supply of oxygen, which can be managed by mechanically coupling the airflow actuator with the manipulated fuel flow actuator). A possible disturbance is the variation in the temperature of the input material. The following transfer functions describe the process dynamics, with time constant units in minutes. First, the process variable or temperature in the oven Y changes with respect to the combustible flow U is given by:

$$G_p(s) = \frac{Y(s)}{U(s)} = \frac{0.6}{2s+1} \tag{9.7}$$

The way the oven temperature is affected by the input material temperature D is described by:

$$G_d(s) = \frac{Y(s)}{D(s)} = \frac{0.5}{2.5s+1} \tag{9.8}$$

The controller in a simple feedback scheme is based on Y(s)/U(s) and is proposed as a PID algorithm with $k_c = 1/0.6$, $\tau_i = 2$, and $\tau_d = 0$.

The basic system is represented by the block diagram of Fig. 9.4, and its performance during unit set point and disturbance changes are shown in Fig. 9.5.

The feedforward function required to compensate for the disturbance effect is given by:

$$G_{FF}(s) = -\frac{G_D(s)}{G_p(s)} = -\frac{0.5}{0.6}\left(\frac{2s+1}{2.5s+1}\right) \tag{9.9}$$

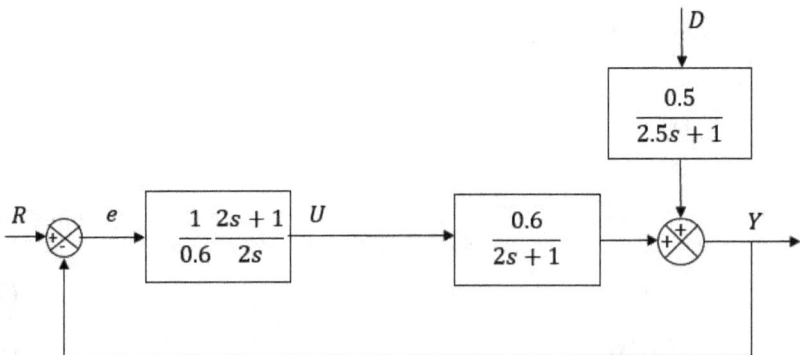

Fig. 9.4. Block diagram of the feedback control system. **Signals:** set point (*R*), disturbance (*D*), process output (*Y*), error (*e*), and controller manipulation (*U*).

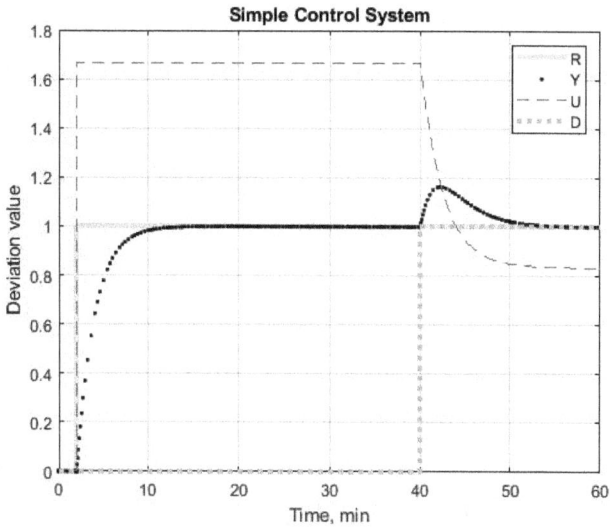

Fig. 9.5. Performance of basic control system: no feedforward calculation to compensate disturbance D that occurs at time 40.

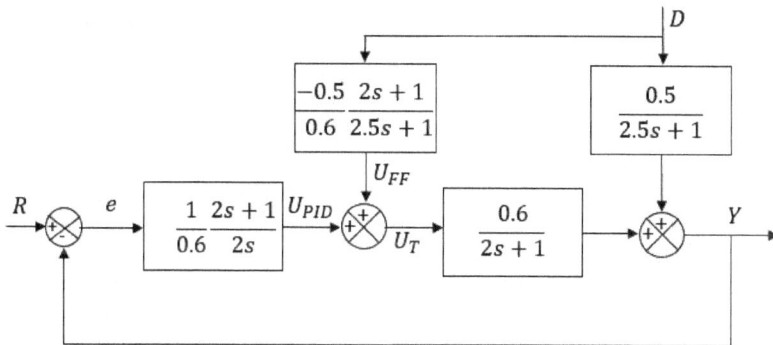

Fig. 9.6. Block diagram of a feedforward control scheme. The dynamic compensation $G_{FF}(s)$ is used.

The use of the feedforward function is shown in Fig. 9.6. The compensation would be perfect, as demonstrated in Fig. 9.7. However, an imperfect compensation often renders an important performance improvement. Consider a static feedforward function, which may be an appropriate choice if modeling errors are present or if a simplified and easy-to-implement function is desired:

$$G_{FF}(s) \approx -\frac{G_D(s)}{G_p(s)} = -\frac{0.5}{0.6} \tag{9.10}$$

The use of only a feedforward gain causes the performance shown in Fig. 9.8.

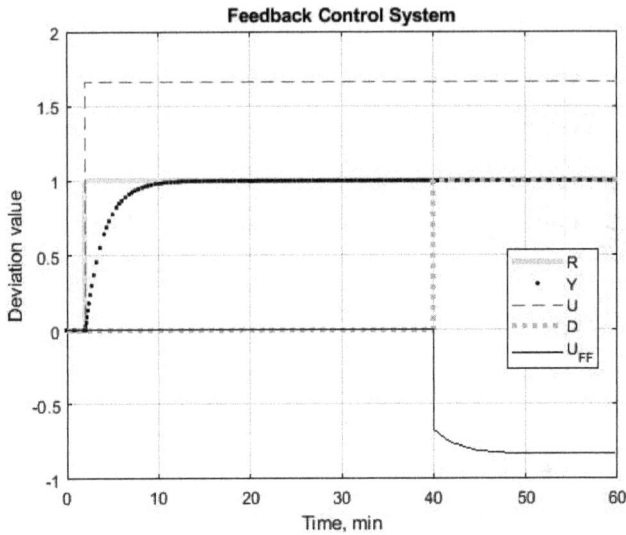

Fig. 9.7. Performance of feedforward control system: feedforward calculation U_{FF} fully compensates disturbance D that occurs at time 40, avoiding a deviation in process variable Y.

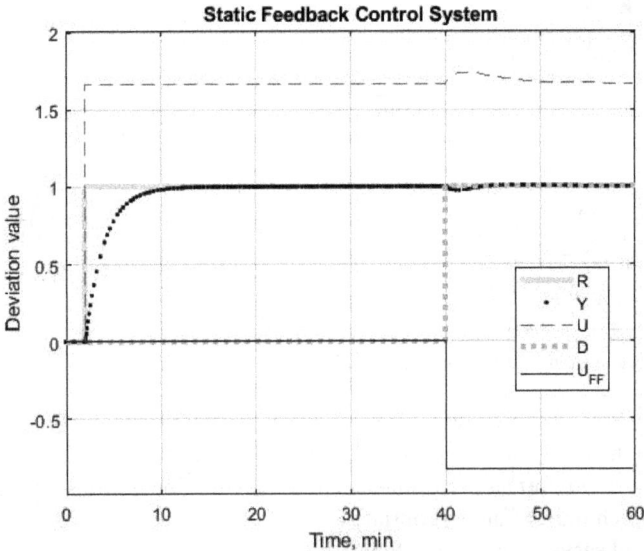

Fig. 9.8. Performance of control system with static feedforward: static D feedforward calculation U_{FF} partially compensates disturbance D that occurs at time 40, allowing a small deviation of process variable Y.

3. Cascade Control

The key to suggesting a cascade strategy is to divide the process into two subsystems in a series connected by an intermediate measurable variable (Wilamowski, 2018). The addressed disturbance impacts first on this intermediate variable (output of

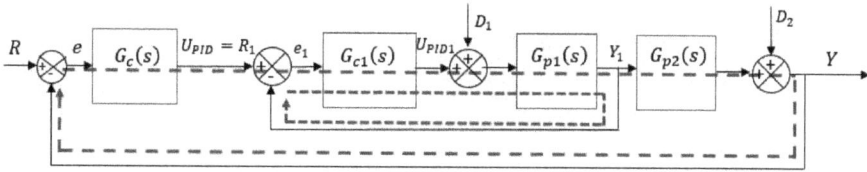

Fig. 9.9. Cascade control scheme: two nested loops are created based on each process part.

the first subsystem) before causing an effect on the process variable (output of the second subsystem). The measurement of this connecting variable would reflect an early effect of the disturbance, which can be used to start adjusting the manipulation before the process variable deviates.

The linking variable between the two parts or subsystems of the process is the auxiliary variable of the cascade strategy. To improve the performance when the disturbance happens, the deviation on the auxiliary variable needs to be corrected through an internal controller (that receives feedback from the first subsystem). If this correction is made fast enough, the effect on the process variable or main variable will cause a small error that can be eliminated by an external controller (whose feedback comes from the second subsystem). The cascade strategy consists of nested loops, as shown in Fig. 9.9.

Notice that in the cascade strategy, the output of the main controller or external controller is the set point for the internal or auxiliary controller and that the actuator is the same one that would be used in the simple loop-control scheme without the process separation in the two subsystems. The cascade configuration suggests that the internal controller could be tuned for set-point changes because its function is to track the set-point changes specified by the external controller, and the velocity of the response of the internal loop assures the attenuation of the effect on the external loop. The external controller could be tuned with a regulatory objective to favor the reaction to other disturbances impacting the second subsystem of the process since the internal controller already handles those in the first subsystem.

For the process in an oven discussed above, consider the fuel transport through the actuator as the first subsystem in a series, with a second subsystem constituted by the heating process that produces a specific temperature. Another type of disturbance D_1, different from the one considered in feedforward compensation designated from this point as D_2, is given by a deviation of flow that affects fuel input through the installed actuator. Disturbance D_1 can be managed by an internal flow controller in order to diminish its effect on the main temperature variable Y.

The transfer function of the first subsystem relates flow Y_1 to changes in actuator position U:

$$\frac{Y_1(s)}{U(s)} = \frac{0.2}{0.1s + 1} \tag{9.11}$$

The transfer function of the second subsystem is given by the ratio of the oven temperature Y with respect to fuel flow Y_1:

$$\frac{Y(s)}{Y_1(s)} = \frac{3}{2s+1} \tag{9.12}$$

Notice that the second subsystem causes the main dynamic content and that the product of the two subsystems approximates the overall transfer function Y/U managed before:

$$\frac{Y(s)}{U(s)} = \frac{0.2}{0.1s+1} \frac{3}{2s+1} \approx \frac{0.6}{2s+1} \tag{9.13}$$

The simple feedback system uses a single PID controller with the parameters $k_c = 1/0.6$, $\tau_i = 2$, and $\tau_d = 0$ (as proposed before). Figure 9.10 shows the simple feedback control system and the process configuration as a series of two subsystems with their corresponding disturbances. The performance of this feedback system when unit disturbances D_1 and D_2 occur is shown in Fig. 9.11.

The cascade strategy uses two controllers: the internal PID controller is proposed with $k_{c1} = 1/0.2$, $\tau_{i1} = 0.1$ and $\tau_{d1} = 0$, and the external PID controller has $k_c = 1/3$, $\tau_i = 2$ and $\tau_d = 0$. The same tests applied to the simple feedback control system are applied to the cascade strategy achieving superior performance in the case of the disturbance D_1 (Figs. 9.12 and 9.13).

Feedforward and cascade strategies can be combined for improved performance for both types of disturbances: the controllers obtained above are used together with the dynamic feedforward function:

$$G_{ff}(s) = -\frac{G_D(s)}{\dfrac{Y_1(s)Y(s)}{U(s)Y_1(s)}(s)} = -\frac{0.5}{3}\left(\frac{2s+1}{2.5s+1}\right) \tag{9.14}$$

where $\dfrac{Y_1(s)}{U(s)}\dfrac{Y(s)}{Y_1(s)}$ is the process dynamics 'observed' by the main or external controller, which includes the internal closed loop transfer function $Y_1(s)/U(s)$

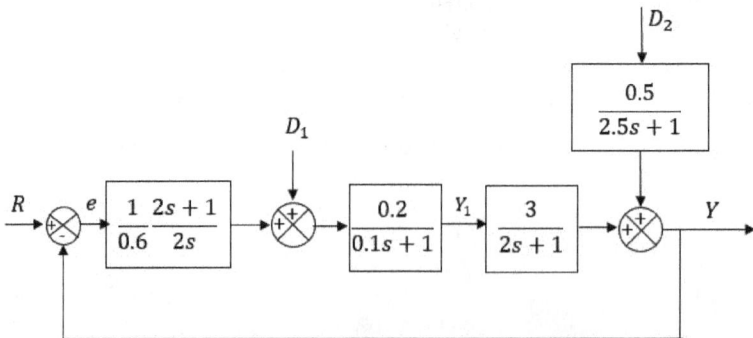

Fig. 9.10. Block diagram of feedback control system showing process division and incidence point of disturbances D_1 and D_2.

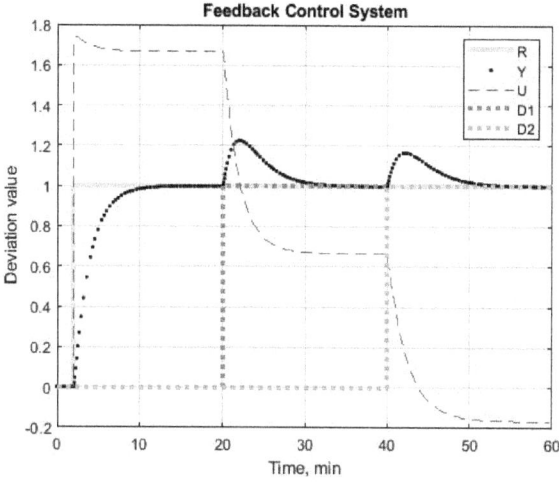

Fig. 9.11. Feedback control system: set point input at the time of 2 min, unit disturbance D_1 (on the first subsystem) at 20 min, and unit disturbance D_2 at time of 40 min.

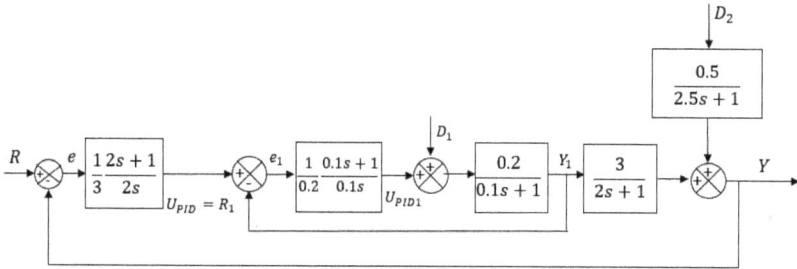

Fig. 9.12. Block diagram of the cascade control system to attenuate disturbance D_1.

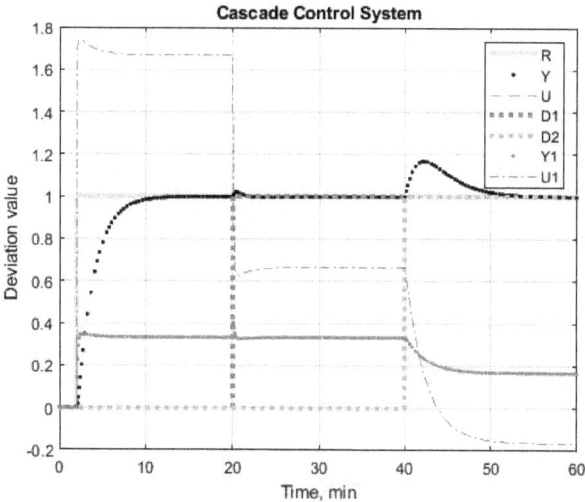

Fig. 9.13. Cascade control system: attenuation of unit disturbance D_1; same performance for a set-point change and disturbance D_2 than the performance obtained with the feedback control system.

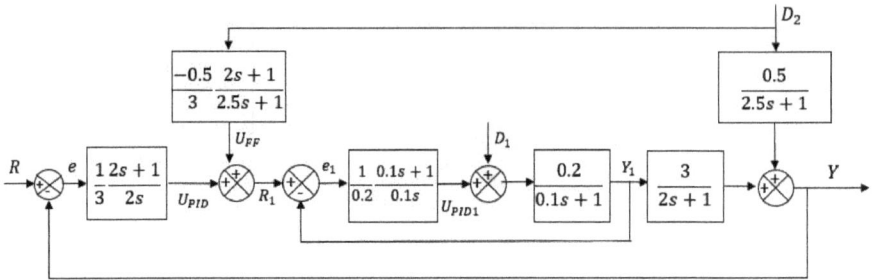

Fig. 9.14. Block diagram of feedforward-cascade control system.

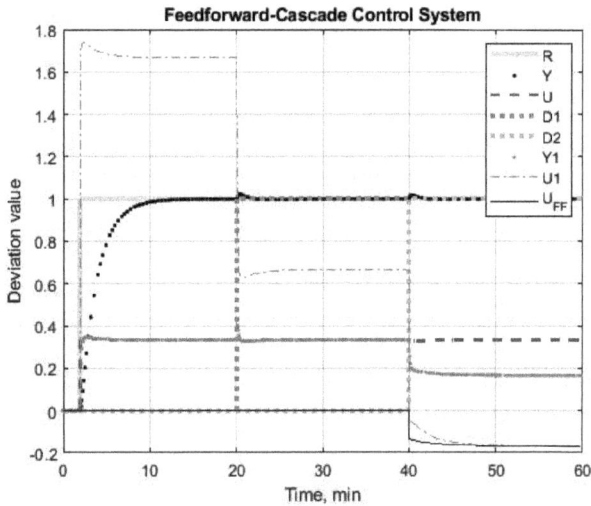

Fig. 9.15. Feedforward-cascade control system: disturbances D_1 and D_2 (of unit magnitude and applied at 20 and 40 min, respectively) are managed with clear advantage with respect to the simple feedback system.

or $Y_1(s)/R_1(s)$ in series with open-loop transfer function of the second subsystem $Y(s)/Y_1(s)$. The configuration of the feedforward-cascade control system is shown in Fig. 9.14.

The performance of the feedforward-cascade control system is shown in Fig. 9.15, which shows the expected drastic reduction of the process variable deviation before both types of disturbances.

4. Ratio Control

In some processes, the ratio of variables is important for proper operation (Seborg et al., 2016). The input of a reactor often requires a stoichiometric proportion between reactants or a particular excess percentage with respect to the demand of a limiting reactant.

Combustion processes typically illustrate the need for a specific proportion between oxygen and fuel. Consider the balanced chemical equation:

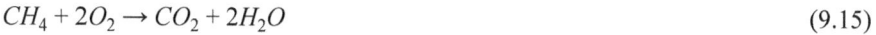

$$CH_4 + 2O_2 \rightarrow CO_2 + 2H_2O \tag{9.15}$$

For each unit of CH_4 fuel, 2 oxygen units are needed for complete combustion. Oxygen is provided by air, which is 21% oxygen. Therefore, a volumetric flow ratio of 10 air units to one of fuels is desired.

Different ratio control schemes can be used: series, parallel, and cross-limits. The following explanations consider that the flow dynamics of the fuel is faster than the one for the air. This corresponds to the assumption of fuel transport through a system of oil ducts at high pressure and air supply by a system of ventilators and dampers at low pressures.

The different ratio control schemes are demonstrated by varying the fuel demand arbitrarily, with the proper adjustment of the air flow. However, in the context of an application such as the temperature control problem in an oven discussed in this chapter, the fuel flow set point would correspond to the manipulation of the temperature controller; that is, the ratio control scheme would be subordinated to the temperature control. In the previous sections, the manipulation of the fuel flow would be mechanically coupled with an airflow actuator for proper combustion, which allows no further consideration in the control strategy. Nevertheless, precise control of the combustion process can be achieved by the ratio control techniques explained here, where the controls of both fluids, fuel, and air, are determined explicitly.

For the simulation of the ratio control schemes, the following flow dynamics are considered:

$$\frac{F(s)}{U_f(s)} = \frac{1}{s+1} \tag{9.16}$$

$$\frac{A(s)}{U_a(s)} = \frac{1}{10s+1} \tag{9.17}$$

where F is fuel flow, A is airflow, U_f is the manipulation of the fuel actuator, and U_a is the manipulation of the air actuator. The time constant for the fuel transport process is 1 s, and the time constant for the air flow process is 10 s assuming fuel is supplied at a much higher pressure than air and fuel flow changes can be produced much faster.

The controllers of the strategy are G_{cf} and G_{ca} for fuel and air regulation, respectively. The PI controller parameters for the fuel flow process are $k_{cf} = 1$ and $\tau_{if} = 1$. For the airflow control, $k_{ca} = 1$ and $\tau_{ia} = 10$ are used for a PI algorithm too.

4.1 Parallel Ratio Control

The ratio between two flows can be achieved by scaling the set point of the control loop of one flow based on the set point of the control system of the other flow. The disadvantage is that the desired ratio is only achieved at a steady state once the respective set-point values have been reached. Figure 9.16 shows the parallel arrangement of the control systems for each flow.

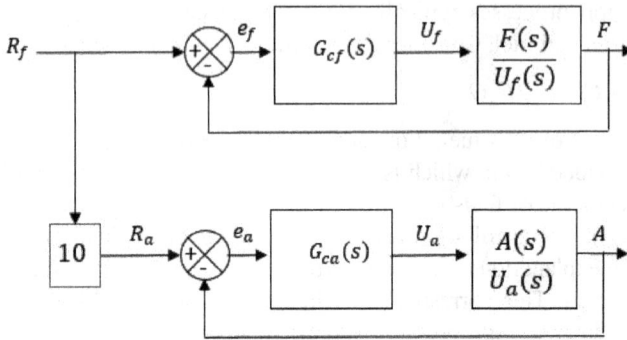

Fig. 9.16. Block diagram of a parallel scheme for ratio control: desired air to fuel ratio is 10.

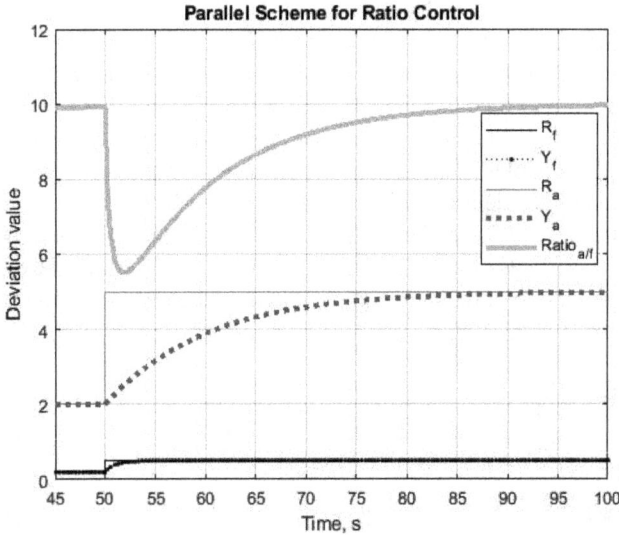

Fig. 9.17. Performance of parallel scheme for ratio control: desired air-to-fuel ratio is 10; fuel demand changes from 0.2 to 0.5 flow units.

The parallel ratio control strategy is demonstrated by an increment in the combustion demand from 0.2 to 0.5 fuel flow units (Fig. 9.17). For simplicity, this change is applied in the fuel flow set-point value, which is considered an external input to the system specified by the user. Fuel and air set-point changes are applied simultaneously; in the transitory, the air to fuel ratio drops 4.5 units and recovers as fast as the air control system, which is the slowest, stabilizes.

4.2 Series Ratio Control

In the series scheme, the output of one substance flow control loop is multiplied by the desired ratio or its inverse to obtain the desired flow of the other substance (Fig. 9.18). This suggests that two variations can be managed depending on the choice of the leading flow. In this figure, subscript *lead* designates the system's

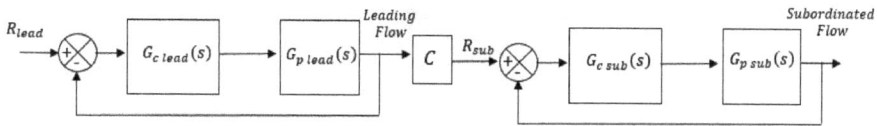

Fig. 9.18. Block diagram of the series scheme for ratio control.

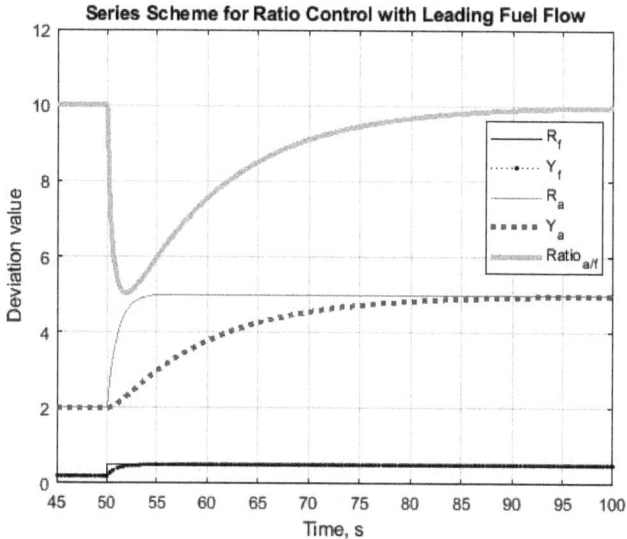

Fig. 9.19. Performance of series scheme for ratio control with fuel as leading flow: desired air-to-fuel ratio is 10; fuel demand changes from 0.2 to 0.5 flow units.

reference variable R, or controller and process elements (Gc and Gp) for controlling the leading fluid. Subscript *sub* designates variables or elements of the system for controlling the subordinated fluid. The constant C may be the desired air to fuel ratio if the leading flow is the fuel or the inverse of the desired air to fuel ratio if the leading flow is air.

In the combustion case, if the leading flow is the fuel, the actual air-to-flow ratio may deviate from the value of 10 in a transitory due to the slower dynamics of the subordinated air flow (the maximum deviation from the desired ratio is almost 5 units in Fig. 9.19). On the other hand, if the leading flow is the air, which changes more slowly than fuel, the air-to-fuel ratio may be maintained closer to the ideal value of 10 in the transition from one operation point to the other (the ratio experiences a maximum change of one unit in Fig. 9.20).

4.3 Cross-Limit Ratio Control

The cross-limit ratio control is based on the series scheme and has the flexibility of alternating the designation of the leading flow.

If an air-to-fuel ratio slightly higher than 10 is considered safer than slightly lower than 10, the following scheme is proposed. In the case fuel demand increases, the leading flow is the air, and the set point of its control system is determined first,

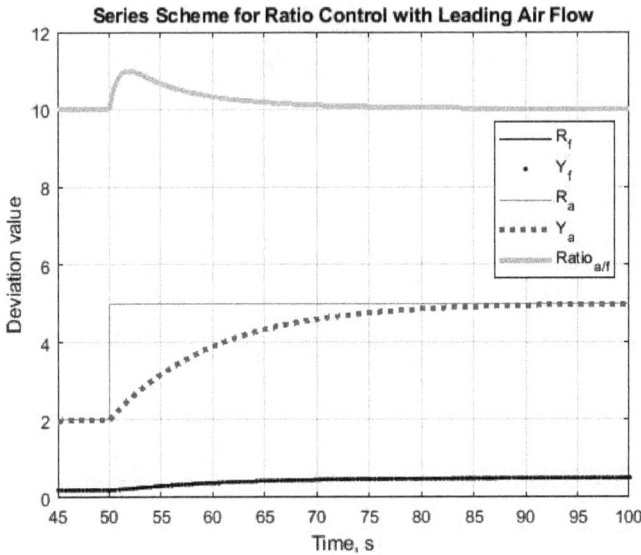

Fig. 9.20. Performance of series scheme for ratio control with air as leading flow: desired air-to-fuel ratio is 10; fuel demand changes from 0.2 to 0.5 flow units.

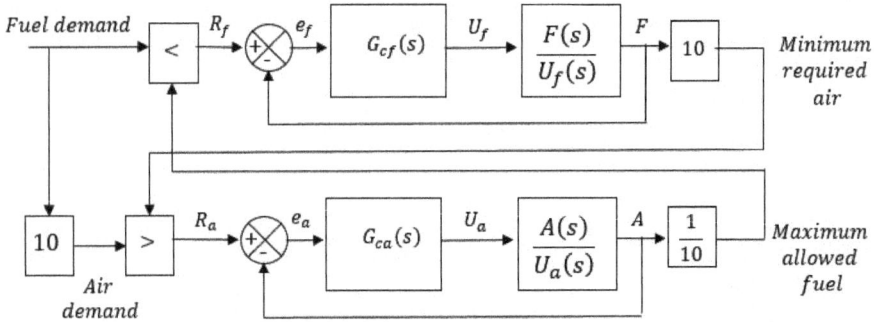

Fig. 9.21. Block diagram of the cross limit ratio control scheme.

and the fuel set point is dynamically obtained by dividing the current airflow by 10. If the fuel demand decreases, now the leading flow is the fuel, and its set point is directly applied, and the set point of the air is adjusted based on the fuel flow actual value, making the air the subordinated flow. Since the airflow is changed at a lower velocity, the actual ratio would always be greater or equal to 10, as wanted.

In other words, the set point of the fuel is equal to whatever is lower between the current fuel set point or the fuel that can be managed by the current airflow, that is, the fuel value calculated by dividing the airflow by 10. The set point of the air is the higher between the air required by the current flue flow, or fuel multiplied by 10, and the air set point or fuel set point multiplied by 10. Figure 9.21 shows the configuration of the cross-limit ratio control scheme.

This cross-limit strategy assures the desired ratio by allowing the choice of the leading and subordinated flows and may surpass the performance of the series

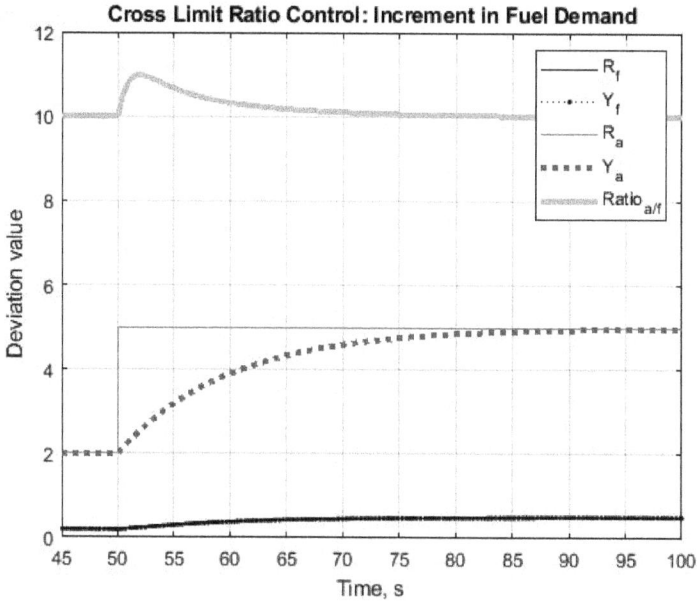

Fig. 9.22. Performance of cross limit ratio control scheme. Fuel demand increment from 0.2 to 0.5 flow units.

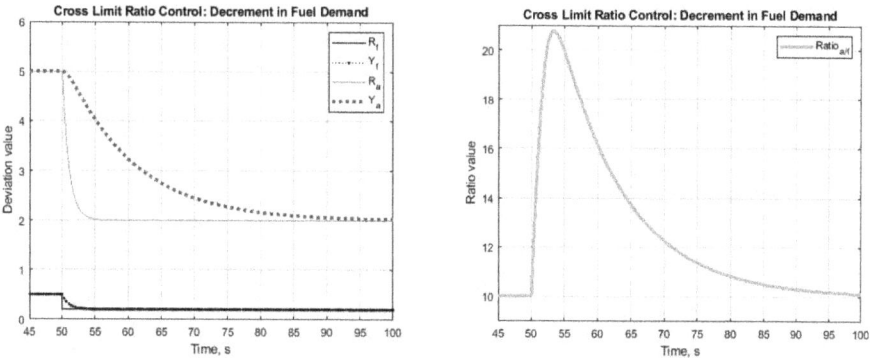

Fig. 9.23. Performance of cross limit ratio control scheme. Fuel demand decrement from 0.5 to 0.2 flow units. (Left: fuel set point and actual value, and air set point and actual value. Right: air-fuel ratio.)

scheme with air leading flow in assuring a safe ratio higher than the minimum desired value during transitory changes. Figure 9.22 shows the performance of the control scheme when the fuel demand increases: a set-point step change is applied on the airflow controller, while the fuel flow set point is adjusted gradually; the air-fuel ratio increases by 1 unit and recovers in approximately 25 seconds.

Figure 9.23 shows the behavior when the fuel demand decreases: the set point's sudden and direct change is applied on the fuel control system to assure a high air-fuel ratio; since the air transport process is much slower, the ratio increases considerably (more than a 100%) and recovers the normal value in approximately 50 seconds.

This performance is preferable to decreasing the air-fuel ratio below the established value of 10, even for a short transitory.

5. Two Degree of Freedom PID

The degree of freedom of a control system is defined as the number of closed-loop transfer functions that can be adjusted independently. The design of control systems is a multi-objective problem; for this reason, a two degree of freedom (2DOF) control system is expected to have advantages over a one degree of freedom (1DOF) control system (Araki and Taguchi, 2003). This idea was first studied by Horowitz (1963); later, it attracted engineers' attention to exploit the advantages of the 2DOF structure for PID control systems in industrial applications (Hiroi, 1986). 2DOF PID controllers have the capability of reducing overshoot in set-point tracking while they can react fast to reject disturbances; this is, they constitute a suitable control scheme for applications that require servo-control and regulatory control.

5.1 Feedforward 2DOF PID Configuration

A conventional 2DOF PID controller configuration is equivalent to a feedforward controller plus a PID controller, as shown in Fig 9.24.

The 2DOF PID controllers are based on set-point weighting on the proportional and derivative terms. In the feedback configuration, the controller, $G_c(s)$, is a conventional PID controller, and the feedforward controller, $G_{ff}(s)$, is a PD controller with different weights, α, and β, on the proportional and derivative terms, respectively.

$$G_c(s) = \frac{U(s)}{E(s)} = k_c\left[1 + \frac{1}{\tau_i s} + \tau_d s\right] \qquad (9.18)$$

$$G_{ff}(s) = \frac{U(s)}{R(s)} = -k_c\left[\alpha + \beta\tau_d s\right] \qquad (9.19)$$

The difference equation for the 2DOF PID implementation of the feedforward configuration can be expressed as follows:

$$u_{1,n} = u_{1,n-1} + k_c\left[(e_n - e_{n-1}) + \frac{T}{\tau_i}e_n + \frac{\tau_d}{T}(e_n - 2e_{n-1} + -e_{n-2})\right] \qquad (9.20)$$

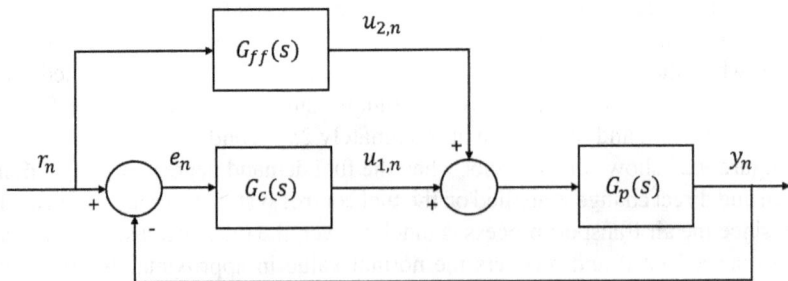

Fig. 9.24. Feedforward 2DOF PID configuration.

$$u_{2,n} = k_c \left[\alpha r_n + \beta \frac{\tau_d}{T} (r_n - r_{n-1}) \right] \tag{9.21}$$

$$u_n = u_{1,n} + u_{2,n} \tag{9.22}$$

5.2 Feedback 2DOF PID Configuration

An alternative configuration comprises a PID controller and a feedback controller, as shown in Fig. 9.25.

In the feedback configuration, the controller, $G_c(s)$, is a PID controller with weights, $(1-\alpha)$ and $(1-\beta)$, on the proportional and derivative terms, respectively, and the feedback controller, $G_{fb}(s)$, is a PD controller with weights α on the proportional term, and β on the derivative term.

$$G_c(s) = \frac{U(s)}{E(s)} = k_c \left[(1-\alpha) + \frac{1}{\tau_i s} + (1-\beta)\tau_d s \right] \tag{9.23}$$

$$G_{fb}(s) = \frac{U(s)}{Y(s)} = k_c \left[\alpha + \beta \tau_d s \right] \tag{9.24}$$

The difference equation for the 2DOF PID implementation of the feedback configuration can be expressed as follows:

$$u_{1,n} = u_{1,n-1} + k_c \left[(1-\alpha)(e_n - e_{n-1}) + \frac{T}{\tau_i} e_n + (1-\beta)\frac{\tau_d}{T} (e_n - 2e_{n-1} + e_{n-2}) \right] \tag{9.25}$$

$$u_{2,n} = k_c \left[\alpha(y_n - y_{n-1}) + \beta \frac{\tau_d}{T} (y_n - 2y_{n-1} + +y_{n-2}) \right] \tag{9.26}$$

$$u_n = u_{1,n} + u_{2,n} \tag{9.27}$$

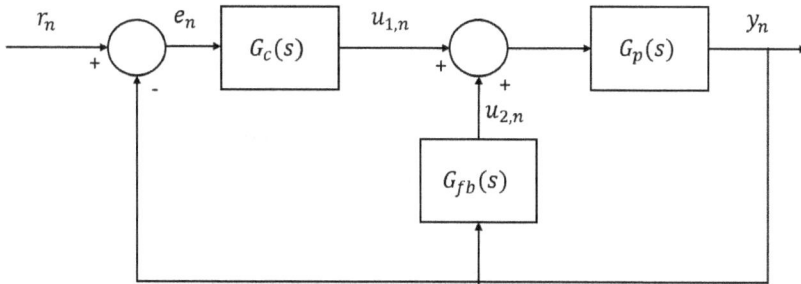

Fig. 9.25. Feedback 2DOF PID configuration.

5.3 Method for Tuning 2DOF PID

Model Reference Control 2DoF IMC-PID structures have two transfer functions that can be adjusted independently and have been used for effective disturbance rejection and set point tracking (Araki, 2003). The simple tuning method is based on two steps: first, adjustment of basic parameters, k_c, τ_i, τ_d, with the conventional direct synthesis or IMC-based tuning; second, adjustment of the weights, α and β, that are the parameters introduced by the 2DOF configuration. The parameter β is decreased to reduce the overshoot under set-point changes, thereby increasing the controller's gain to improve the disturbance rejection. These parameters must range from 0 to 1 to keep the system stable. This tuning approach has the advantage that the classical PID tuning can be utilized.

6. Case Study: Reheating Furnace

Industrial plants continuously improve process automation to increase productivity, product quality, and energy efficiency. Steel-making companies have energy intensive consumption processes and are committed to developing actions to decrease fossil fuel consumption to mitigate greenhouse gas emissions. Modern steel rolling mills produce large quantities of thin sheet metal but consume significant energy, particularly in the reheat furnaces. The operation of reheating furnaces is relevant to the steel-making process. The fuel consumption is up to 15% of the operational cost of a rolling mill process. The reheat furnace capacity often defines the production rate for the rollers. Therefore, it is a critical process to achieve the required production.

6.1 Reheat Process Description

A reheat furnace is a direct heating process with burners providing heat through combustion reactions. The reheating furnace is a process required before rolling mills, where steel billets, plates, or blocks are usually heated from ambient temperature to the required temperature, so the billets can be hot rolled to achieve specific metallurgical, mechanical, and dimensional properties. Optimal operation requires stable and safe temperature control to get the steel billet thermal soaking while reducing fuel consumption. Figure 9.26 depicts a reheat furnace and the multiple process and control variables related to the operation of the furnace.

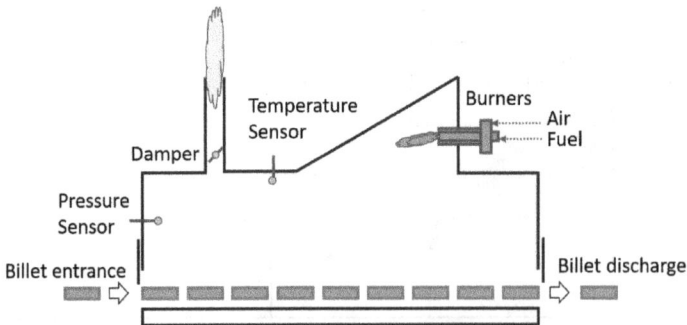

Fig. 9.26. Steel reheat furnace.

6.2 Furnace Control Systems

Furnace Temperature Control

Heat transfer to the steel billets depends upon the temperature control of the furnace; the temperature control is often implemented in a cascade strategy with the flame control of the burners, which is achieved by controlling the fuel and air flows (see

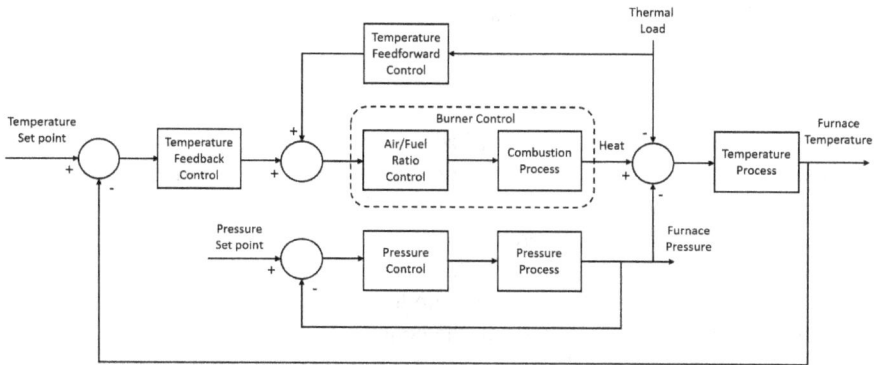

Fig. 9.27. Furnace temperature and pressure control.

Fig. 9.27). The criterion for controlling the temperature in a furnace is based on fixing the temperature set point so that pieces loaded into the furnace, as they move within it in a given cycle time, are heated to a desired temperature.

The furnace temperature is primarily affected by the number of billets loaded into the furnace, representing a variable heat load. The furnace temperature is also affected by some disturbances like the operating furnace pressure and other conditions, such as the opening and closing of the billet charge and discharge doors. A feedforward controller, as shown in Fig. 9.27, must compensate for the variations that occur by the dynamic heat load and the furnace pressure considering these fluctuations as external disturbances to the temperature process.

From the point of view of energy transfer, there is an opportunity to balance better the amount of energy supplied to the furnace with the heat required by the billets. The feedforward controller is a predictive control action that detects load variations. The feedforward controller requires less fuel to control the required temperature when a reduction in the heat load is detected and slightly limits the fuel consumption by an increase in the heat load; therefore, if well-tuned, it contributes to smoother temperature control.

Furnace Pressure Control

The furnace pressure control is also very important. If the furnace pressure is too low, air entering the furnace with oxygen potentially causes scale formation, impacting quality and productivity. If the furnace operation pressure is too high, fuel consumption will increase. Pressure control is achieved by an actuated damper installed on the flue gas stack. Furnace pressure may be controlled independently, as shown in Fig. 9.27.

Burners Air-to-Fuel Ratio Control

The efficiency of the combustion ratio is improved with the airflow to fuel flow ratio control. This can be implemented with a cross-limit ratio control strategy to operate and achieve efficient combustion reactions. The air-to-fuel ratio control must be considered fuel injection for a safe and efficient operation and control of combustion reactions. The burners need to be instrumented with air and gas flow control valves and with adequate flow sensors.

Additional Automation Functions

All the discussed control strategies (feedforward, cascade, and ratio control) can be implemented in programmable logic controllers (PLC). Additionally, other automation functions are implemented, such as the automatic detection of operation delays, where the furnace temperature may be lowered to a delay temperature set point to avoid the overheating of the billets. The criterion may be to regulate a lower furnace temperature value. While there is no load or when the load flow has stopped completely, or if the process allows it, the control loops can be switched to manual mode in the event of greater delays to keep the burners with a minimum flame. Some other diverse control actions can be implemented to balance energy use with the processed material to save fuel when the furnace is not at its maximum operational capacity or when it is in a delay condition, as long as the safe and productive operation is maintained.

7. Conclusions

The presented strategies cover two control problems: recovering an operation point from deviations caused by disturbances and changing reference values to maintain a proper ratio between two process variables.

Improving the controlled response to disturbances requires an analysis of the incidence point of such perturbations. When the disturbance directly affects the process output and the disturbance model can be estimated, the feedforward control strategy can be applied by adding a trajectory through a compensating function. Even with modeling errors or a simplified feedforward function, the process response benefits from an action anticipated to the transitory and static effect of the disturbance. On the other hand, if the disturbance is manifested in an auxiliary intermediate variable, that is, in a variable that experiences an earlier effect than the process output variable, a cascade control strategy can be used. For cascade control, the process is divided in two parts in series, such that an inner loop is defined with the first part, and this interior loop in series with the second part of the process interacts with an outer controller. The interior loop may reject the disturbance before observing an important effect on the main process variable. The bigger the difference in the speed of response of the auxiliary and main variables, the greater the benefit of the cascade strategy to attenuate the internal disturbance.

The ratio control is illustrated by the need for an appropriate proportion between air and fuel flows in a combustion process. The simplest way to control the air-fuel ratio is the parallel scheme, where the desired ratio is managed by the set point variables. The parallel ratio control is ineffective in maintaining the proper ratio between the flow variables during transitory responses. The series ratio control achieves a better actual air-fuel ratio by determining the set point of the fuel flow from the slower measured air flow; in such a scheme the air is the leading flow. The cross-limit ratio control interchanges the leading and the subordinated flows as necessary to keep a safe actual ratio value, assumed to be greater or equal to 10 for the air-fuel ratio.

The feedforward, cascade, and ratio strategies can be combined to cover the control needs of a process. The knowledge of these control schemes constitutes a basis for advanced control techniques. These control schemes are conceptually proposed through block diagrams and continuous functions but can be implemented on discrete platforms.

The automation and control of an industrial reheat furnace was presented as a case study to show how the furnace temperature and pressure control requires different multi-loop control strategies such as cascade, feedforward, and ratio control schemes for controlling the multiple coupled process variables.

References

Araki, M. 1985. Two-degree-of-freedom control system: Part I. *Systems and Control* 29: 649–656.

Araki, M. and Taguchi, H. 2003. Tutorial paper two-degree-of-freedom PID controllers. *International Journal of Control, Automation, and Systems* 1(4).

Shah, S. and MacGregor, J.F. (Vol. eds.). 2004. Dynamics and Control of Process Systems *IPV –IFAC Proceedings*. Elsevier IFAC Publications.

Hiroi, K. 1986. Two-degree-of-freedom PID control system and its application. *Instrumentation* 29: 39–43.

Horowitz, I.M. 1963. *Synthesis of Feedback Systems*. Academic Press.

Seborg, D.E., Edgar, T.F., Mellichamp, D.A. and Doyle III, F.J. 2016. *Process Dynamics and Control*. 4th Edition. Wiley.

Wilamowski, B. and Irwin, J.D. 2018. *Control and Mechatronics*. CRC Press.

Chapter 10
Digital Control Design

1. Introduction

Digital controllers are conveniently implemented in microcontrollers, programmable controllers, and computers. Control software algorithms are developed based on the difference equations of discrete-time controllers.

The controller design must be properly based on the characteristics of the process (Kuo, 2002). Such characteristics can be expressed as parameters of a continuous low-order model: gain, time constant and dead time for a first-order system, or gain, time constant, damping ratio, and dead time for a second-order system. Third or higher order continuous models are not parametrized and are often reduced by neglecting non-dominant poles directly by eliminating polynomial factors in the denominator. Typically, a PID controller can be proposed based on continuous process models and subsequently discretized for implementation.

There are several methods available for the design of digital controllers: (1) model the plant or process in continuous time, discretize a continuous-time controller (for example, express the PID as a difference equation), and then tune up the controller with the parameters of the continuous-time process; (2) model the plant or process in continuous time, discretize the process model to obtain a transfer function in z-domain, synthesize or design the digital controller directly in z-domain and obtain the difference equation; and (3) identify the system directly in discrete time (obtain a transfer function in z-domain), design the digital controller in z-domain and obtain the difference equation.

Methods 2 and 3 imply the design of the controller in discrete time without reference to a continuous PID (Dorf and Bishop, 2004). In the case of discrete process models, even low order parametric models lack a precise physical description, and the parameters are considered numerical and particular to the specific magnitude of the data and the sample time; that is, the parameters do not characterize separately a particular effect, such as sensitivity or velocity of response, but all the process traits influence all the parameters. Model order reduction is possible for discrete systems but implies recalculating all the parameters (not only direct elimination of denominator factors). Nevertheless, the discrete controller design requires procedures dissociated from the design of an equivalent PID (Ogata, 1995).

This chapter explains a basic approach for the design of digital controllers based on the discrete model of the process, that is, the whole design procedure is carried out in the discrete-time domain (z-domain). In Section 2, a general design procedure is described. Different controller laws are possible. However, three specific controller algorithms are reviewed and discussed: dead-beat in Section 3, Kalman in Section 4, and Dahlin in Section 5. Section 6 presents the derivation of the PID difference equation from its specification in the z-domain and Section 7 provides the conclusions.

2. General Design Method

The discrete control system is depicted in Fig. 10.1, showing the basic elements and variables in the domain of variable z. The open loop transfer function $HG_p(z)$ represents the instrumented process, that is, the set of the actuator, physical process, and sensor. $HG_p(z)$ models the effect of the manipulated variable, $U(z)$, on the process variable, $Y(z)$. The process dynamics with respect to the disturbance variable, $D(z)$, is expressed as $G_d(z)$, giving the possibility to be similar or more disadvantageous (slower, for example) than the dynamics with respect to $U(z)$. The digital controller, $G_c(z)$, processes the error, $e(z)$, or difference between $Y(z)$ and its desired or reference value, $R(z)$.

An algebraic procedure is applied in the design of digital controllers based on the analysis of the closed-loop transfer function given by:

$$\frac{Y(z)}{R(z)} = \frac{G_c(z)HG_p(z)}{1+G_c(z)HG_p(z)} \tag{10.1}$$

The only unknown variable in the closed loop transfer equation (9.1) is $G_c(z)$. Open-loop transfer functions are determined previously by identification procedures. However, the process transfer function, $HG_p(z)$, with respect to $U(z)$, is the only one required. Therefore, the model coefficients are assumed to be available, for example:

$$HG_p(z) = \frac{b_1 z^{-1-d} + b_2 z^{-2-d}}{1+a_1 z^{-1}} \tag{10.2}$$

On the other hand, the ratio $\dfrac{Y(z)}{R(z)}$ in (9.1) should be considered the design specification, that is, it should be defined for proper control performance. The closed-loop response should present no offset and a desirable faster response than the open loop. Therefore, equation (10.1) can be solved for $G_c(z)$:

$$G_c(z) = \frac{1}{HG_p(z)} \frac{\dfrac{Y(z)}{R(z)}}{1-\dfrac{Y(z)}{R(z)}} \tag{10.3}$$

Equation (10.3) is the general design equation for the plant dynamics cancellation in digital controllers (Isermann, 1981). This controller explicitly considers the

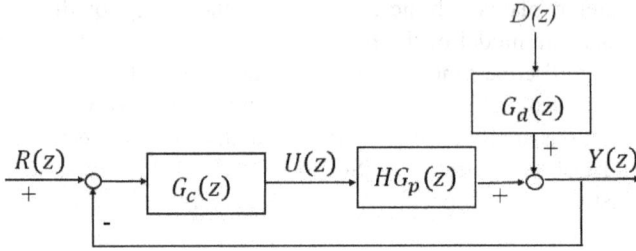

Fig. 10.1. Discrete closed loop control system.

inverse of the plant model (Widrow and Walach, 2008). The cancellation effect can be explained by substituting the controller function in the closed-loop transfer function equation (10.1) to reduce the right-hand side expression to $Y(z)/R(z)$, as in the left side of (10.1), by canceling $HG_p(z)$ algebraically. Such cancellation is easily observed in the controller-plant product used in (10.1):

$$G_c(z)HG_p(z) = \frac{1}{HG_p(z)} \frac{\dfrac{Y(z)}{R(z)}}{1-\dfrac{Y(z)}{R(z)}} HG_p(z) = \frac{\dfrac{Y(z)}{R(z)}}{1-\dfrac{Y(z)}{R(z)}} \tag{10.4}$$

The disturbance performance with controller (9.3) can be analyzed by the corresponding transfer function based on the block diagram of Fig. 10.1:

$$\frac{Y(z)}{D(z)} = \frac{G_d(z)}{1+G_c(z)HG_p(z)} \tag{10.5}$$

Substituting (10.4) in (10.5):

$$\frac{Y(z)}{D(z)} = \frac{G_d(z)}{1+\dfrac{\dfrac{Y(z)}{R(z)}}{1-\dfrac{Y(z)}{R(z)}}} = G_d(z)\left(1-\frac{Y(z)}{R(z)}\right) \tag{10.6}$$

The closed-loop transfer function (10.6) with respect to the disturbance input clearly includes the open-loop function $G_d(z)$ as one of its factors. Consequently, the open-loop dynamics prevails and contributes to the dominant closed-loop poles. This indicates that when a disturbance occurs, the closed loop will exhibit a response speed similar to the open-loop response, which is expected to be slow, with no other advantage than the final return of the process response to the set-point value. Therefore, the cancellation effect of the controller, according to (9.4), is produced only under a reference change or servocontrol response, imposing the desired dynamics represented by the specification $Y(z)/R(z)$.

The design of a cancellation controller requires proper choices of sample time and set point closed-loop response specification. As for the sample time, it is important to notice that the zeros of the plant are poles of the controller since the inverse of the plant transfer function is a factor in the controller equation (10.3). For a digital system to be stable, all its z-poles should have a magnitude less than 1. Then, the zeros of the plant, which are part of the poles of the controller, should satisfy this condition (as a minimum-phase plant model). The location of the zeros of the plant depends on the sample time, and the criterion of the zeros' magnitude being less than 1 should guide its selection. For example, for the plant given by (10.2), the sample time adequacy is guaranteed by this relationship between the model numerator coefficients:

$$|z_{zero}| = \left|\frac{b_2}{b_1}\right| < 1 \tag{10.7}$$

Moreover, the magnitude of the poles coming from the plant zeros should be very small to favor a fast response or not to dominate over the closed-loop specification dependent factor in (10.3). Therefore, it is more convenient to choose a sample time such that:

$$|z_{zero}| = \left|\frac{b_2}{b_1}\right| \to 0 \tag{10.8}$$

Inverse plant modeling techniques are currently being investigated to develop controllers (Widrow and Walach, 2008), considering even non-minimum-phase plants and including the effect of disturbances and noise. The concept of the inverse of the plant model has been explored for feedforward control schemes (Devasia, 2002). Research also covers feedback control with online inverse plant model approximation without prior knowledge of the plant dynamics (Waegeman et al., 2012). However, here the basic strategy of feedback for closed-loop control and inverting an available plant model are considered.

As for the design specification $Y(z)/R(z)$, the gain of this closed-loop function should be one to avoid an offset error. The analysis of this requirement can be made assuming a unit step reference input to the system:

$$R(z) = \frac{1}{1 - z^{-1}} \tag{10.9}$$

The response of the closed-loop system would be:

$$Y(z) = \frac{Y(z)}{R(z)} \frac{1}{1 - z^{-1}} \tag{10.10}$$

The final steady state value theorem can be applied on the response function:

$$\lim_{k \to \infty} Y(k) = \lim_{z \to 1} (1 - z^{-1}) Y(z) = \lim_{z \to 1} (1 - z^{-1}) \frac{Y(z)}{R(z)} \frac{1}{1 - z^{-1}} = \frac{Y(z)}{R(z)}\bigg|_{z=1} \tag{10.11}$$

For the response to reach the reference value of 1, and the steady state error to be null, the final response value should also be 1. The design condition is stated as:

$$\frac{Y(z)}{R(z)}\bigg|_{z=1} = 1 \tag{10.12}$$

Additionally, the poles of $Y(z)/R(z)$ should be stable and produce a faster response than the open-loop system or $HG_p(z)$. It is also possible for the specification $Y(z)/R(z)$ not to have poles at all, which means that the response can reach the reference in a certain number of sample times after the dead time has elapsed. Therefore, the fastest response that can be specified is:

$$\frac{Y(z)}{R(z)} = z^{-1-d} \tag{10.13}$$

where d is the integer number of sample periods covered by the dead time of the process, and $1 + d$ is the minimum number of sample periods to make the response equal to the reference after the latter has been changed stepwise. Different particular designs are presented in the following sections.

3. Design 1: Dead-Beat Controller

The simplest representation of the closed-loop response and the strictest specification is given by equation (10.13). In one sample time, after the dead time of the process, the output variable is expected to be at the desired value. For this specification to be realizable, the sample time is chosen to be relatively big in magnitude, considering, additionally, the zeros of the plant to be within the unit circle.

If the plant, $HG_p(z)$ is given by (10.2), and the desired closed-loop response is (10.13), then the controller calculated with (10.3) results:

$$G_c(z) = \frac{1 + a_1 z^{-1}}{z^{-1-d}(b_1 + b_2 z^{-1})} \frac{z^{-1-d}}{1 - z^{-1-d}} = \frac{1 + a_1 z^{-1}}{b_1 + b_2 z^{-1} - b_1 z^{-1-d} - b_2 z^{-2-d}} \tag{10.14}$$

Since $G_c(z) = U(z)/e(z)$, the controller recursive equation needed for its implementation or programming is:

$$u(k) = \frac{1}{b_1}(e(k) + a_1 e(k-1) - b_2 u(k-1) + b_1 u(k-1-d) + b_2 u(k-2-d)) \tag{10.15}$$

To illustrate the use of the dead-beat controller, consider a process described as the following continuous first-order system with dead time:

$$G_p(s) = \frac{Y(s)}{U(s)} = \frac{3e^{-2s}}{8s+1} \tag{10.16}$$

The process dynamics before a disturbance input is considered the same as with respect to the manipulation. That is, for simplicity (with no fault to the generality), $G_d(s) = G_p(s)$.

The discrete model with sample time, T, of 10 time units (approximately the time constant of the process, in this case) is:

$$HG_p(z) = z^{-1} \frac{1.8964 + 0.2441 z^{-1}}{1 - 0.2865 z^{-1}} \tag{10.17}$$

The *d* value is zero since the sample time is greater than the dead time. Therefore, the dead-beat design specification is:

$$\frac{Y(z)}{R(z)} = z^{-1} \tag{10.18}$$

The controller calculated with (10.3) or (10.14) is:

$$G_c(z) = \frac{1 - 0.2865z^{-1}}{1.8964 - 1.6522z^{-1} - 0.2441z^{-2}} \tag{10.19}$$

3.1 Dead-Beat Controller Nominal Performance

The implementation of the closed-loop system with the corresponding recursive equations for the plant and the controller can be done in a spreadsheet. The closed-loop performance of the dead-beat controller is shown in Fig. 10.2. A reference unit step change is applied at time zero, and a disturbance step input with an amplitude of 0.5 is introduced at time 50 (or time 60 in Figs. 10.5 and 10.6). The physical range for the manipulation is assumed to be [–0.6, 0.6]; that is, numerical manipulations out of the range have no other effect than the exceeded limit value.

The dead-beat controller achieves a fast response when the set point is changed since the process variable takes one sample time to reach the desired value. However, it can be observed that the manipulation becomes steady after two sample periods. This indicates that it is possible for the process variable to oscillate between sample times as a consequence of the variation of the process input or manipulation. That is, the stabilization of the process variable is apparent but not real because the manipulation is still moving. This oscillation is hidden due to the big sample time used. Analysis of equations (10.18) and (10.19) supports this observation: the closed-loop response given by (10.18) only has z^{-1} in the numerator, which indicates that the response equals the reference with a delay of only one sample time; the controller equation (10.19) has the most negative power of z^{-2} in the denominator, which makes the manipulation dependent of the same manipulation variable two samples times behind.

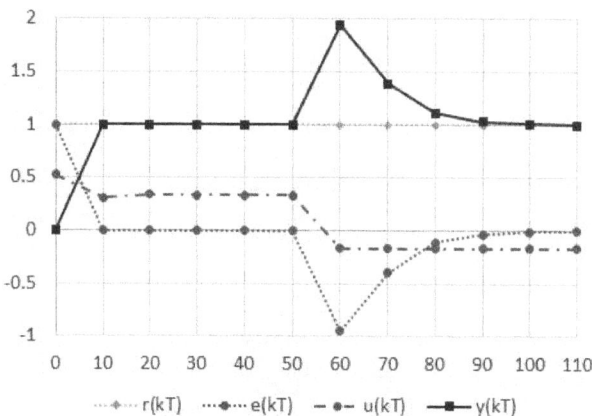

Fig. 10.2. Control loop performance with the dead-beat controller. Plant with nominal parameters.

When a step disturbance of magnitude 0.5 occurs, the process response shows a transitory of 40 time units and a maximum deviation of 1 unit. The response time is considered slow because it is approximately equal to the open-loop response of 34 times units, or four times the first-order time constant plus the dead time of the process, according to (10.16). The disturbance is effectively rejected because the response returns to the set-point value, but that happens with the open-loop speed. This is explained by recognizing that the dynamic cancellation effect of the controller is complete only when a set-point change is applied but not when a disturbance change is considered, as stated by (10.6).

3.2 Dead-Beat Controller Sensitivity and Saturation

The algebraic cancellation of the process dynamics depends on the accuracy of the model used in the calculation of the controller. The sensitivity to modeling errors can be explored by modifying the plant parameters without changing the calculated controller. For example,if the process model used for the controller calculation is given by (10.16) with an error of -20% in the process gain, then the real process would be

$$G_p'(s) = \frac{3.6e^{-2s}}{8s+1} \qquad (10.20)$$

and the process would be simulated with the corresponding discrete model with a sample time of 10 units:

$$HG_p'(z) = z^{-1}\frac{2.276+0.2929z^{-1}}{1-0.2865z^{-1}} \qquad (10.21)$$

To cover the opposite variation, an error of $+20\%$ in the process gain for the controller calculation can be simulated from the following equations (sample time of 10 units):

$$G_p''(s) = \frac{2.4e^{-2s}}{8s+1} \qquad (10.22)$$

$$HG_p''(z) = z^{-1}\frac{1.517+0.1953z^{-1}}{1-0.2865z^{-1}} \qquad (10.23)$$

Figures 10.3 and 10.4 show the effect of errors in the estimation of the process gain when using a dead-beat controller. In Fig. 10.3, when the process gain is higher than expected according to the process model used for the calculation of the controller, the process responds with more sensitivity, and an overshoot is produced. In Fig. 10.4, the actual process gain is smaller with respect to the one used for the controller design. Therefore, the response is less sensitive, and the process variable does not rise above the set-point value. In both cases, the response deviates from the dead-beat characteristic, taking two sample times to stabilize instead of one. Similar simulations can be worked to determine the effect of errors on the modeling of the time constant and the dead time of the process. These are left as an exercise for the reader.

Another issue with implementing digital controllers, in general, is the use of calculated manipulation values out of the physical range in subsequent evaluations of the controller equation. The current manipulation value may be above or below the maximum or minimum physical values, respectively; that is, the manipulation is saturated. An attempt to apply this manipulation will have the effect of applying

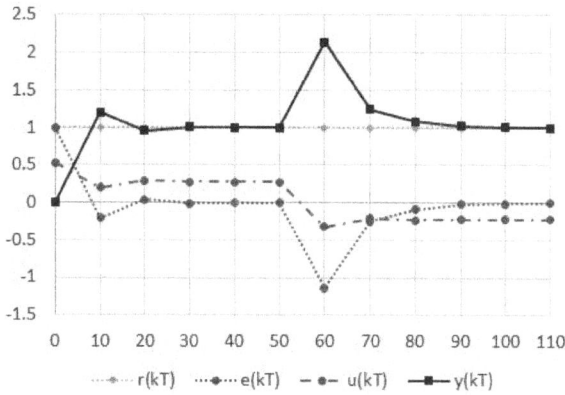

Fig. 10.3. Effect of underestimation of process gain on control loop performance with the dead-beat controller. (Plant simulated with gain 20% greater than the nominal value. Controller calculated assuming a lower process gain, using the nominal value.)

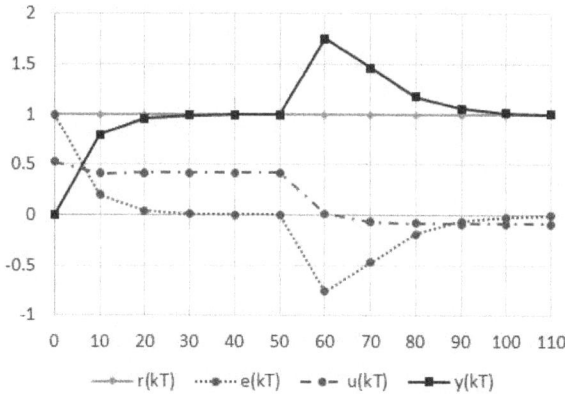

Fig. 10.4. Effect of overestimating the process gain on control loop performance with the dead-beat controller. (Plant simulated with gain 20% lower than the nominal value. Controller calculated assuming higher process gain, using the nominal value).

a physical limit value. Nevertheless, in the next work cycle of the controller, the manipulation just calculated becomes the previous manipulation, and two choices are numerically possible: either to work the value directly obtained before or use the value limited to the upper physical value for greater manipulations, or to the inferior physical value for lower manipulations. The latter option is called saturation feedback. This may be favorable or not depending on the controller, which is something to determine by simulation. Table 10.1 clarifies the difference between using and omitting saturation feedback.

In the case of the dead-beat controller, Figs. 10.5 and 10.6 show that it is better to use saturation feedback to avoid a response overshoot. The fast speed of the controller produces a rapid increment in the manipulation value, especially if its numerical value is not limited, and the effective manipulation stays at the physical maximum longer (Fig. 10.5). On the other hand, if the saturation of the manipulation

Table 10.1. Preparation of Evaluation of Controller Equation for the Next Work Cycle: f is the discrete function of the controller, u_{send} is the value within the physical range of the manipulation $[u_{min}, u_{max}]$ to be applied to the actuator of the process, and k is the discrete time.

With Saturation Feedback	Without Saturation Feedback
Controller difference equation: $u(k) = f(e(k), e(k-1),\dots, u(k-1), u(k-2),\dots)$	Controller difference equation: $u(k) = f(e(k), e(k-1),\dots, u(k-1), u(k-2),\dots)$
Preparation of manipulation value to apply $u_{send}(k) = u(k)$ *if* $u_{send}(k) > u_{max}$, *then* $u_{send}(k) = u_{max}$ *if* $u_{send}(k) < u_{min}$, *then* $u_{send}(k) = u_{min}$ *apply* $u_{send}(k)$	Preparation of manipulation value to apply $u_{send}(k) = u(k)$ *if* $u_{send}(k) > u_{max}$, *then* $u_{send}(k) = u_{max}$ *if* $u_{send}(k) < u_{min}$, *then* $u_{send}(k) = u_{min}$ *apply* $u_{send}(k)$
Shifting u values in preparation for next cycle: $u(k) = u_{send}(k)$ $\dots$ $u(k-2) = u(k-1)$ $u(k-1) = u_{send}(k)$	Shifting u values in preparation for next cycle: $u(k)$ is not limited $\dots$ $u(k-2) = u(k-1)$ $u(k-1) = u(k)$

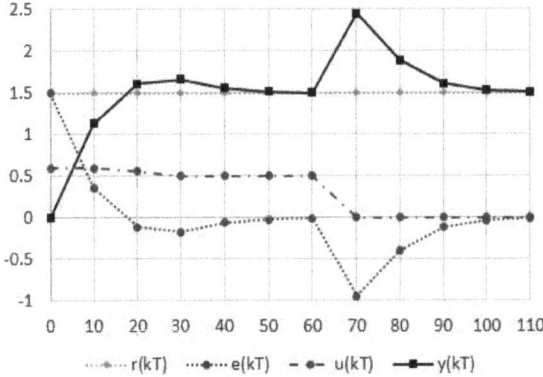

Fig. 10.5. Control loop performance with dead-beat controller and no output saturation feedback. Plant with nominal parameters. (Manipulation values are not limited to the physical limits for subsequent controller evaluations.)

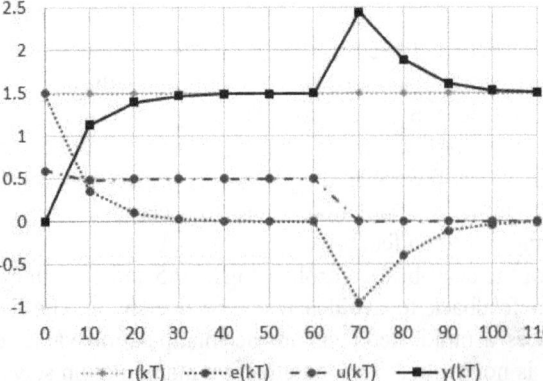

Fig. 10.6. Control loop performance with dead-beat controller and output saturation feedback. Plant with nominal parameters. (Manipulation is limited to the physical limits for subsequent controller evaluations.)

is fed back, by eliminating the excess over the maximum value for the evaluation of the controller, the manipulation is restricted for calculations, no oscillation is produced, and the system stabilizes sooner (Fig. 10.6).

4. Design 2: Kalman Controller

The design specification for the Kalman controller is based on the process model. The variables related by the transfer function $HG_p(z)$ are restated by algebraic operations to obtain the desired closed-loop function from the numerator. First, on the left side of the equation, $Y(z)$ and $U(z)$ are divided by $R(z)$:

$$HG_p(z) = \frac{Y(z)}{U(z)} = \frac{Y(z)/R(z)}{U(z)/R(z)} = \frac{b_1 z^{-1-d} + b_2 z^{-2-d}}{1 + a_1 z^{-1}} \tag{10.24}$$

Then, on the right-hand side of the equation, the numerator and denominator are divided by $b_1 + b_2$, which is the gain of the numerator polynomial:

$$HG_p(z) = \frac{Y(z)/R(z)}{U(z)/R(z)} = \frac{(b_1 z^{-1-d} + b_2 z^{-2-d})/(b_1 + b_2)}{(1 + a_1 z^{-1})/(b_1 + b_2)} \tag{10.25}$$

One way to satisfy the above equation is by making the numerators of the left and right sides equal and, therefore, equating also the denominators. From the numerators, the design specification is obtained:

$$\frac{Y(z)}{R(z)} = \frac{b_1 z^{-1-d} + b_2 z^{-2-d}}{b_1 + b_2} \tag{10.26}$$

This closed-loop function has unit gain and no poles. The response is expected to reach the reference value in $2 + d$ sample times, with no offset.

From the denominators, it can be observed that the manipulation takes one sample time to stabilize once the reference value has been changed:

$$\frac{U(z)}{R(z)} = \frac{1 + a_1 z^{-1}}{b_1 + b_2} \tag{10.27}$$

Since the manipulation will stabilize faster than the response, the apparent stabilization of the process variable will be real, and no hidden oscillations would occur (as in the case of the dead-beat controller).

To achieve a stable response in a few sample times $(2 + d)$, the sample period is recommended to be big, similar to the sample time used for the dead-beat controller.

The Kalman cancellation controller is then calculated as:

$$G_c(z) = \frac{1 + a_1 z^{-1}}{z^{-1-d}(b_1 + b_2 z^{-1})} \frac{z^{-1-d}(b_1 + b_2 z^{-1})/(b_1 + b_2)}{1 - z^{-1-d}(b_1 + b_2 z^{-1})/(b_1 + b_2)} = \frac{1 + a_1 z^{-1}}{b_1 + b_2 - b_1 z^{-1-d} - b_2 z^{-2-d}} \tag{10.28}$$

with the corresponding difference equation:

$$u(k) = \frac{1}{b_1 + b_2}(e(k) + a_1 e(k-1) + b_1 u(k-1-d) + b_2 u(k-2-d)) \tag{10.29}$$

For the same process considered before (gain 3, time constant 8, and dead time 2) with the same sample time of 10, the discrete process model is given by (10.17), and the Kalman controller is:

$$G_c(z) = \frac{1 - 0.2865z^{-1}}{2.1405 - 1.8964z^{-1} - 0.2441z^{-2}}$$ (10.30)

4.1 Kalman Controller Nominal Performance

The same simulation tests are applied to the Kalman control system: unit step reference at time zero, 0.5 step size of a disturbance at time 50 (or time 60 in Fig. 10.10 and 10.11), and manipulation range of [–0.6, 0.6]. Figure 10.7 shows the characteristic response of the control loop closed with a Kalman controller: with a sample time chosen such that $d = 0$, the response reaches the new reference value in two sample periods, and the manipulation acquires a constant value at the first sample period, which assures that the response does not move between sampled values that coincide with the set point. The Kalman controller corrects the deviation of the response due to the disturbance but with the velocity of the open-loop dynamics, as expected for cancellation controllers.

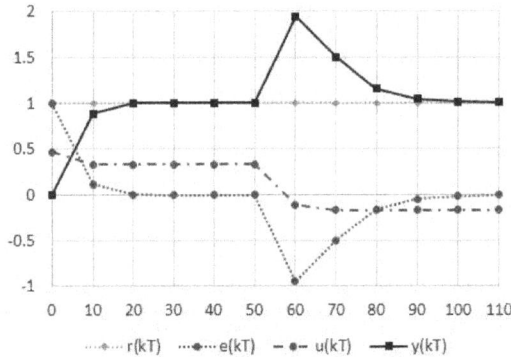

Fig. 10.7. Control loop performance with Kalman controller. Plant with nominal parameters.

4.2 Kalman Controller Sensitivity and Saturation

The robustness of the calculated Kalman controller can be tested by changing the process model according to (10.21) and (10.23), which correspond to the increment and decrement of the process gain of 20%, respectively. Figure 10.8 shows that when the controller is designed with a process model with a gain lower than the real one, the process is more sensitive than expected and an overshoot appears under the set-point change or servocontrol test, and a bigger deviation is obtained under the disturbance test. Particularly, the overshoot of the servocontrol response is clearly smaller with respect to the overshoot observed with the dead-beat controller in the same simulation conditions, as it can be seen in Fig. 10.3. When the Kalman controller is calculated with a process model with a gain higher than the one of the real process, the response, shown in Fig. 10.9, is slower than in the case with no modeling errors, and without overshoot when the reference is changed, similar to the response obtained with the dead-beat controller in Fig. 10.4. In the performed simulations, the Kalman controller tends to be more robust than the dead-beat controller.

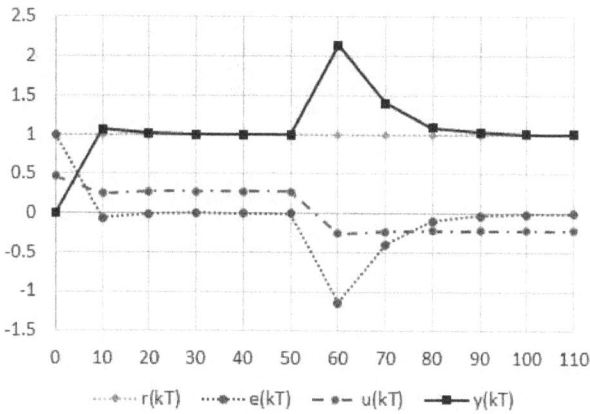

Fig. 10.8. Effect of underestimation of process gain on control loop performance with Kalman controller. (Plant simulated with gain 20% greater than the nominal value. Controller calculated with the nominal process gain.)

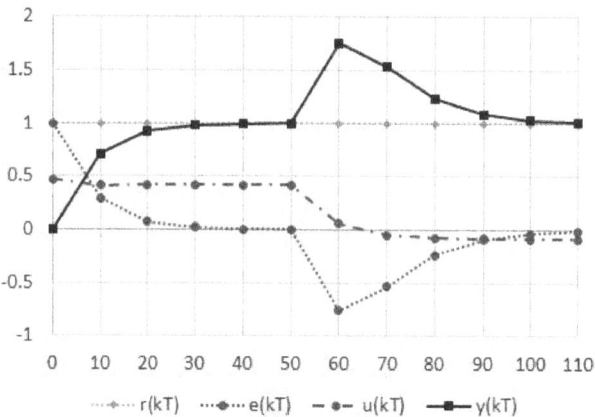

Fig. 10.9. Effect of overestimating process gain on control loop performance with Kalman controller. (Plant simulated with gain 20% lower than the nominal value. Controller calculated with the nominal process gain.)

The saturation feedback effect is explored with the Kalman controller in Figs. 10.10 and 10.11, where no modeling errors are considered. Again the set-point change is increased to 1.5 units, such that the initially calculated manipulations are higher than the physical limit of 0.6 units. Figure 10.10 shows the loop response when no saturation feedback is applied, or the numerical value of the manipulation is not restricted for the evaluation of the controller equation. In this case, an overshoot appears, and the stabilization time takes longer than two sample times. The disturbance response is the same as can be observed in Fig. 10.7 because the manipulation does not saturate. In contrast, when saturation feedback is used, the response shown in Fig. 10.11 does not present an overshoot, and the stabilization is faster under the set-point change. Therefore, saturation feedback is recommended for the Kalman controller, too, as in the case of the dead-beat controller.

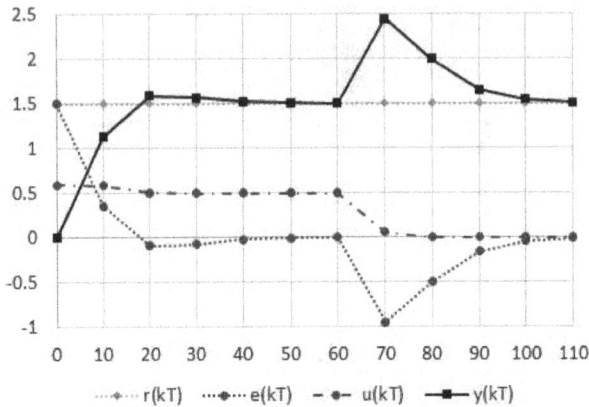

Fig. 10.10. Control loop performance with Kalman controller and no output saturation feedback. Plant with nominal parameters.

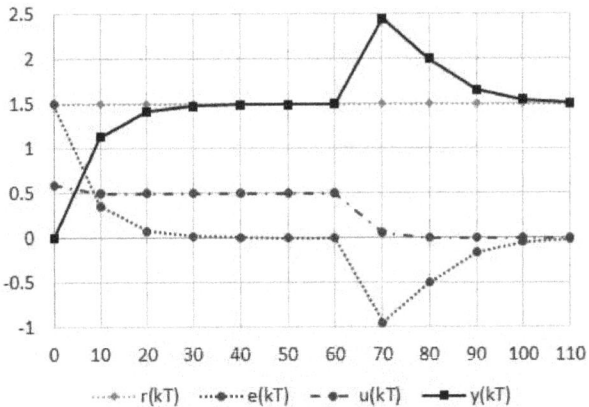

Fig. 10.11. Control loop performance with Kalman controller and output saturation feedback. Plant with nominal parameters.

5. Design 3: Dahlin Controller

The lambda tuning technique, also known as the Dahlin controller, is a dead-time compensator developed in discrete time that uses a first order plus dead time (FOPDT) model of the plant; this controller is based on the direct synthesis method and also requires the tuning of one parameter, the desired closed-loop response is a FOPDT with unit static gain (Chen and Seborg, 2002).

The Dahlin cancellation controller aims to approximate the continuous control of the process by achieving a smooth response over time. This requires using a small sample time and a specification of the desired response speed in terms of a closed-loop time constant and not in terms of stabilization in a certain number of sample periods (as in the case of the two previous controllers).

The design specification is stated in the continuous domain as a first-order transfer function. The closed-loop gain must be unitary to avoid an offset error. The dead time of the closed loop should be minimum; therefore, it is set equal to the dead time of the process. Finally, the closed-loop time constant should be less than the one of the open loop to produce a speed advantage in the response by using the controller. The design specification is then:

$$\frac{Y(s)}{R(s)} = \frac{e^{-\theta s}}{\lambda s + 1} \tag{10.31}$$

where θ is the dead time of the process, and λ is the closed-loop time constant.

For the design of the discrete controller, the desired closed-loop transfer function needs to be discretized with a small sample time.

$$\frac{Y(z)}{R(z)} = \frac{d_1 z^{-1-d} + d_2 z^{-2-d}}{1 + c_1 z^{-1}} \tag{10.32}$$

The process model, $HG_p(z)$ given by (10.2) for a first-order dynamics, is calculated with the same sample time to be used to control the process.

The general design equation (10.3) is evaluated with (10.32) and the process model to calculate the Dahlin controller:

$$G_c(z) = \frac{1 + a_1 z^{-1}}{z^{-1-d}(b_1 + b_2 z^{-1})} \frac{\dfrac{z^{-1-d}(d_1 + d_2 z^{-1})}{1 + c_1 z^{-1}}}{1 - \dfrac{z^{-1-d}(d_1 + d_2 z^{-1})}{1 + c_1 z^{-1}}} = \frac{(1 + a_1 z^{-1})(d_1 + d_2 z^{-1})}{(b_1 + b_2 z^{-1})(1 + c_1 z^{-1} - d_1 z^{-1-d} - d_2 z^{-2-d})} \tag{10.33}$$

$$G_c(z) = \frac{d_1 + (a_1 d_1 + d_2)z^{-1} + a_1 d_2 z^{-2}}{b_1 + (b_1 c_1 + b_2)z^{-1} + b_2 c_1 z^{-2} - b_1 d_1 z^{-1-d} - (b_1 d_2 + b_2 d_1)z^{-2-d} - b_2 d_2 z^{-3-d}} \tag{10.34}$$

The following recursive equation is used for the implementation:

$$\begin{aligned}
u(k) = \frac{1}{b_1}[&d_1 e(k) + (a_1 d_1 + d_2)e(k-1) + a_1 d_2 e(k-2) - (b_1 c_1 + b_2)u(k-1) \\
&- b_2 c_1 u(k-2) + b_1 d_1 u(k-1-d) + (b_1 d_2 + b_2 d_1)u(k-2-d) \\
&+ b_2 d_2 u(k-3-d)]
\end{aligned} \tag{10.35}$$

To continue with the illustration of this controller design, the same process with gain 3, time constant 8, and dead time 2 is considered, and a sample time of 1 unit is chosen. The discrete model of the process is now:

$$HG_p(z) = \frac{0.3525 z^{-3}}{1 - 0.8825 z^{-1}} \tag{10.36}$$

The desired closed-loop time constant is specified as one fourth of the one of the open-loop process. The design specification is first given in the continuous domain and then discretized:

$$\frac{Y(z)}{R(z)} = \frac{e^{-2s}}{1 + 2s} \tag{10.37}$$

$$\frac{Y(z)}{R(z)} = \frac{0.3935z^{-3}}{1-0.6065z^{-1}} \tag{10.38}$$

The Dahlin controller is given by:

$$G_c(z) = \frac{0.3935 - 0.3473z^{-1}}{0.3525 - 0.2138z^{-1} - 0.1387z^{-3}} \tag{10.39}$$

5.1 Dahlin Controller Nominal Performance

Figure 10.12 shows the simulation of the closed-loop system with a unit step reference change at time zero as a servocontrol test, a 0.5 step disturbance at time 30 as a regulatory test, and a manipulation interval of [−1.2, 1.2]. The design specification is clearly satisfied, achieving stabilization in 10 time units under the set-point change (four times the closed-loop time constant of 2, plus the dead time of 2). The disturbance is managed by returning the process variable to the set point value but in approximately 34 time units, which is the open-loop stabilization time. The smaller sample time used allows the maximum deviation under the disturbance to be noticeably reduced with respect to the ones observed with the other controllers.

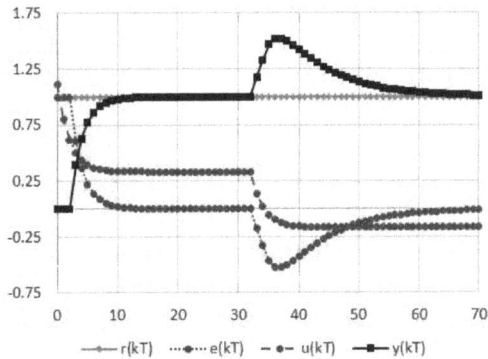

Fig. 10.12. Control loop performance with Dahlin controller. Plant with nominal parameters.

5.2 Dahlin Controller Sensitivity and Saturation

The Dahlin controller performance under modeling errors is shown in Figs. 10.13 and 10.14. The plant model is altered with variations of ±20% in the gain value. The plant models with increased and decreased gain are given by equations (10.40) and (10.41), respectively, with the sample time of 1:

$$HG_p'''(z) = z^{-3}\frac{0.423}{1-0.8825z^{-1}} \tag{10.40}$$

$$HG_p^{iv}(z) = z^{-3}\frac{0.282}{1-0.8825z^{-1}} \tag{10.41}$$

When the process gain is underestimated and equation (10.40) is used, the system with the Dahlin controller (10.39) responds faster, causing the peak values in both the servocontrol and the regulatory tests to be slightly increased, as illustrated in Fig. 10.13. In fact, the small overshoot (1.9%) in the set-point change can be

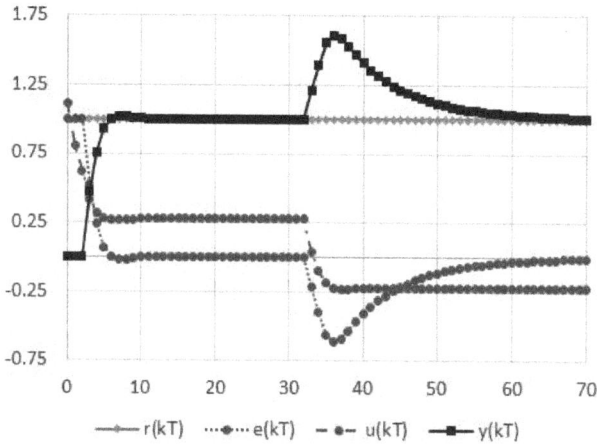

Fig. 10.13. Effect of underestimation of process gain on control loop performance with Dahlin controller. (Plant simulated with gain 20% greater than the nominal value. Controller calculated with the nominal process gain.)

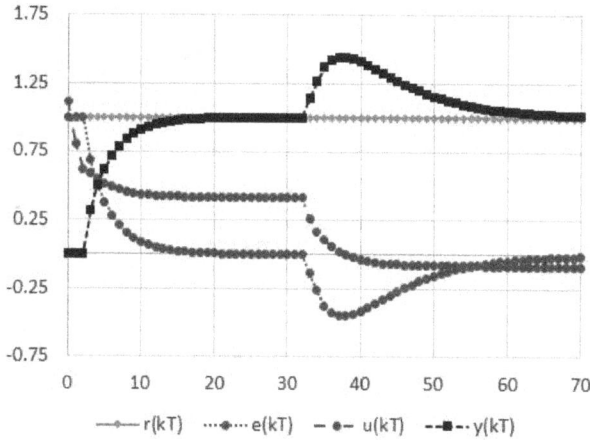

Fig. 10.14. Effect of overestimation of process gain on control loop performance with Dahlin controller. (Plant simulated with gain 20% lower than the nominal value. Controller calculated with the nominal process gain.)

considered within a reasonably tight stabilization response band (±2%) and, therefore, neglected to affirm that the expected first-order response is still obtained. On the other hand, Fig. 10.14 shows that when the process gain is overestimated by using equation (10.41) without changing the controller, the servocontrol response is slower, no oscillation is observed, and the maximum deviation in the regulatory test decreases a bit and is eliminated with the same long stabilization time.

Figures 10.15 and 10.16 address the aspect of saturation of the manipulation at the value of 0.6. Figure 10.15 shows that no output saturation feedback (or unlimited numerical values for the previous manipulations when calculating the current controller output) reduces the rising time and produces a small overshoot

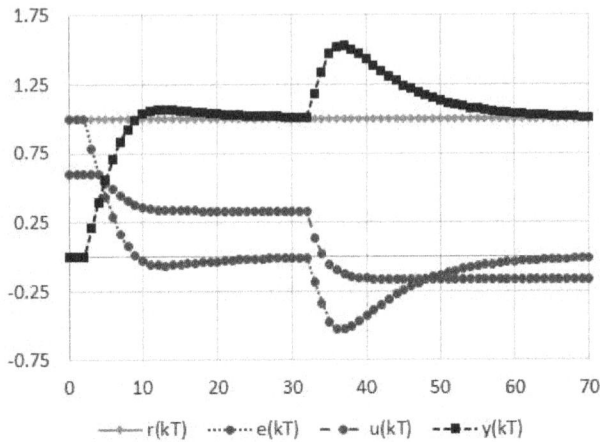

Fig. 10.15. Control loop performance with Dahlin controller and no output saturation feedback. Plant with nominal parameters.

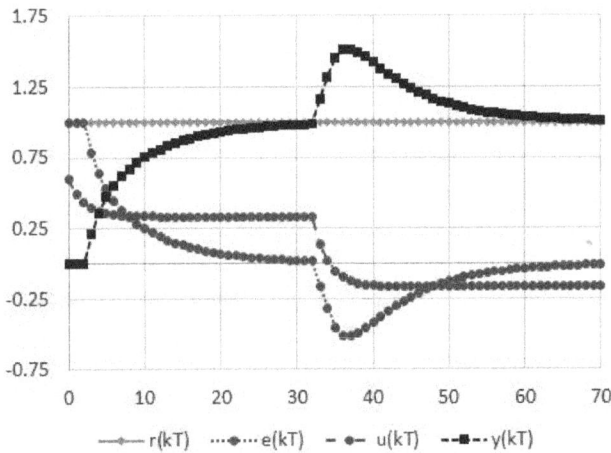

Fig. 10.16. Control loop performance with Dahlin controller and output saturation feedback. Plant with nominal parameters.

compared to the ones obtained with a dead-beat and Kalman controllers. Lack of saturation feedback tends to maintain the manipulation saturated longer with respect to Fig. 10.16. When saturation feedback is applied, a sluggish response is obtained for reaching the new set point, as shown in Fig. 10.16. If the overshoot is neglected, it can be preferable not to apply saturation feedback when using the Dahlin controller.

6. PID Controller in z-Domain

The design of discrete time PID controllers can be done using z-transform methods. For most applications, the poles and zeros of the transfer function of the controller are at low frequencies compared to the sampling frequency (or execution time), and, therefore, the controller design can be done in the s-domain. In some cases, and depending upon the implementation platform, like an embedded controller or Field Programmable Gate Arrays (FPGAs), or when the poles and zeros of the transfer function are close to the sampling frequency, the systems can be best modeled in the z-domain.

The basic transfer functions to obtain the components of a PID controller directly specified in the z-domain are:

Proportional term:

$$G_c(z) = k_p \tag{10.42}$$

Integral term:

$$G_c(z) = \frac{k_i z}{z-1} = \frac{k_i z^{-1}}{1-z^{-1}} \tag{10.43}$$

Derivative term:

$$G_c(z) = \frac{k_d(z-1)}{z} = k_d(1-z^{-1}) \tag{10.44}$$

6.1 PI Controller

The proportional and integral terms are added to obtain the PI transfer function:

$$G(z) = \left(k_p + \frac{k_i z}{z-1} \right) = \frac{(k_p + k_i)z - k_p}{z-1} \tag{10.45}$$

$$G(z) = \frac{U(z)}{E(z)} = \frac{(k_p + k_i) - k_p z^{-1}}{1-z^{-1}} \tag{10.46}$$

To derive the control law in discrete time, express it as a two polynomial side equation:

$$(1 - z^{-1})\, U(z) = ((k_p + k_i) - k_p z^{-1})E(z) \tag{10.47}$$

Then, solve for $U(z)$:

$$U(z) = z^{-1}\, U(z) + (k_p + k_i)E(z) - k_p z^{-1}E(z) \tag{10.48}$$

Apply inverse z-transform to obtain u_n:

$$u_n = u_{n-1} + (k_p + k_i)\, e_n - k_p\, e_{n-1} \tag{10.49}$$

$$u_n = u_{n-1} + k_p(e_n - e_{n-1}) + k_i e_n \tag{10.50}$$

6.2 PD Controller

The proportional and derivative terms are added to obtain the PD transfer function:

$$G(z) = \left(k_p + \frac{k_d(z-1)}{z} \right) = \frac{(k_p + k_d)z - k_d}{z} \tag{10.51}$$

$$G(z) = \frac{U(z)}{E(z)} = (k_p + k_d) - k_d z^{-1} \qquad (10.52)$$

Solve for $U(z)$ to derive the control law in discrete time:

$$U(z) = (k_p + k_d)E(z) - k_d z^{-1} E(z) \qquad (10.53)$$

Apply inverse z-transform to obtain u_n:

$$u_n = (k_p + k_d)\, e_n - k_d\, e_{n-1} \qquad (10.54)$$

$$u_n = k_p\, e_n + k_d\,(e_n - e_{n-1}) \qquad (10.55)$$

6.3 PID Controller

The proportional, integral, and derivative terms are added to obtain the PID transfer function:

$$G_c(z) = \left(k_p + \frac{k_i z}{z-1} + \frac{k_d(z-1)}{z}\right) = \frac{k_p z(z-1) + k_i z^2 + k_d(z-1)^2}{z(z-1)} \qquad (10.56)$$

$$G_c(z) = \frac{k_p z^2 - k_p z + k_i z^2 + k_d(z^2 - 2z + 1)}{z^2 - z} \qquad (10.57)$$

$$G_c(z) = \frac{(k_p + k_i + k_d)z^2 + (-k_p - 2k_d)z + k_d}{z^2 - z} \qquad (10.58)$$

$$G_c(z) = \frac{U(z)}{E(z)} = \frac{(k_p + k_i + k_d) + (-k_p - 2k_d)z^{-1} + k_d z^{-2}}{1 - z^{-1}} \qquad (10.59)$$

For the PID controller, $G_c(z)$ of Eq. 10.59 can have two complex conjugate zeros or two real zeros. For a stable closed-loop response, all system poles must be within the unit circle of the z-plane. However, zeros can be outside of the unit circle. The dominant poles closer to the origin, the faster the system response, and the greater in magnitude the manipulation of the controller.

To derive the control law in discrete time, solve for $U(z)$:

$$(1 - z^{-1})\, U(z) = [(k_p + k_i + k_d) + (-k_p - 2k_d)z^{-1} + k_d z^{-2}]E(z) \qquad (10.60)$$

$$U(z) = z^{-1} U(z) + [(k_p + k_i + k_d) + (- k_p - 2k_d)z^{-1} + k_d z^{-2}]E(z) \qquad (10.61)$$

Apply inverse z-transform to obtain u_n:

$$u_n = u_{n-1} + (k_p + k_i + k_d)e_n - (k_p + 2k_d)\, e_{n-1} + (k_d)\, e_{n-2} \qquad (10.62)$$

$$u_n = u_{n-1} + k_p(e_n - e_{n-1}) + k_i\, e_n + k_d(e_n - 2e_{n-1} + e_{n-2}) \qquad (10.63)$$

The discrete-time control law requires the proportional integral and derivative gains to be specified. The disadvantage of design in the z-domain is that the meaning of the parameters of the physical systems is not directly visible in the difference equation, compared to the case of s-domain methods and discretization techniques.

The execution time interval affects the controller's response; gain adjustments must be taken into consideration whenever the execution time interval is changed. For a direct equivalence to a PID in discrete time and to include the sampling interval explicitly, the gain parameters are:

$$k_p = k_c \tag{10.64}$$

$$k_i = k_c \left(\frac{T}{T_i} \right) \tag{10.65}$$

$$k_d = k_c \left(\frac{T_d}{T} \right) \tag{10.66}$$

Therefore, the control law can be expressed as:

$$u_n = u_{n-1} + k_c (e_n - e_{n-1}) + k_c \left(\frac{T}{T_i} \right) e_n + k_c \left(\frac{T_d}{T} \right) (e_n - 2e_{n-1} + e_{n-2}) \tag{10.67}$$

$$u_n = u_{n-1} + k_c \left[(e_n - e_{n-1}) + \left(\frac{T}{T_i} \right) e_n + \left(\frac{T_d}{T} \right) (e_n - 2e_{n-1} + e_{n-2}) \right] \tag{10.68}$$

This last equation is equivalent to the difference equation obtained by discretizing a continuous time PID equation using the backward Euler method. This conversion process can be applied to any PID controller described as a transfer function in z-domain.

7. Conclusions

In this chapter, a basic approach to digital design has been presented. The qualitative behavior of the cancellation controllers is consistent regarding effective cancellation of process dynamics only before set-point changes, rejection of disturbance with open-loop stabilization time, faster velocity or higher transitory slopes of response with no saturation feedback, and when the process gain is underestimated.

Quantitative differences in the controller performance are summarized in Table 10.2. Although the dead-beat and Kalman controllers are calculated with a wide sample time, they can be considered fast controllers. Table 10.2 shows that these controllers are more sensitive to modeling errors regarding the process gain than the Dahlin controller. Dead-beat and Kalman controllers should not use manipulation saturation feedback to avoid important oscillation. In contrast, the Dahlin controller can be favored in velocity of response by the use of saturation feedback, especially if the closed-loop time constant is closer to the open-loop time constant.

The concept of cancellation controllers can be applied for varied closed-loop specifications beyond the cases of dead-beat, Kalman, and Dahlin controllers. For instance, second-order dynamics could be defined for the closed loop expected

Table 10.2. Comparison of cancellation controller performance (* apparent stabilization; stabilization tolerance band of ±2% of final set-point value).

Aspect	Test	Indicator	Dead-beat	Kalman	Dahlin
Nominal performance	Servocontrol test (Figs. 10.2, 10.7, 10.12)	Overshoot (%)	0	0	0
		Settling time	10 *	20	10
	Regulatory test (Figs. 10.2, 10.7, 10.12)	Maximum deviation	0.948	0.948	0.524
		Settling time	40	50	35
Effect of underestimation of process gain	Servocontrol test (Figs. 10.3, 10.8, 10.13)	Overshoot (%)	20	6.33	0
		Settling time	30	20	6
	Regulatory test (Figs. 10.3, 10.8, 10.13)	Maximum deviation	1.138	1.138	0.608
		Settling time	40	50	35
Effect of overestimation of process gain	Servocontrol test (Figs. 10.4, 10.9, 10.14)	Overshoot (%)	0	0	0
		Settling time	30	30	15
	Regulatory test (Figs. 10.4, 10.9, 10.14)	Maximum deviation	0.758	0.758	0.447
		Settling time	50	60	37
No saturation feedback	Servocontrol test (Figs. 10.5, 10.10, 10.15)	Overshoot (%)	11.15	5.97	6.3%
		Settling time	50	40	24
Use of saturation feedback	Servocontrol test (Figs. 10.6, 10.11, 10.16)	Overshoot (%)	0	0	0
		Settling time	30	30	30

performance, with the flexibility of adjusting two parameters instead of one: damping ratio and time constant, since the unit gain and minimum dead time are prerequisites.

Direct design of controllers in the discrete domain is a powerful possibility in front of the discretization of PID type controllers, in terms of flexibility and variability of controller equations: the process can be modeled by a high order transfer function with or without dead time, and the controller equation may have a different number of terms with different delays. However, the disadvantage of discrete time designed controllers is that they need to be recalculated in the case of an important variation of process dynamics. In contrast, the response of the PID-based controllers can still be adjusted by the PID parameters.

References

Chen, D. and Seborg, D.E. 2002. PI/PID controller design based on direct synthesis and disturbance rejection. *Industrial & Engineering Chemistry Research* 41(19): 4807–4822.

Devasia, S. 2002. Should model-based inverse inputs be used as feedforward under plant uncertainty? *IEEE Transactions on Automatic Control* 47(11).

Dorf, R. and Bishop, R. 2004. *Modern Control Systems*. Prentice Hall.

Isermann, R. 1981. Cancellation controllers. *In: Digital Control Systems*. Berlin, Heidelberg. Springer.

Kuo, B. 2002. *Automatic Control Systems*. Wiley and Sons.

Ogata, K. 1995. *Discrete-Time Control Systems*. Australia, Sydney. Prentice Hall.

Waegeman, T., Wyffels, F. and Schrauwen, F. 2012. Feedback control by online learning an inverse model. *IEEE Trans Neural Netw. Learn. Syst.* 23(10): 1637–1648.

Widrow, B. and Walach, E. 2008. Inverse plant modeling. *In*: *Adaptive Inverse Control, Reissue Edition: A Signal Processing Approach*. Wiley-IEEE Press.

Chapter 11

PLCs and Sequential Logic Control

1. Introduction

Automated systems, where inputs and output variables are mostly discrete-binary, require the implementation of both combinational and sequential logic control functions. Combinational logic is used in digital electronics and logic control to perform Boolean algebra on input signals. Sequential logic is used to automate operation sequences. Practical logic control applications are a mixture of combinational and sequential logic.

Many automated systems, where inputs and output variables are mostly binary discrete, can be modeled as discrete event systems. The state of the system may change because new conditions have been detected (event-driven changes) or because a certain amount of time has elapsed (time-driven changes). In a washing machine, the agitation cycle is set to operate for a specific time. By contrast, filling the tub is event-driven because it starts when the user pushes a button and stops when the high level is reached and indicated by a sensor (Groover, 2008). Both event-driven and time-driven events can be simulated within the programmable controllers or in external computers connected to the electronic control units to validate the control logic functions (Vieira et al., 2017).

This chapter reviews basic concepts and methodologies to address combinational and sequential logic control problems and presents a basic approach for simulating discrete event systems. Section 2 reviews combinational and sequential control logic and their basic representation using logic gates and logic functions. Section 3 describes the Programmable Logic Controllers (PLC) and their programming languages, particularly the popular ladder logic. Section 4 presents a control logic design method based on time diagrams to deduce logic functions for their implementation as ladder logic. Section 5 presents another design methodology based on sequenced or sequential latching functions and their implantation in ladder logic. Section 6 presents a basic approach for discrete event process simulation required to validate sequential logic implemented in PLCs. Section 7 reviews the conclusions.

2. Combinational and Sequential Logic

In automata theory, combinational logic is a type of logic where the output at any moment depends exclusively on the values of the inputs. In sequential logic, the outputs are activated based on the present value of the input signals and memory conditions generated by the previous activation of the inputs. Sequential logic has states (or memory functions), while combinational logic does not.

2.1 Combinational Logic Control

In combinational logic, the binary output signals depend only on the present value of the binary inputs. Binary variables can take only two values: ON/OFF or 1/0. Combinational logic functions have no memory and are time independent; as soon as inputs are changed, the information from the previous inputs is lost.

Combinational logic is implemented by logical Boolean functions. The basic logical operators NOT, AND, OR, NAND, and NOR and their truth o logical tables are shown in Fig. 11.1. Practical design of combinational logic control systems may require some delay in the logic evaluation since binary sensors or actuator elements have their own response time.

Combinational logic control can be implemented using the basic logic functions AND, OR, and NOT. For example, consider a tank operated with two solenoid valves, one for filling the tank ($SV1$) and the other for discharging the tank ($SV2$). There are binary level switches to automatically close the valves: one to detect the tank is full ($LS1$ normally open switch), and the other to detect the tank is empty ($LS2$ normally open switch). The filling and discharging operations are activated with a three-position selector. Left position, $SW1$, to fill, and right position, $SW2$, to discharge. The logic functions required with the basic operators are:

$$SV1 = SW1 \cdot \overline{LS1} \tag{11.1}$$

$$SV2 = SW2 \cdot LS2 \tag{11.2}$$

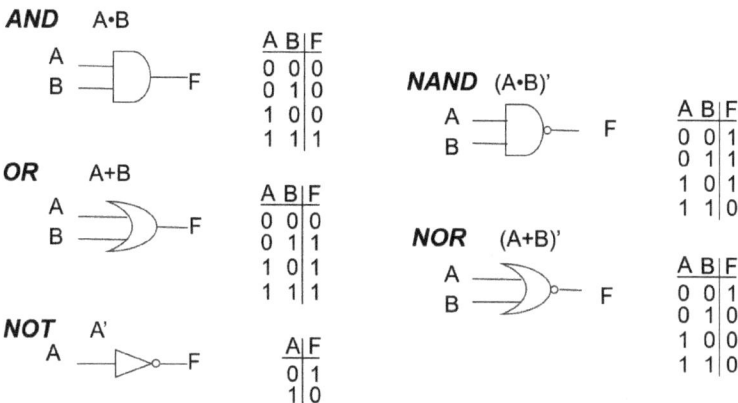

Fig. 11.1. Basic logic functions.

2.2 Sequential Logic Control

In sequential logic, the outputs depend not only on the present inputs but also on the history of the inputs. In this sense, sequential logic has memory elements (Fig. 11.2). Sequential logic control is used to control the execution of event- and time-driven changes. Sequential logic also uses internal timing devices to determine when to initiate changes in output variables.

Sequential logic is typically required for the automation of sequential processes. The concept of latching is very relevant to implement sequential logic functions; this implies the use of memory elements or the feedback of the outputs as inputs to the logic circuit, as shown in Fig. 11.3. The truth table considers the state of the output to solve the logical function for the same output.

The logical function for the (set/reset) latching function is:

$$Mem = (Mem + Set) \cdot \overline{Rst} \tag{11.3}$$

A simple application example of the latching function is the start/stop operation of a motor using "start and stop" push buttons that a user manually activates.

$$MtrRun = (MtrRun + StartPB) \cdot \overline{StopPB} \tag{11.4}$$

Sequential Logic

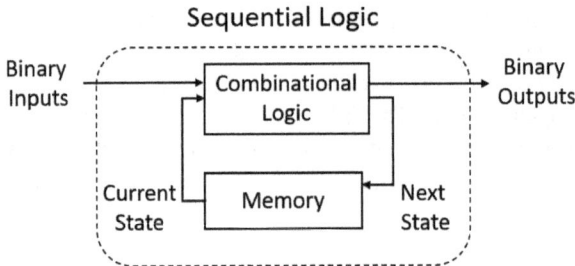

Fig. 11.2. Sequential logic components.

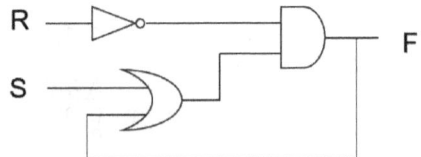

S	R	F	F
0	0	0	0
0	0	1	1
0	1	0	0
0	1	1	0
1	0	0	1
1	0	1	1
1	1	0	0
1	1	1	0

Fig. 11.3. Truth table and logic gates for set/reset latching function.

Another simple example is the starting and stopping of a pump to fill a tank. In this example, the binary inputs come from sensor signals. The pump starts with the low level switch and stops with the high level switch. Consider low and high level switches are both normally open. To start the pump, the empty tank is detected with the low level switch deactivated, and this logical condition starts the pump. The high level switch is activated when the tank is full, and this condition stops the pump.

$$PumpStart = (PumpStart + \overline{LLS}) \cdot \overline{HLS} \tag{11.5}$$

Now consider that a system must be started and stopped with the same push button. To solve this problem, a memory variable can be used to generate the start-and-stop logic functions. These two logical conditions correspond to set-and-reset inputs to generate a latching function, that is, the memorized condition of the system: "on or off" status.

$$Set = PB \cdot \overline{SysOn} \tag{11.6}$$

$$Reset = PB \cdot SysOn \tag{11.7}$$

$$SysOn = (SysOn + Set) \cdot \overline{Reset} \tag{11.8}$$

2.3 Mixing and Heating Process Example

Recall the Mixing and Heating Process reviewed in Chapter 2, also shown in Fig. 11.4. The process consists of a tank, two feeding valves SVA and SVB, one discharge valve SVC, and three binary level switches: LS1 to sense high level, LS2 for intermediate level, and LS3 for low level or empty tank. The motor to drive the mixer is energized with contactor M1 and operates while there is a level detected by LS2 either during filling or discharging.

Latching functions can implement the required sequential logic: one for activating the first filling valve, another for the second filling valve, and a third one for the discharge valve. The mixer motor can be started with one latching condition

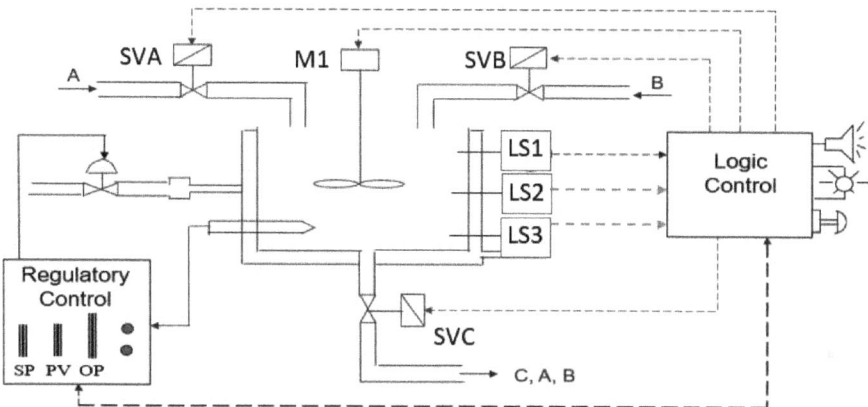

Fig. 11.4. Mixing and heating process.

and stopped with another, assuming one latching condition corresponds to one step in the logic control sequence. However, in this case, the mixer is active if the tank level is greater or equal to the intermediate level. The corresponding logic functions are:

$$SVA = (SV1 + \overline{LS3}) \cdot \overline{LS2} \tag{11.9}$$

$$SVB = (SV2 + LS2) \cdot \overline{LS1} \tag{11.10}$$

$$SVC = (SV1 + LS1) \cdot LS3 \tag{11.11}$$

$$M1 = LS2 \tag{11.12}$$

Logic control algorithms contain a mixture of combinational and sequential logic. Combinational and sequential logic can be implemented in microcontrollers, PLC, and digital electronic circuits. Many sequential control logic problems can be addressed with sequential activation of latching functions. A sequenced latching methodology is detailed in Section 6.

3. Programmable Logic Controllers

The programmable logic controller, PLC, is a microcomputer based controller that uses stored instructions in programmable memory to implement logic, sequencing, timing, counting, and arithmetic functions through digital or analog modules to control machines and processes in process industries and discrete manufacturing. PLCs are industrial-grade controller units with greater reliability, easier maintenance, and communication capability for connecting to computers and information systems.

PLC was introduced to replace hard-wired electromechanical relay panels. PLC can be reprogrammed and can perform a greater variety of control functions. The application of PLC for industrial automation started with the invention of the first MODICON 084 programmable controller (MOdular DIgital CONtroller) in 1969 (Dunn, 2009). The PLC, in conjunction with sensors and actuators equipped with electronic conditioning, allowed to implement more precise and repetitive control systems both in the manufacturing industry (metalworking, automotive, electronics, etc.) and in the process industries (iron and steel, chemical, petrochemical, cement, etc.) (Hugh, 2010).

3.1 PLC Operation

In a PLC, the processor unit senses inputs and executes logic and sequencing functions to determine the appropriate output signals. The memory unit contains the logic and sequential programs, and I/O operations. In a small PLC, the inputs and outputs are integrated into the same equipment. Larger PLCs use inputs and outputs modules for connection to field devices. PLCs are powered with 120 VAC or 24 VDC power supplies, see Fig. 11.5.

The operating cycle of a PLC, which is also called a scan cycle, is as follows: (1) input scan – inputs are read by the processor and stored in memory; (2) program scan – control program is executed, input values stored in memory are used in the control logic calculations to determine values of outputs; (3) output scan – output

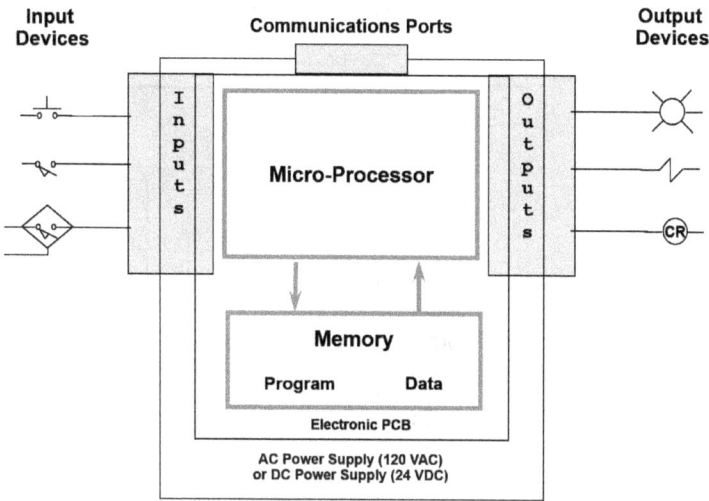

Fig. 11.5. Architecture of PLC.

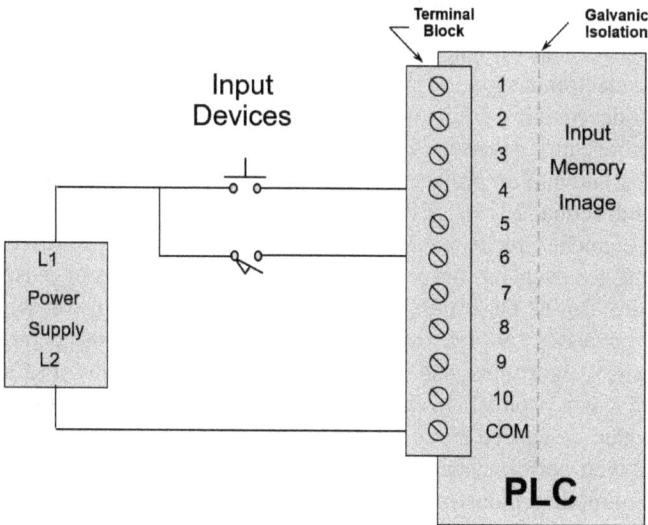

Fig. 11.6. A PLC input module.

values are updated with the calculated values, and (4) communication functions, internal system verification, and scan time calculation. The time to perform the scan cycles depends on the capacity of the PLC: it can be less than 1 ms for high performance PLCs and around 10–20 ms for more compact (micro or mini) PLC.

In PLC systems, there are dedicated modules both for inputs and outputs. The input module is connected to the CPU, which processes all the input data. An input module, see Fig. 11.6, detects the status of input signals such as push buttons, switches, temperature sensors, etc.

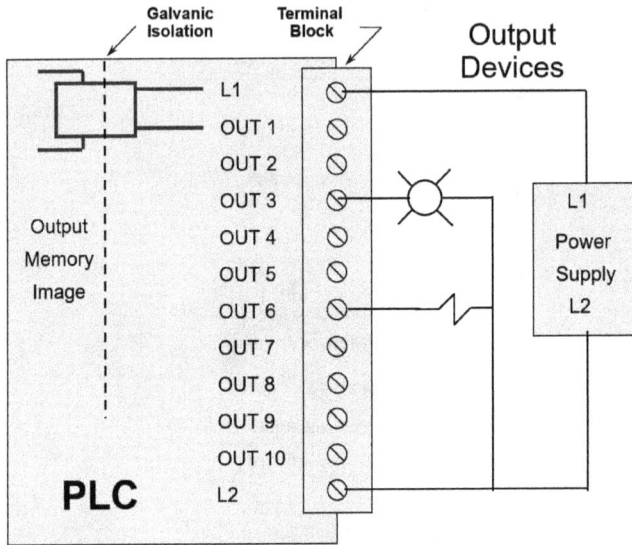

Fig. 11.7. A PLC output module.

The CPU generates output data to the output module. The output module provides an electrical/electronic signal to the output device. The signals can be anything like activating or deactivating output devices. An output module controls devices such as relays, solenoids, motor starters, lights, etc. (see Fig. 11.7).

Complex automation projects are supported by the software resources of PLC (Ranjeeta and Verma, 2018). A PLC can execute combinational logic, timing, and sequencing elements, for example, logic functions, timers to sequence operations or for waiting for events to occur, and counters to count events or pulses received. Analog control can be performed using PID algorithms available on some PLC for continuous processes. A PLC can execute arithmetic functions, permitting more complex control algorithms, and perform basic signal processing. PLCs also serve as devices for data acquisition, data logging, and reporting. The PLC concept has evolved; modern advanced PLCs are called Programmable Automation Controllers (PACs), and their performance can equal or exceed the processing and connectivity capabilities of modern industrial computers.

3.2 PLC Programming Languages

If possible, a PLC can be programmed using graphical languages such as ladder logic diagrams, the most widely used PLC programming language, or function block diagrams, which are instructions composed of operation blocks that transform input signals. Another popular language is sequential function charts composed of steps and transitions from one state to the next. There are also text-based languages like instruction list, which is a low level computer language resembling assembler language, or structured text, which is a high level computer language.

The international standard IEC 61131-3 (International Electrotechnical Commission, 2013) seeks to establish specific guidelines in instruction list programming, structured text, function block diagrams, ladder diagrams, and sequential function charts. Each specified language has advantages and disadvantages and must be selected depending on the automation application. The development of this standard is an example of the emphasis on programming but not on logic control design (Music et al., 2002).

Ladder Diagram or Ladder Logic

The more popular PLC programming language is Ladder Diagram (LD), also called Ladder Logic. It is the most used PLC language because it was designed to substitute hardwired control systems implemented with electromechanical relays. It is a graphical programming language that resembles schematic diagrams of relay logic, allowing easier training of maintenance personnel with electro-technical skills. LD is recommended as a first language to learn PLC programming because it is easy to understand and because most PLCs provide the capability to debug LD programs while the system is online.

The LD is a diagram in which various logic elements, conditional statements, and other components are displayed along horizontal rungs connected on either end to two vertical rails (Fig. 11.8). An LD program is interpreted from left to right and from top to bottom in a continuous execution loop. All conditional instructions on the left must be in logical 'true' condition to activate the output coils. The elements and components on an LD diagram are: (1) contacts, logical inputs for binary sensors, for example, limit switches and photo-detectors; (2) coils as electrical loads for binary outputs, for example, motors, lights, alarms, and solenoids; (3) timers to implement

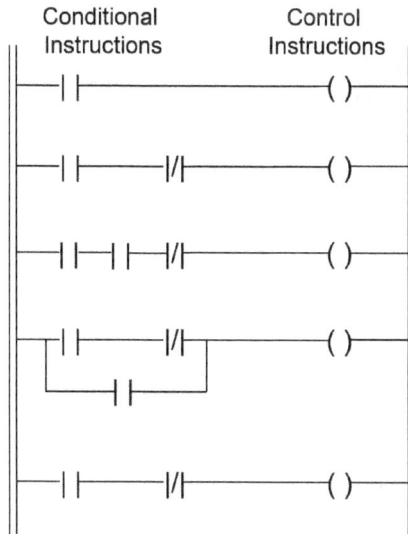

Fig. 11.8. Ladder logic.

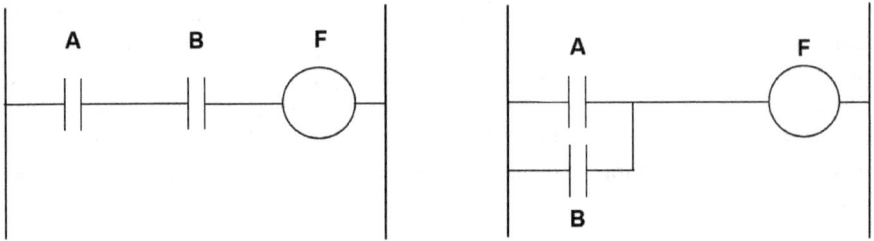

Fig. 11.9. Basic logic functions in ladder logic: AND (left), OR (right).

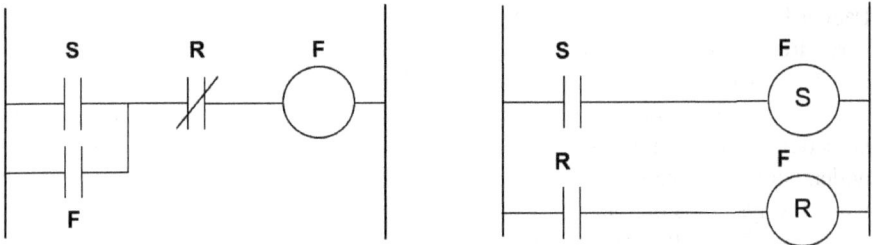

Fig. 11.10. Ladder Logic: (a) latching rung (left), (b) with set and reset coils (right).

timing functions required in the sequential logic, and (4) counters to count discrete events and pulses received from digital inputs or to perform sequencer operations.

Basic logic functions can be easily implemented in PLCs with LD. Normally open contacts in series implement the logical AND, normally open contacts in parallel implement the logical OR, and the NOT function uses the normally closed contact, see Fig. 11.9.

In LD, the latching function is implemented by using normally open contacts of the coil elements or by using memory bits with set and reset coil functions (Fig. 11.10). The easy implementation of sequential logic functions is one of the reasons for the broad application of PLCs in the industrial automation.

LD presents limitations for complex applications because only basic functions are provided, but most control systems requiring logic automatisms can be developed with LD. Some PLCs provide the capability to implement functions and therefore, the ladder code can be encapsulated for reuse.

Structured Text

Structured Text (ST) is another popular language for PLC programming. ST looks very similar to BASIC or C programming and is the best option for programming control systems that require mathematics, such as regulatory control. Complex calculations and data processing can be implemented easier than in LD. Because ST is not a graphical language, it also runs faster and requires less memory.

In ST-based programming, the basic logical and latching functions are easily implemented with IF, THEN, ELSE sentences, and by using logical operators in the conditional expressions, the programming details depend on the programming language, for example, in C:

```
Bool A, B, F;
// Logical AND
if (A && B) then { F=true; } else { F=false; }
// Logical OR
if (A || B) then { F=true; } else { F=false; }
// Inputs
Bool Set, Rst;
Bool Mem;
// Latching logic
if ((Mem || Set )&& (!Rst)) then { Mem=true; } else { Mem=false; }
```

Function Block Diagram

Function Block Diagram (FBD) is another popular PLC programming language. In FBD, program blocks are connected to create a program. Typical logical, timing, and counting functions used in LD are also available in FBD. An added benefit of using FBD is that many lines of code can be incorporated into a single block (Wcislik et al., 2015); therefore, FBD is easier to read and conceptualize.

FBD is not often available as a PLC programming resource; depending on the size of the PLC program or the application, FBD may or may not be the optimal choice. If the system requires motion trajectory control or repetitive tasks, then FBD could be the best approach. However, if the application requires specific code development, additional effort is required to develop and organize the function block diagrams.

Sequential Function Chart

Sequential function chart (SFC) is a visual programming language derived from GRAFCET (GRAphe Fonctionnel de Commande Etapes/Transitions) that was originally designed for the computational implementation of Petri Nets. IEC61131-3 also defines the SFC standard.

The components of SFC are steps or stages with associated control actions, transitions with associated logic conditions, and directed links between steps and transitions. The execution of the SFC language is that step blocks are active until the transition below it is activated and generates a step change. The transition contains all the conditions that must be met for the next step block to activate. The advantage of using the SFC is that large automation programs can be organized into sequential steps where only some code is executed based on the system's current state.

Instruction List

Instruction List (IL) is a programming language consisting of lines of code resembling the assembly language. This is a low level with a single instruction per line, and each line is executed sequentially. IL uses mnemonic codes such as Load (LD), Store (ST), etc. The IL is a programming tool that allows creating more compact, optimal, and time-critical code when it is required by the application. Large programs coded in IL are complicated to organize and debug.

4. Logic Design by Time Diagram

PLC is a widely used automation platform; however, their programming is not only subject to the mastery of their programming languages but also to the practices and styles of each programmer. This situation prevails despite recently proposed programming methodologies (Vieira et al., 2017; Darvas et al., 2015; Music et al., 2002).

A popular logic design method for PLC programming is based on time diagrams. In this section, the context of a sequential problem to be solved is established by an example of a pick-and-place machine. The analysis of time and motion (timing diagram) is described and formulated in logic functions and then converted to ladder logic that can be programmed in a PLC.

4.1 Pick-and-Place System Description

A pick-and-place (P&P) machine is a material transfer system composed of a traverse crane and a cylinder-based mechanism with electromagnets. It transports metallic materials from a receiving (load) band to a discharge band that feeds a manufacturing cell (Fig. 11.11). The automatic operation mode is activated and deactivated with a

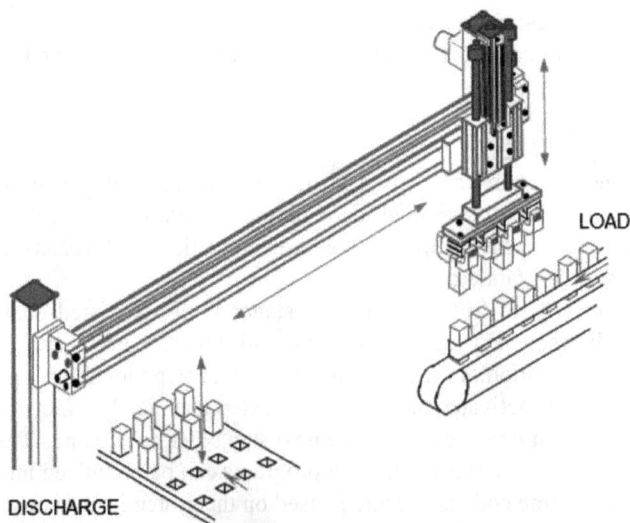

LOAD

DISCHARGE

Fig. 11.11. Pick-and-place crane of a manufacturing process.

simple selector switch. The ZS1 proximity sensor detects the presence of material in the load band.

The crane has a double-effect cylinder to lower and lift the electromagnets and a motor for horizontal movement. The cylinder has extended (down, ZSB) and retracted (up, ZSA) position sensors and is operated with the solenoid valves SVA upwards (retraction) and SVB downwards (extension). The positions on the bands are detected with the sensors ZSC (on the load band), and ZSD (on the discharge band), and the motor is operated by energizing the contactors M1 for movement towards the discharge band and M2 for movement towards the loading band. The electromagnets are activated with the M3 contactor.

The sequence to be implemented for the transport operation consists of these steps: lower the cylinder in the loading area to reach the material that consists of a set of pieces, activate the electromagnets to hold the pieces, lift the pieces and go to the unloading area, lower the parts, deactivate the electromagnets to release the parts, and lift the electromagnets, and return to the loading area to repeat the operation. In this transfer operation, it is important to consider a time delay to fully activate and deactivate the electromagnets and ensure the correct holding and releasing of the parts, respectively.

4.2 Pick-and-Place Control with Time Diagrams

The timing diagram in Fig. 11.12 shows the evolution of the sequence. Initially, the cylinder is up in the loading area, and the motor and magnets are de-energized. The lines show the binary state of the variables at different times. For the cylinder, the binary positions of the top (ZSA), bottom (ZSB), discharge side (ZSD), and load side (ZSC) are indicated. The vertical bars highlight the instants in which specific

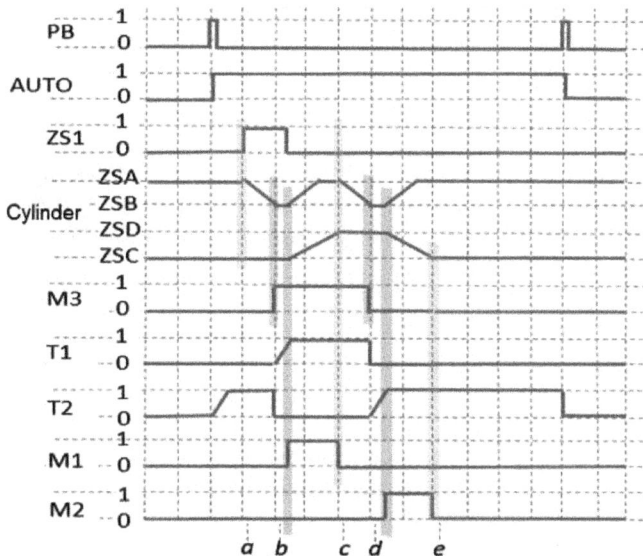

Fig. 11.12. Time diagram.

control actions take place: once in the automatic mode and the material is detected, the cylinder is lowered (bar at instant a), the magnet is activated (M3, at instant b), the cylinder is raised (retracted), and the contactor M1 is activated (just after b), the cylinder goes down in the discharge zone (instant c), the magnet contactor M3 is deactivated (instant d), the cylinder is raised, and the M2 contactor returns it to the loading area (a little after instant d). The timers (T1 and T2) introduce delays that allow the correct material handling.

Firstly, the PB push button controls the automatic sequence's activation and deactivation. The AUTO mode is activated by momentarily pressing the button when the AUTO function is deactivated; it remains activated when the button is released. In any other case, the mode of operation is not automatic but manual so that the AUTO function can be expressed as:

$$AUTO_{set} = PB \cdot \overline{AUTO} \tag{11.13}$$

$$AUTO_{reset} = PB \cdot AUTO \tag{11.14}$$

$$AUTO = (AUTO_{set} + AUTO) \, \overline{AUTO_{reset}} \tag{11.15}$$

The energization of the electromagnet to hold the material requires the mechanism in the loading position and the cylinder extended or down (instant b). However, these conditions are not maintained since the electromagnet must remain energized until the material is deposited in the discharge area (time d), so the logic of the electromagnet function can be defined as follows:

$$M_{3,set} = ZSB \cdot ZSC \tag{11.16}$$

$$M_{3,reset} = ZSB \cdot ZSD \tag{11.17}$$

Typically, the definition of bistable variables in a PLC only require indicating the activation and deactivation conditions. However, the detailed sequential function can be derived directly as:

$$M_3 = (ZSB \cdot ZSC + M_3)(\overline{ZSB} + \overline{ZSD}) \tag{11.18}$$

Two timers are used to guarantee the correct holding and releasing of the pieces by the electromagnet. As soon as it is activated (instant b), the preset time in timer 1 (T1) is measured, and when it is turned off (instant d), the preset time in timer 2 (T2) is measured:

$$T1 = M_3 \tag{11.19}$$

$$T2 = \overline{M_3} \tag{11.20}$$

If the system is in automatic mode, the vertical cylinder is in the loading position, and the material is detected in the receiving band, the command to lower the cylinder is produced (instant a). On the discharge side, the cylinder is only lowered if the magnet is energized (instant c):

$$SVB = ZSC \cdot ZS1 \cdot AUTO + ZSD \cdot M_3 \tag{11.21}$$

To raise the cylinder, it is required that the magnet has already been properly energized or de-energized, which is ensured by waiting a time after generating the command to hold the load (after instant b) and a time after ordering the release of the load (after instant d), respectively:

$$SVA = ZSC \cdot T1 + ZSD \cdot T2 \qquad (11.22)$$

The M1 motor is activated in the loading area once the magnet has been energized long enough to ensure the load is attached correctly (after time b). This causes the mechanism to move towards the discharge area, where its detection deactivates the M1 motor (instant c):

$$M_{1, set} = ZSC \cdot T1 \qquad (11.23)$$

$$M_{1, reset} = ZSD \qquad (11.24)$$

$$M_1 = (ZSC \cdot T1 + M_1) \overline{ZSD} \qquad (11.25)$$

Similarly, motor 2 must be activated in the discharge zone once the magnet has been de-energized, releasing the material thoroughly, which is ensured with the use of timer T2 (after time d). Motor 2 will transport the machine to the loading area (it stops at time e). When automatic operation stops, M2 activates to return the system to the loading area:

$$M_{2, set} = ZSD \cdot T2 \qquad (11.26)$$

$$M_{2, reset} = ZSC \qquad (11.27)$$

$$M_2 = (ZSD \cdot T2 + M_2) \overline{ZSC} \qquad (11.28)$$

The ladder diagram in Fig. 11.13 implements the solution derived from the timing diagram. The set and reset functions are used to simplify the PLC program,

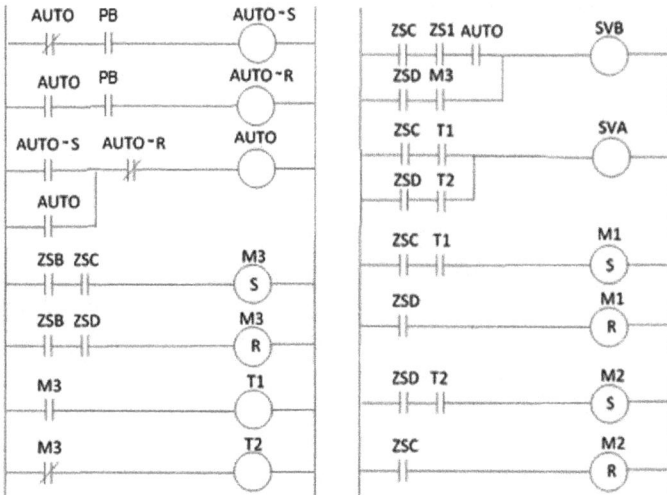

Fig. 11.13. Ladder diagram program of the solution derived from the analysis of the time diagram.

however, the equations for the implementation using only contacts are also provided: (11.18), (11.25), and (11.28). This solution does not include manual operation.

5. Logic Design by Sequenced Latching

In PLC programming, different implementations can be used (Wcislik et al., 2016). Sequential control can also be implemented using a methodology of sequenced latching. In this section, this methodology is addressed with the pick-and-place problem of the previous section, but now by defining and then programming the logic functions in LD.

5.1 Sequential Latching Method

A basic approach for sequential logic control is using sequential latching functions. The sequent operation can be specified in steps directly from the operation description of the system to be automated. Sequenced latching defines the equation of each state or step with a logic function with the following structure:

$$Step_i = (Step_i + Step_{i-1} \cdot Set_i) \cdot \overline{Rst_i} \tag{11.29}$$

This sequence considers an initial step, intermediate steps, and final step; the activation of the next step implies the deactivation of the current step:

$$Step_1 = (Step_1 + Set_1) \cdot \overline{Step_2} \tag{11.30}$$

$$Step_2 = (Step_2 + Step_1 \cdot Set_2) \cdot \overline{Step_3} \tag{11.31}$$

...

$$Step_n = (Step_n + Step_{n-1} \cdot Set_n) \cdot \overline{Step_{n+1}} \tag{11.32}$$

The set function of the first step must be carefully defined, it can include the required run interlocks. The condition to enable automatic operation (i.e., a mode selector) can be added in all the logic functions to ensure the sequence operates exclusively in automatic mode.

The sequence can be cycled; under this scenario, the deactivation of the last step requires the activation of the first step:

$$Step_n = (Step_n + Step_{n-1} \cdot Set_n) \cdot \overline{Step_1} \tag{11.33}$$

5.2 Pick-And-Place Control with Sequenced Latching

Consider the P&P machine as an example. The logical functions for automating the pick-and-place crane can be deduced by applying the sequential latching method corresponding to the following steps, see Table 11.1.

Outputs may be directly activated with the step signals or by the combinational logic of more than one step, as indicated in Table 11.2.

Table 11.1. Sequenced latching.

Sequence Step	Logic Function
Lower Cylinder, SVB	$Step_1 = Auto \cdot (Step_1 + Set_1) \cdot \overline{Step_2}$
Energize Magnets, M3	$Step_2 = Auto \cdot (Step_2 + Step_1 \cdot Set_2) \cdot \overline{Step_3}$
Lift Cylinder, SVA	$Step_3 = Auto \cdot (Step_3 + Step_2 \cdot Set_3) \cdot \overline{Step_4}$
Travel to discharge position, M1	$Step_4 = Auto \cdot (Step_4 + Step_3 \cdot Set_4) \cdot \overline{Step_5}$
Lower Cylinder, SVB	$Step_5 = Auto \cdot (Step_5 + Step_4 \cdot Set_5) \cdot \overline{Step_6}$
De-energize Magnets, M3	$Step_6 = Auto \cdot (Step_6 + Step_5 \cdot Set_6) \cdot \overline{Step_7}$
Lift Cylinder, SVA	$Step_7 = Auto \cdot (Step_7 + Step_6 \cdot Set_7) \cdot \overline{Step_8}$
Travel to charge position, M2	$Step_8 = Auto \cdot (Step_8 + Step_7 \cdot Set_8) \cdot \overline{Rst_8}$

Table 11.2. Outputs activation.

Output Energize	Logic Function
Lower Cylinder, SVB	$SVB = Auto \cdot (Step_1 + Step_5) \cdot \overline{EStop}$
Lift Cylinder, SVA	$SVB = Auto \cdot (Step_3 + Step_7) \cdot \overline{EStop}$
Travel to discharge position, M1	$M1 = Auto \cdot Step_4 \cdot \overline{EStop}$
Travel to charge position, M2	$M2 = Auto \cdot Step_8 \cdot \overline{EStop}$
Energize Magnets, M3	$M3 = (M3 + Step_2) \cdot \overline{Step_6}$

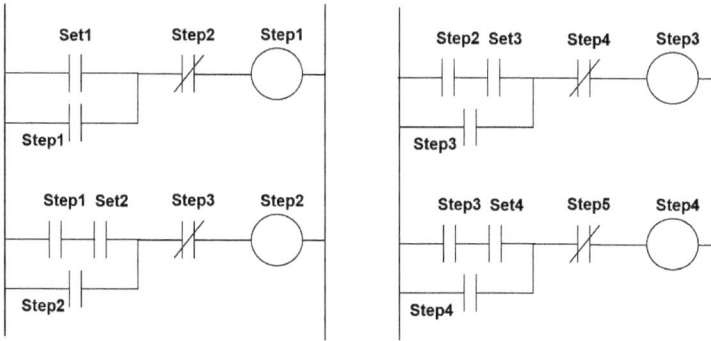

Fig. 11.14. Ladder diagram for the sequential latching of first four steps of the pick-and-place sequence.

Figures 11.14, 11.15, and 11.16 show the implementation of the sequential latching solution in an LD. The sequential latching functions of the operation's steps are shown in Figs. 11.14 and 11.15, and the logic functions to energize the outputs are shown in Fig. 11.16.

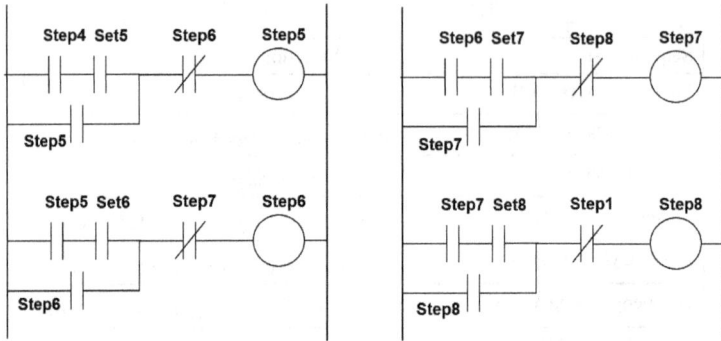

Fig. 11.15. Ladder diagram for the sequential latching of the last four steps in the pick-and-place sequence.

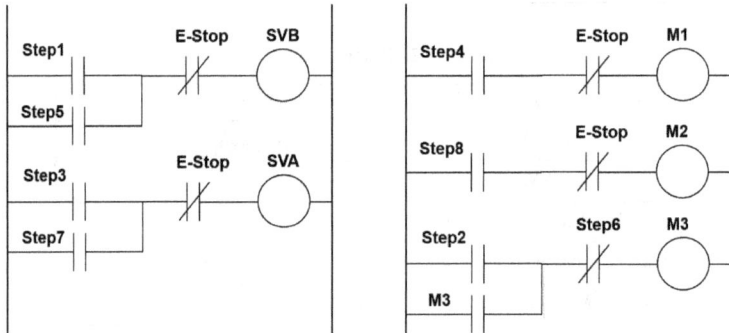

Fig. 11.16. Ladder diagram for the activation of outputs.

6. Discrete Event Systems Simulation

Hardware in the loop and software in the loop simulation schemes require process simulation, including sensors and actuators, to close control loops within the control units to validate the programmed automatic operation sequences. There are two possible approaches for process simulation: one basic approach uses timers to generate the discrete events, and a second approach uses integrators of speed or flow process variables to produce position or volume, respectively, to simulate the sensor with compare instructions.

6.1 Actuators and Sensors Simulation with Timers

The simulation approach with timers can be implemented using basic timer functions typically available in a PLC. The basic idea is to activate the timer functions with the actuators, and when the configured time is elapsed, a related sensor is then activated. Consider that a double-effect cylinder can be simulated with timers (see Fig. 11.17) The timers to extend and retract the cylinder are activated by the output variables for the solenoid valves. When the configured time is elapsed, the corresponding limit sensors for the extended or retracted positions are activated.

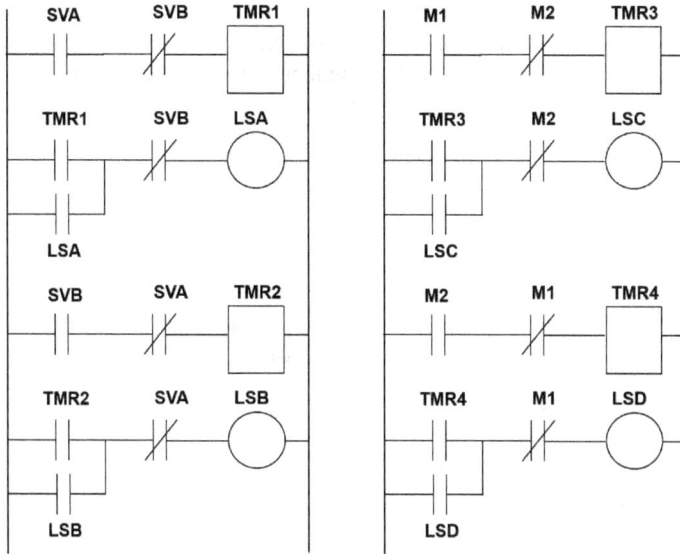

Fig. 11.17. Simulation with timers in ladder logic.

6.2 Actuators and Sensors Simulation with Integrators

A second process simulation approach is by programming a numerical simulation. Actuators can be approached by implementing basic models with integration functions and then by using the COMPARE instructions for the simulated process variables with respect to specific ranges or thresholds.

In processes with hydraulic and pneumatic cylinders, the volumetric flow can be integrated in discrete time to obtain a certain volume quantity that can be directly proportional to displacement. The retracted position limit switch can be energized when the integration value reaches the minimum volume value. The extended position limit switch can be energized when the volume value reaches the maximum. In processes using electric motors, the speed commanded by an actuator can be integrated in discrete time to compute the position. This simulates an analog position that can be converted to counts such as those generated by a high speed counter integrating pulses from quadrature encoders.

For example, consider the numerical simulation for the pick-and-place machine. The pneumatic cylinder and the crane can be simulated with integration equations triggered by the binary actuators, and the sensor can be simulated with comparison functions.

The pneumatic cylinder can be simulated by considering the solenoid valves ON state, enable the equations to integrate flow into volume.

$$if\ (SVA\ and\ V_k < V_{max})\ then\ V_k = V_{k-1} + T_s\ f_s \tag{11.34}$$

$$if\ (SVB\ and\ V_k > V_{min})\ then\ V_k = V_{k-1} - T_s\ f_s \tag{11.35}$$

The integrated volume can be considered to be directly proportional to the displacement of the cylinder rod. The limit switches can be activated by comparing specific maximum and minimum volume values. The logic to simulate the retracted and extended limit switches is as follows:

$$if\,(V_k \geq V_{max})\ then\ LSB = TRUE\ else\ LSB = FALSE \tag{11.36}$$

$$if\,(V_k \leq V_{min})\ then\ LSA = TRUE\ else\ LSA = FALSE \tag{11.37}$$

Motion control applications and processes or machines using speed and position variables can also be simulated with integrators. For example, for the same pick-and-place machine, the crane displacement can be simulated as a speed-and-position process. The equations to simulate the position of the crane are

$$if\,(M1\ and\ z_k < z_{max})\ then\ z_k = z_{k-1} + T_s\,v_k \tag{11.38}$$

$$if\,(M2\ and\ z_k > z_{min})\ then\ z_k = z_{k-1} - T_s\,v_k \tag{11.39}$$

As for the pneumatic cylinder, the comparison logic is applied to minimum and maximum position values to simulate the proximity switches that are located over the load and discharge conveyors:

$$if\,(Z_k \geq Z_{max})\ then\ LSC = TRUE\ else\ LSC = FALSE \tag{11.40}$$

$$if\,(Z_k \leq Z_{min})\ then\ LSD = TRUE\ else\ LSD = FALSE \tag{11.41}$$

7. Conclusions

Automated systems where inputs and output variables are mostly binary discrete require the implementation of both combinational and sequential logic control functions. Combinational logic is to perform Boolean algebra on input signals, while sequential logic is used to automate operation sequences. Practical logic control applications are a mixture of combinational and sequential logic. Sequential logic control is designed as a collection of logic functions obtained by design methodologies. Basic combinational and sequential logic functions can be represented with truth tables and implemented with logic gates and relay logic. Sequential logic problems can be solved with set/reset (latching) functions or binary memories.

A PLC is the workhorse controller of industrial automation. PLCs are industrial computing control units that sense inputs and execute logic and sequencing functions to determine the required output signals. Ladder logic is considered the most popular PLC programming language and consists of a diagram in which combinational and sequential logic functions are implemented. Sequential logic implies using input values stored in memory in the control logic calculations to determine the values of outputs. Methodological approaches for PLC programming allow designing better the sequential logic required to automate the event- or timed-based sequential logic automation problems. Time diagrams or sequential latching methods can approach the solution to automation sequences.

There are two possible approaches for the basic simulation of discrete event systems. The first approach uses timer instructions enabled by the actuators to generate the corresponding activation of sensors. The second approach involves a numerical simulation using integrators and first-order linear systems to simulate cylinders and their related position sensors. With basic process simulation, it is possible to perform hardware- or software-in-the-loop simulations to validate sequential logic programs.

References

Darvas, D., Blanco, E. and Majzik, I. 2015. A formal Specification Method for PLC-based Applications. *15th International Conference on Accelerator and Large Experimental Physics Control Systems*, Melbourne, Australia, pp. 91–94.

Dunn, A. 2009. The father of invention: Dick Morley looks back on the 40th anniversary of the PLC [White paper]. Manufacturing Automation. https://www.automationmag.com/855-the-father-of-invention-dick-morley-looks-back-on-the-40th-anniversary-of-the-plc/.

Groover, M.P. 2008. *Automation, Production Systems, and Computer-Integrated Manufacturing.* Third Edition, Prentice Hall.

International Electrotechnical Commission, IEC 61131-6, Programmable Controllers. Part 3: Programming Languages, 2013.

Music, G., Matko, D. and Zupancic, B. 2002. Model Based Programmable Control Logic Design. *15th Triennial World Congress*, Barcelona, Spain.

Ranjeeta, S. and Verma, H. 2018. Development of PLC-based controller for pneumatic pressing machine in engine-bearing manufacturing plant. *Procedia Computer Science* 125: 449–458.

Vieira, A.D., Portela, E.A. and De Queiroz, M.H. 2017. A method for PLC implementation of supervisory control of discrete event systems. *IEEE Transactions on Control Systems Technology* 25(1): 175–191.

Wcislik, M., Suchenia, K. and Laskawski, M. 2015. Programming of sequential control systems using functional block diagram language. *IFAC-PapersOnLine* 48(4): 330–335.

Wcislik, M., Suchenia, K. and Laskawski, M. 2016. Method of programming of sequential control systems using LabVIEW environment. *IFAC–PapersOnLine* 49(25): 476–481.

Chapter 12

Logic Control with State Machines

1. Introduction

Programmable logic controllers (PLCs) or programmable controllers (PCs) are essential industrial automation components because they constitute a versatile platform for process automation and operation. For this reason, it is important to explore PLC programming methods and techniques further since the lack of them results in poorly structured programs that are difficult to maintain and transfer. The PLC programming languages, such as instruction list, structured text, and function block diagrams (according to the IEC 61131-6 standard), provide basic and advanced tools to program the control logic but not a method to design it. Sequential function chart (SFC) is a graphical language that allows the design and implementation of sequential control to be addressed in a structured way. A similar tool is GRAFCET (GRAphe Fonctionnel de Commande Etapes/Transitions). SFC and GRAFCET, as advanced languages, are only available in PLCs with superior computing and programming capabilities and higher cost.

Formal sequential design methods that allow the development of structured programs for automatic systems are often used in academic and research fields but are less used in industry. These formal methods help to reduce the additional cost of libraries, allowing the implementation with basic programming resources. The finite state machine is a mathematical model used in computer science and allows the derivation of logical equations from the sequential system.

This chapter presents methods to develop and implement sequential logic control with Finite State Automata. Section 2 on Modeling Discrete Event Systems describes the discrete events systems (DESs) and the use of finite state automata (FSA) for their modeling. The advantages of State Machines for Logic Control are presented in Section 3. Section 4, Ladder Logic Synthesis of State Machines, presents the ladder logic for implementing state machines and events charts. Section 5 on Concurrent DES Process Modeling with Petri Networks deals with DESs with concurrent processes and their modeling and control with Petri Networks, particularly to handle simultaneous processing problems. Section 6 discusses the Conclusions.

2. Modeling Discrete Event Systems

Among the tools developed for modeling and analysis of discrete systems are timed automation, process algebra, Petri nets, and state machines (Popescu and Martinez, 2010). Timed automation is essentially based on individual automata for each device, with transitions produced by timers and synchronization channels, which demand a large memory space and computational time. Process algebra describes a system using algebraic or axiomatic expressions. There are different types of process algebra (for the calculation of communication systems and sequential processes, for example), and they differ in the definition of their operators. The process algebra formalism does not handle actions in parallel; rather, it allows an arbitrary order to events that should occur simultaneously. Petri nets allow the graphical representation of a sequential system using places, transitions, arcs, and tokens (marks at each place); the tokens regulate the flow of the sequence: they activate the transitions, and these are triggered by performing a task, eliminating the input tokens and producing tokens in each output place, allowing the simultaneity of actions. The definition of the tokens has a mathematical representation. The finite state machine emerges as a computational model for the design of programming languages (Wagner, 2006).

2.1 Finite States Machines Description

Finite state machines (FSMs) are used in computer sciences to design programming languages and computer programs. They are also helpful for the design of sequential logic circuits. The FSM is an abstract machine that has a finite number of states. The machine is active in only one state at a time; this active state is called the current or actual state. The machine can change from one state to another with a set of transitions triggered by events or other conditions. A particular FSM is defined by a list of its states, the set of events, and the transition functions.

The Moore FSM uses only inputs and the state of the system to define transitions, and the outputs depend only on the current state. The sequence of input changes required to determine the state machine behavior is stored in an internal state variable (Fig. 12.1). The advantage of the Moore model is its simplicity.

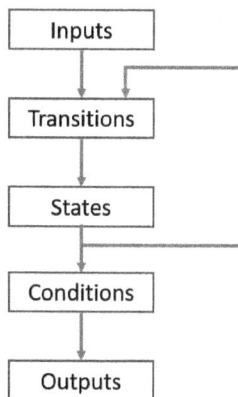

Fig. 12.1. Moore finite state machine.

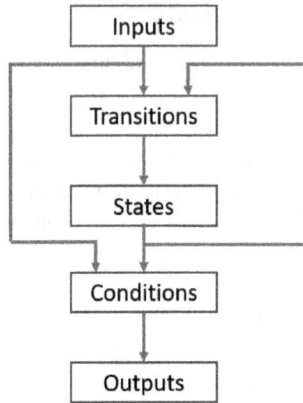

Fig. 12.2. Mealy finite state machine.

The Mealy FSM can use inputs directly to activate outputs. Both the state transition and output conditions are functions of inputs and states. The use of a Mealy FSM often reduces the number of states (see Fig. 12.2).

State machines are a particular type of Petri nets, distinguished by a simplification of the sequence representation using linear structures, reducing the graphic elements to circles for states (to which actions are associated), and arrows for transitions (associated with consequences of actions of the previous state and at the same time requirements for the next state). The concurrency of events can be solved with the combination of state machines.

The analysis and design of a discrete-event system can be done in a structured way using state machines. Typically, in the industrial environment, this tool is not applied (except by some original equipment manufacturers, OEMs), and the programs are created without any formal methodology that would facilitate their revision, modification, or transfer. In other words, the software developed is often inflexible and not very understandable by other programmers, so the maintenance of the systems frequently requires their re-elaboration.

2.2 *Finite States Machines Representation*

The finite automata or finite state machine is an abstract model of digital computation but has a practical approach that can ease the implementation of sequential logic programs. The state machine describes step-by-step the sequence or process to be implemented, including the actions to be carried out and the conditions that allow determining that a step has finished and that it is possible to continue with the next one (transitions between stages or states). A graphical or tabular representation of the state machine may be used. The implementation of the system is directly derived from this representation: logic equations for programming states and outputs based on the active states. In this way, the state machine assists both the design and programming of the sequential control system. The disadvantage of the state machine can be the large number of states that have to be defined for a very long process. Several works have proposed methods to simplify or reduce the state machine (Yue and Yan, 2022).

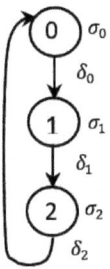

Current State	Action	Transition	Next State
0	σ_0	δ_0	1
1	σ_1	δ_1	2
2	σ_2	δ_2	0

Fig. 12.3. Finite state machine in equivalent graphical and tabular forms.

The finite state machine or finite automaton is defined mathematically with the quintuple: $G = (X, \Sigma, \delta, x_0, X_m)$, where X is the set of states or stages that compose the sequence, Σ is the set of symbols for the actions, δ represents the events or logical functions for the transitions between states, x_0 is the initial state, and X_m is a subset of final states. Transition functions are logical conditions for state change, expressed in terms of input events or conditions for detecting that the current stage actions are completed and the system can go to the next state. In the current active state, actions are performed; that is, output commands from Σ are produced.

A finite state machine can be represented by a graphic or table, as shown in Fig. 12.3. The graphic consists of numbered nodes or circles representing states and arcs or arrows representing transitions. Actions are indicated beside the state where they must be executed. The FSM graph resembles a GRAFCET available in some PLC brands.

3. State Machines for Logic Control

Control logic is often written in a not structured way, and this makes it complicated to analyze and debug. The logical sequence required to automate a process can be programmed using a state machine with several advantages: (1) the state diagram allows a visual aid to the control solution; (2) a complex problem can be split into simpler sections; (3) the state diagram or table can be translated to any programming language. There is a direct relationship between the state machine and the programming of the sequential logical system, which is a difference with respect to traditional algorithm flow diagrams (Ranjeeta and Verma, 2018).

Finite state machines are relevant not only for sequential logic control but for hybrid control of batch processes since the continuous control of certain process variables with specialized techniques can be assigned to one or more states within a sequence that considers from the input of materials to the product discharge. State machines can also be used for a wide variety of implementation technologies. For example, the implementation has been derived for PLC (Darvas et al., 2015; Vieira et al., 2017) and for devices based on FPGA using functional block diagrams (Wcislik et al., 2015), as well as the implementation for programmable automation controllers, PACs, using LabVIEW graphics (Wcislik et al., 2016). These implementations use binary memories. Here we propose a structured programming style for PLC using memory registers and compare it against typical practices to demonstrate the higher degree of flexibility obtained. Concurrence or simultaneity of states is also discussed and solved systematically.

3.1 Sequential Problem: Drilling Station

Consider a drilling station example. Drilling is the process of cutting holes in solid materials with drill bits using rotating cutting tools. The drilling station has a storage pile of work materials that are fed to the drill area with a pneumatic cylinder, C1. The drill is lowered with another cylinder, C2, to reach the workpiece, drill a hole, and retract the drill bit when the drilling time has elapsed. A third cylinder, C3, ejects the workpiece from the drill area to a discharge bin. The pneumatic cylinders are double acting, operated with solenoid valves to extend and retract the rod from the fully retracted and fully extended positions, which are sensed with electromechanical limit switches. Figure 12.4 shows the drilling station with the pneumatic cylinders.

One of the first tasks in designing a control solution for the sequential logic program is elaborating a complete list of PLC inputs and outputs (IO list) or connected instrumentation. It is desirable to identify which devices are for control (sensors and actuators) and which are for process operation, such as push buttons, selectors, and pilot lights. The IO list corresponding to the sensors and actuators to control the drilling station is shown in Table 12.1. Symbols or tags are defined in this IO list.

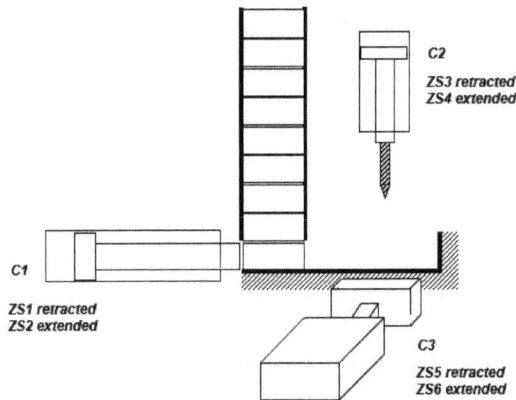

Fig. 12.4. Drilling Station. Double acting cylinders C1, C2, and C3. Sensors: ZS1, ZS2, ZS3, ZS4, ZS5, ZS6 (sensors with odd numbers for retracted positions; sensors with even numbers for extended positions).

Table 12.1. PLC inputs (from sensors) and outputs (to actuators).

PLC Inputs	Symbol	PLC Outputs	Symbol
Cylinder 1 retracted sensor	ZS1	Cylinder 1 extend SV	SV1
Cylinder 1 extended sensor	ZS2	Cylinder 1 retract SV	SV2
Cylinder 2 retracted sensor	ZS3	Cylinder 2 extend SV	SV3
Cylinder 2 extended sensor	ZS4	Cylinder 2 retract SV	SV4
Cylinder 2 retracted sensor	ZS5	Cylinder 3 extend SV	SV5
Cylinder 2 extended sensor	ZS6	Cylinder 3 retract SV	SV6
		Run Drill Motor	CR1

Table 12.2. Control system push buttons and pilot lamps.

PLC Inputs	Symbol	PLC Outputs	Symbol
System Start Button 1	PB1	System On pilot light	PL1
System Stop Button 2	PB2	System Fault	PL2
System Fault Reset	PB3		

The IO list corresponding to push buttons and pilot lights to operate the drilling station is shown in Table 12.2.

Once the complete IO list is defined, it is desirable to write down a functional description of the required sequential logic. It can be described with steps indicating the actions and events related to the inputs and outputs using the defined symbols or tags:

Step 1. Start the system when PB1 is pressed, and stop when PB2 is pressed. Energize PL1 when the system is started.

Step 2. Energize SV1 to extend cylinder 1 until it is fully extended, as detected by ZS2.

Step 3. Energize SV3 to extend cylinder 2 until it is fully extended (ZS4). Energize and latch CR1 to run the drill motor.

Step 4. Start a timer of 2 seconds to keep drilling the piece.

Step 5. When the timer is done, energize SV4 to retract cylinder 2 until it reaches the ZS3 position.

Step 6. Unlatch CR1 to stop the drill motor and energize SV2 to retract cylinder 1 until it is fully retracted (ZS1).

Step 7. Energize SV5 to extend cylinder 3 until it is fully extended, as detected by ZS6.

Step 8. Energize SV6 to retract cylinder 3 until it reaches the ZS5 position. End of the work cycle.

The defined steps clearly indicate which actuators are required to be energized and which sensors or other events indicate the fulfillment of the control actions. The finite state machine can be specified by drawing a state diagram (see Fig. 12.5).

Optionally, a second FSM can be defined to start and stop the system and to detect fault conditions. In this case, the start (PB1) and stop (PB2) buttons activate and deactivate the SysOn status. PL1 is turned on with SysOn to indicate automatic operation. Sys Fault and the indicator PL2 are activated if a Fault Condition is detected, and the fault status is reset with the PB3 button. Figure 12.6 shows the operation FSM.

The state diagram or transition table can be programmed in the available programming language of the control unit. The standard IEC 61131-3 specifies several programming languages, such as ladder logic, instruction list, structured text, and the SFC, that allows direct graphical programming of the FSM. We left the reader with the exercise of implementing the drilling station example in any programming language.

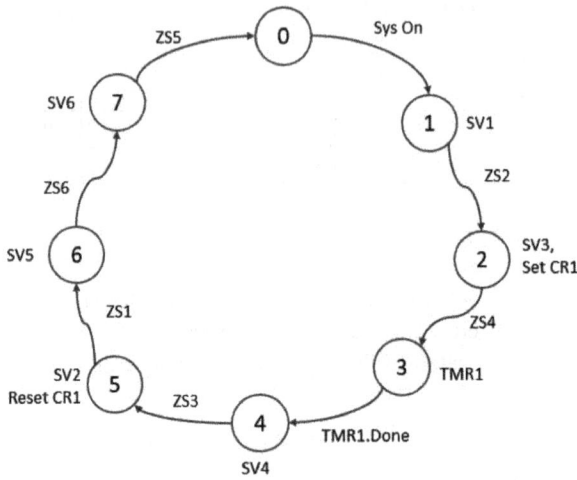

Fig. 12.5. FSM for automated drilling operation.

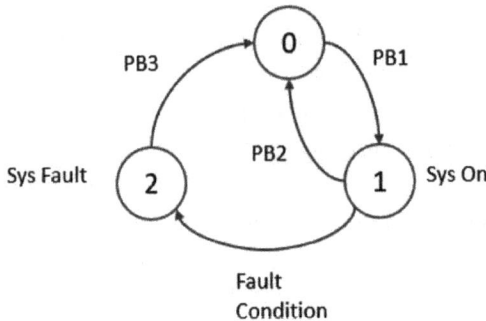

Fig. 12.6. FSM for the operation status.

3.2 Sequential Problem: Pick-and-Place Station

Consider a pick-and-place (P&P) machine example. The P&P transports metallic materials from a receiving (load) band to a discharge band that feeds a manufacturing cell (Fig. 12.7). The operation of the P&P is described with specific actuators and sensors as follows:

- The automatic system mode is activated and deactivated with a simple selector switch.
- The presence of material in the load band is detected by the ZS1 proximity sensor.
- The crane has a motor for a horizontal movement that is operated by energizing the contactors: M1 for movement towards the discharge band and M2 for movement towards the loading band. The positions over the bands are detected with the sensors ZSC (on the load band side) and ZSD (on the discharge band side).

- The crane has a double acting cylinder to lower and lift the electromagnets. The cylinder has extended (down, ZSB) and retracted (up, ZSA) position sensors and is operated with the solenoid valves SVA upwards (retraction) and SVB downwards (extension).

- The electromagnets to pick the metallic materials to be transported are activated with the M3 contactor.

The sequence to be implemented for the transport operation consists of these steps: (1) lower the cylinder in the loading area to reach the material that consists of a set of pieces; (2) activate the electromagnets to hold the pieces; (3) lift the pieces; (4) go to the unloading area; (5) lower the parts and deactivate the electromagnets to release the parts; (6) lift the electromagnets and return to the loading area, and be ready to repeat the operation. In this transfer operation, it is important to consider a time delay to fully activate and deactivate the electromagnets and ensure the correct holding and releasing of the parts, respectively.

For this problem, the automaton is defined with the following elements: $x_0 = 0$ the initial state corresponds to the cylinder raised or retracted and located in the loading area; $X_m = \{x_0\} = \{0\}$ that is, the marker state is simply the initial state to ensure that when the program is activated, the system can start the sequence again; states are defined according to the control sequence as $X = \{x_0, x_1, x_2, x_3, x_5\} = \{0,1,2,3,4,5\}$, the actions $\Sigma = \{\sigma_{0a}, \sigma_{0b}, \sigma_1, \sigma_{2a}, \sigma_{2b}, \sigma_{3a}, \sigma_{3b}, \sigma_4, \sigma_{5a}, \sigma_{5b}\}$, and transition or events symbols $\delta = \{\delta_0, \delta_1, \delta_2, \delta_3, \delta_4, \delta_5\}$. Table 12.3 defines the state machine and details the elements of the sets of states, actions, and events for the case study.

The energization of the electromagnets, or M3 load, is carried out through a memory bit with set and reset coils (σ_{2a} and σ_{5a}). Alternatively, direct activation (as

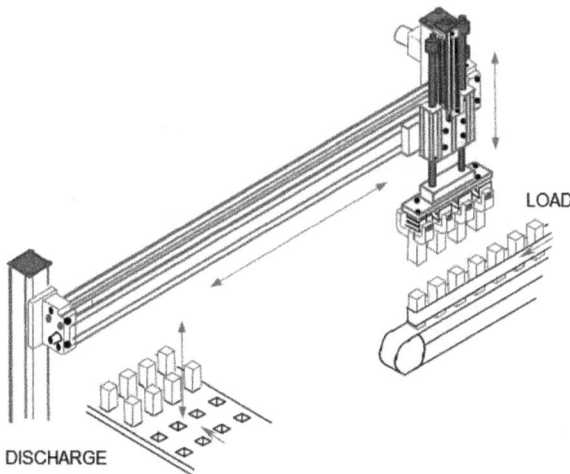

Fig. 12.7. Pick-and-place crane of a manufacturing process.

Table 12.3. Control system push buttons and pilot lamps.

Current State	Actions	Transition Events	Next State
0	$\sigma_{0a} = M_2$ $\sigma_{0b} = SVA$	$\delta_0(x_0, (\sigma_{0a}, \sigma_{0b}))$ $= AUTO \ \& \ ZS1 \ \& \ ZSC \ \& \ ZSA$	1
1	$\sigma_1 = SVB$	$\delta_1(x_1, \sigma_1) = ZSB$	2
2	$\sigma_{2a} = M_{3,set}$ $\sigma_{2b} = T1$	$\delta_2(x_2, (\sigma_{2a}, \sigma_{2b})) = T1_{out}$	3
3	$\sigma_{3a} = SVA$ $\sigma_{3b} = M_1$	$\delta_3(x_3, (\sigma_{3a}, \sigma_{3b}))$ $= ZSA \ \& \ ZSD$	4
4	$\sigma_4 = SVB$	$\delta_4(x_4, \sigma_4) = ZSB$	5
5	$\sigma_{5a} = M_{3,reset}$ $\sigma_{5b} = T2$	$\delta_5(x_5, (\sigma_{5a}, \sigma_{5b})) = T2_{out}$	0

a monostable variable) of M3 could be repeated as an action of steps 2, 3, and 4. Delay times within the sequence are implemented as additional actions to ensure proper handling of the material in the stages corresponding to its collection from the feeding band (stage 2) and its placement in the destination position (stage 5). Coils are assigned with the numerical subscript corresponding to the state number for clarity. States can be defined differently, for example, by separating vertical motion and horizontal motion (σ_{0a} and σ_{0b}, and σ_{3a} and σ_{3b}) into different states. Another possibility is to define state 0 as an idle state, in which no actuation occurs, instead of assigning state 0 to the preparation of the system to start the sequence (by lifting the piston and taking it to the loading area). However, the development of the solutions is based on Table 12.3.

The graphical representation of the finite state machine in Table 12.3 facilitates the modeling of the sequence. The state machine of the operation sequence of the pick-and-place crane is shown in Fig. 12.8.

A detailed verbal description of the stages can be used directly to obtain the graphics of Fig. 12.8:

State 0. Retract (SVA) and move the crane to the loading zone (M2). Wait for the automatic mode (AUTO), the presence of parts (ZS1), and verification of the cylinder position (ZSA, ZSC).

State 1. Lower the crane (SVB) until the detection of the piston stroke end.

State 2. Activate electromagnets to seize the parts (M3 set); wait some time (measured with timer T1) to ensure proper energization and interaction with parts.

State 3. Retract (SVA) and move the crane to the discharge zone (M1) until the desired position is detected (by activation of sensors ZSD and ZSA).

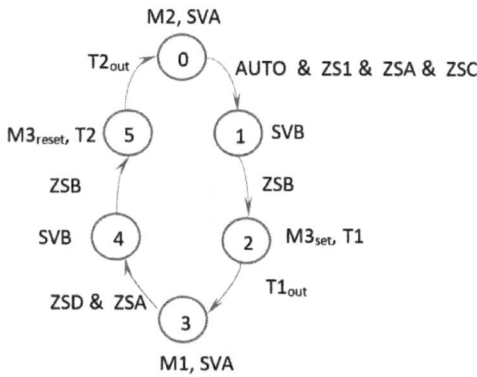

Fig. 12.8. Finite state machine for the sequence of the pick-and-place unit.

State 4. Lower the electromagnets (SVB) until the down position (ZSB) is activated.

State 5. Release parts by deactivating electromagnets (M3 reset) for some time (determined by timer T2).

4. Implementation of Finite State Machines

The finite state machine method is programmed with three possible implementations: using binary variables to represent states, a numerical variable whose values determine the states, and using GRAFCET with the implicit definition of states and their binary representation. In the three programming methods, auxiliary variables are assigned systematically to implement the sequence.

4.1 Finite State Machine Using Binary Variables

The implementation of the state machine can be done based on bistable binary variables, that is, with the activation (set) of the next state and deactivation (reset) of the current state, depending on the occurrence of the corresponding transition events. A bistable variable is defined for each stage: the activation condition (set) will be the logical combination AND between the active status of the previous state and the activation of the transition event to the state in question, and the deactivation condition (reset) will be the logical function AND between the current stage and the transition event to the next state. It is important to note that the combination of the ON status of the current state and transition event is used for both next stage activation and current state deactivation. Figure 12.9 shows the program for state activation and deactivation. Six binary variables are used (M0.0 to M0.5) to represent the six states and are activated one at a time to distinguish the current state.

In the output activation, manual operation is easily incorporated using additional buttons (with labels starting with the prefix MB for manual button) for each load. The automatic mode activated condition is required to attend to the states of the

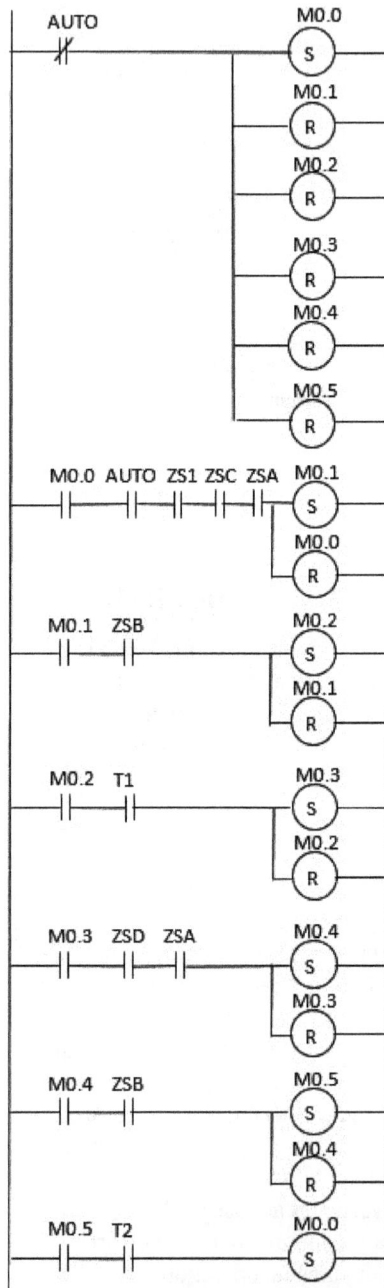

Fig. 12.9. Start of the finite state machine ladder diagram (for the pick-and-place sequence using bistable binary variables with the set (S) and reset (R) commands; logic for state transitions).

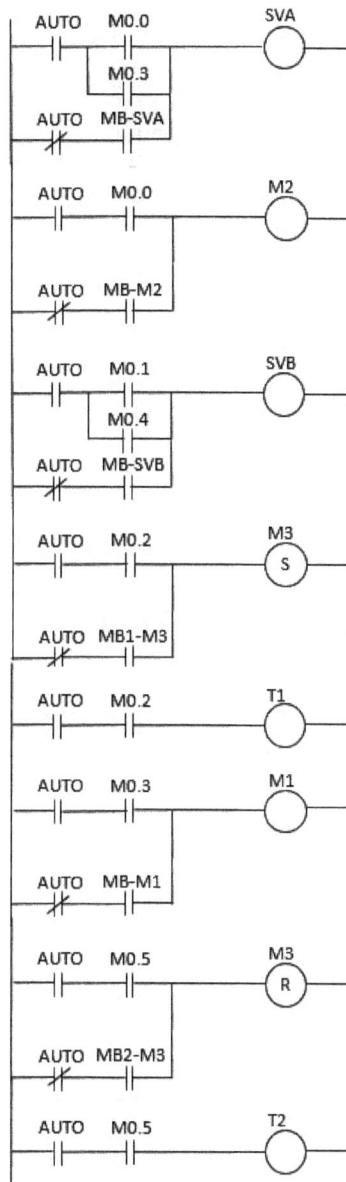

Fig. 12.10. Final part of the finite state machine ladder diagram (for the pick-and-place unit sequence using bistable binary variables; output activation logic).

automaton, and the automatic mode deactivated is necessary to attend to the manual buttons. The addition of the manual mode operation does not imply the redesign of the automata. Figure 12.10 implements the logical activation of the outputs, both automatically and manually.

4.2 Finite State Machine Using Numeric Register

A practical alternative technique for implementing the state machine is to use an integer numerical register of the PLC to control the active stage of the sequence. In this way, a single variable is used to control the sequence instead of using one binary variable per stage. Also, instead of working with activation and deactivation functions for the binary variables, only a different integer value for each stage is assigned to the numeric variable. This programming is shown in Figs. 12.11 and 12.12 for the control of the transport operation sequence.

The auxiliary variable here is a numeric variable whose value changes from 0 to 5. Note that the numerical order of the registry values for the states is not

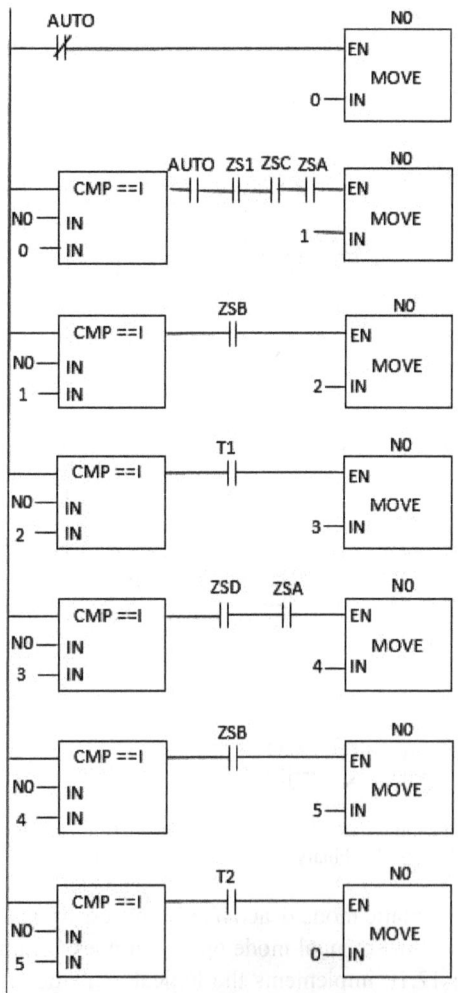

Fig. 12.11. Initial part of the finite state machine ladder diagram (for the pick-and-place sequence using the numerical register (N0); logic of definition of the states; CMP blocks compare the inputs IN (checking for equality); and MOVE blocks assign a value to the register).

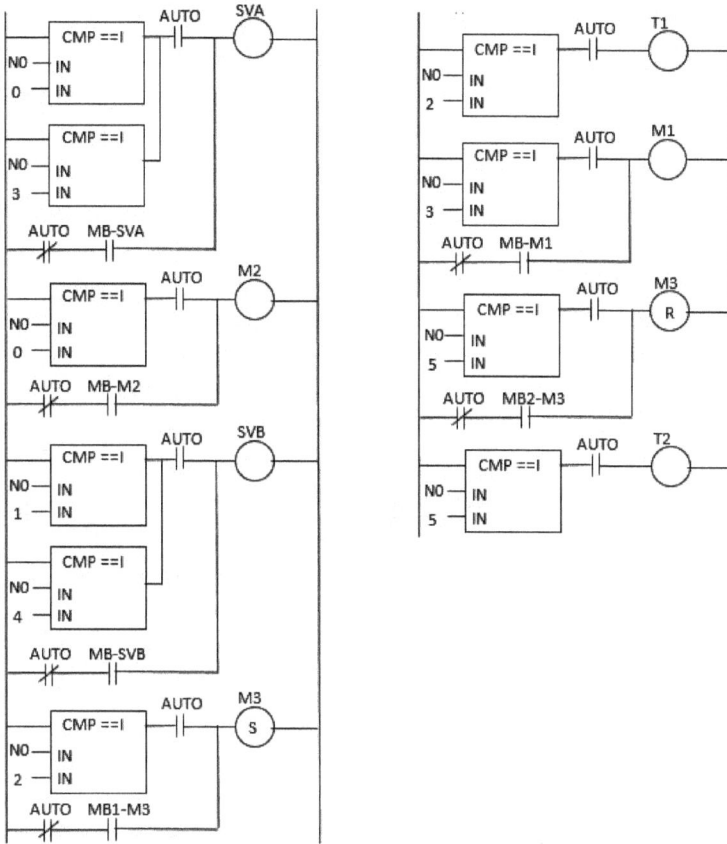

Fig. 12.12. Final part of the finite state machine ladder diagram (for the pick-and-place sequence using a numerical register (N0); programming of the actions corresponding to the states).

required, although it makes the program easier to understand. Using integer numeric registers is typically associated with high-level languages such as structured text, and its versatility in the ladder language is not exploited for PLC programming. The use of a numerical register of the PLC allows a more practical implementation of the stages of the automatic sequence since the value of the register determines the active state. The FSM implementation with registers presents several advantages. In the event of an interruption, it is possible to restart sequences at intermediate stages or to carry out step-by-step tests. The activation of a state implies the deactivation of others, which eases the modification of the program to include more states. The current process step is easily verified by a single variable instead of a set of variables.

4.3 Finite State Machine as a GRAFCET or SFC Program

The GRAFCET is based on the formalism of the state machine to conceptualize the sequence with minimal differences in the graphic elements: boxes for stages or states, 'tees' or crosses for transitions. The solution consists of three sections: the

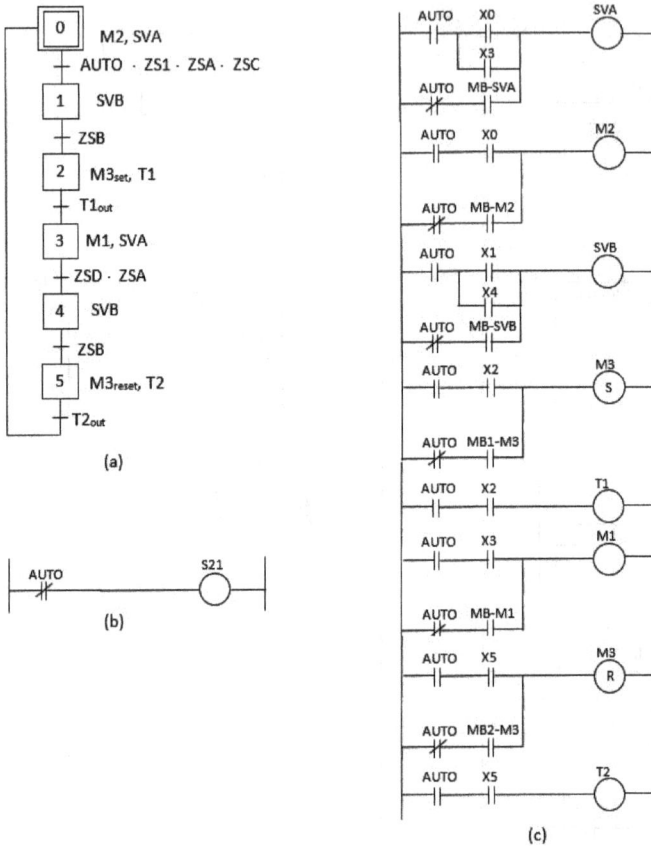

Fig. 12.13. Using GRAFCET to implement the pick-and-place sequence. (a) GRAFCET; (b) Preliminary Section; (c) Posterior Section.

GRAFCET section, where the sequence graph is made; the preliminary section; and the posterior section. The GRAFCET graph is captured in the supporting PLC indicating stage numbers and transitions, and the binary variables representing the stages are automatically defined (X0, X1, ... X5; the number corresponds to the stage number), that is, it is not necessary to implement their logic functions, since they are solved internally. These auxiliary binary variables are directly referenced in the posterior section program for the activation of the outputs or output resolution.

In the preliminary section, the stop of the sequence is programmed. Here a system variable (S21) can be used. It has the function of deactivating any active stage at the moment of stopping and bringing the control of the sequence to the initial stage (stage with double box). Figure 12.13 shows the solution using GRAFCET.

The GRAFCET allows proceeding directly to the programming of outputs using binary variables automatically defined by the PLC system. In addition to the stage variables, system variables can be used to perform special functions upon sequence interruption (such as variable S21 in the preliminary section). These specialized variables hinder the transferability of the program.

5. Concurrence with FSM

In principle, only one state can be active at a time in an FSM to ensure the logical order of the sequence. An FSM can easily model exclusive operation paths, giving flexibility to a system to follow different processing steps, according to a particular selection or different events regarding, for instance, the characteristics of row materials or even safety conditions. On the other hand, simultaneous operation is often necessary to increase productivity, for example, by using identical or different production lines or sequences. The condition of a single active state at any time limits the solution of simultaneous processing stages, which is not a restriction of Petri nets, but can be circumvented. These two sequence possibilities are illustrated for an arbitrary 7-state process in Fig. 12.14.

To be able to propose simultaneous operation paths, the implementation of the FSM can also be done using either bistable variables or numerical registers. When parallel actions are only two, as in the case of the lifting and translation of the pick-and-place crane, both actions can be assigned to a single state. A conditional deactivation of the respective loads can be implemented to avoid energizing each output load longer than necessary. When the simultaneous sequences are more elaborated but have a few steps, using bistable variables can be convenient. Figure 12.15 shows how a state variable and its transition are used to activate more than one state variable at the beginning of the simultaneous paths and how the activated states and transition conditions are used to activate the common subsequent state at the convergence point of the simultaneous sequences, according to the state machine shown in Fig. 12.14b (not related to the pick-and-place station). The output resolution would be made in terms of memory variables M0.0 through M0.6.

If the simultaneous operation paths are long, defining subordinated finite state machines with their own numerical register for controlling each subsequence is advisable. Numerical registers ease the statement of a hierarchical FSM, where inferior levels implement specific sequences as those needed in parallel or simultaneous

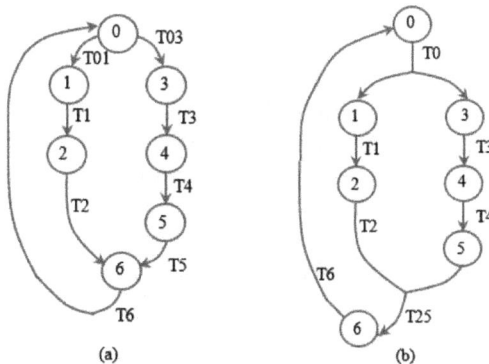

Fig. 12.14. Finite state machines with two exclusive sequence segments (a) and With Two Simultaneous Sequence Segments (b). (States 1 and 2 form a sequence segment, and the other is formed by states 3, 4, and 5. Transitions are indicated by T and related state number(s).)

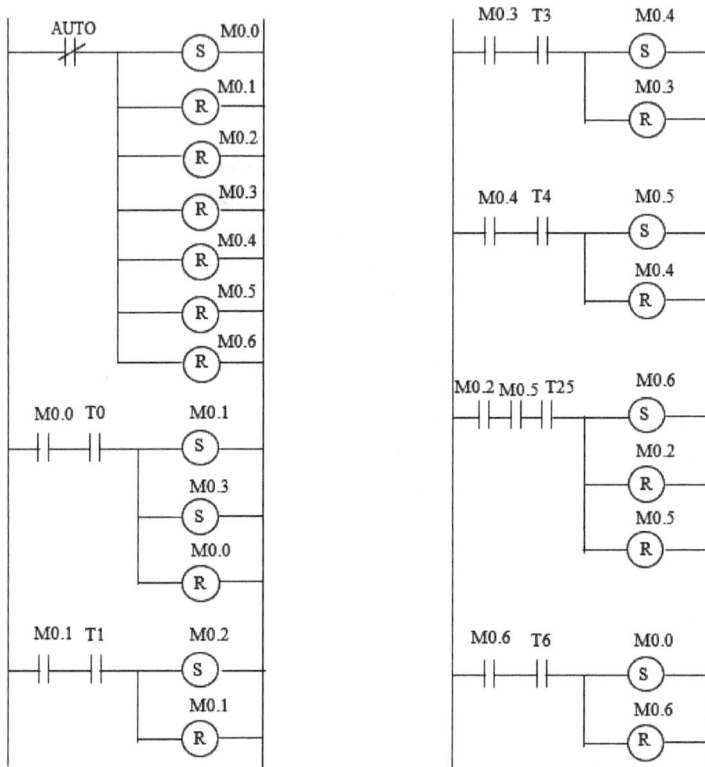

Fig. 12.15. Using bistable variables to activate states for simultaneous sequences (see Fig. 12.14b). (Binary memory variables have the prefix M0. S and R indicate set and reset commands, and transition conditions are represented by T and a number.)

execution trajectories. Figure 12.16 illustrates this type of implementation for the state machine of Fig. 12.14b.

The register corresponding to the general or main sequence is N0 and determines the activation of common states 0 and 6, besides starting simultaneous segments or subsequences. Register N1 is for subsequence formed by states 1 and 2, and register N2 is used for the subsequence with states 3, 4, and 5. N0 is assigned the value of 1 to deactivate state 0 (any value different from 0 and 6 would have the same effect). N1 is assigned the value of 3 to deactivate state 2 (any value different from 1 and 2 could be used). N2 is not assigned values of 1 and 2 to avoid mismatching the numbers of the states defined in its corresponding subsequence. When N2 is made equal to 6, state 5 is deactivated (any value different from 3, 4, and 5 would work the same way). The outputs at each state would be produced with N0 equal to 0 or 6, N1 equal to 1 or 2, and N2 equal to 3, 4, or 5.

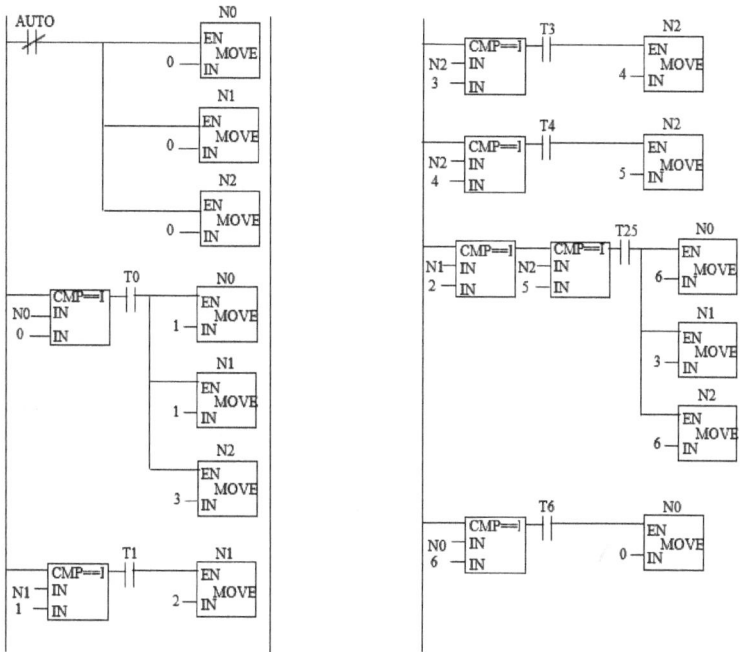

Fig. 12.16. Activation of states for simultaneous sequences (see Fig. 12.14b) Using Numeric Registers. (CMP blocks compare the inputs IN, and MOVE blocks assign a value to the numeric variable.)

6. Conclusions

The finite state machine facilitates the implementation of long and elaborated control sequences because programming can be derived in a direct and structured way. The state machine diagram models the sequence graphically, avoiding the derivation of the detailed logic functions based on considering all input and output variables and introducing minimum auxiliary variables to solve the sequential logic. The FSM tool relies less on the skill of the programmer. Instead, it relies on systematic procedures that guarantee operation and allows its maintenance without needing to redo the programs. The inclusion of the manual mode of operation is done parallelly with the automatic mode, given by the FSM, without interference.

The same state machine diagram for the pick-and-place station is used for alternative implementations, which can be selected depending on the available resources of the type of PLC in use. The implementation with binary variables is a basic and transferable implementation, which can be used even by constructing the logical functions with feedback or interlocking instead of the set and reset functions. Implementation with a numeric variable implies using additional programming resources of a PLC, such as numerical comparators and allocation blocks.

With the use of the register, when the sequence is stopped, the initial stage can be activated by assigning the corresponding specific value to the integer variable.

Immediately, the rest of the stages would be deactivated, regardless of how many had been defined, which is an advantage when adding stages in subsequent modifications. Similarly, it is also possible in the event of an interruption to restart sequences at intermediate stages by implementing more advanced automation functionalities.

GRAFCET is an implicit way of implementing the state machine that makes the state activation logic transparent to the programmer. This advantage has a cost and a limitation regarding the PLCs in which the program can be executed. These issues hinder the operation of processes that require reprogramming and involve an engineering cost.

It can be seen that a greater structure in the automation solution procedure requires more programming resources, but the advantage is that the modifications can be made easily. In the state machine diagram, stages can be added and modified, and the respective changes imply the addition of binary variables for the usual implementation or simply the addition of steps where the new stages are handled with other numerical values without necessarily redefining the values of other stages already programmed in the case of implementation with numerical register. The same advantage, of course, is attributed to GRAFCET.

Using the formalism of the state machines, and maintaining a practical approach, results in programs with guaranteed operation and easy to modify in cases of changes in the operation and equipment of the processes over time. The method is applicable both at the machine control level and at the manufacturing cell control level, that is, for coordinating several machines or stations.

Nowadays, with the more advanced automation requirements for manufacturing cells and lines imposed by the introduction of cyber-physical systems, it is desirable to provide PLC programs to support even more flexible operations or configurations. In this sense, implementing FSM with registers represents a more flexible programming technique. Virtual models of the manufacturing systems will require synchronization with the physical entities and the control system. PLC-based control systems can achieve easier synchronization by the use of registers in the implementation of state machines since the state of the virtual machine can be updated more simply on the real machine and vice versa.

References

Darvas, D., Blanco, E. and Majzik, I. 2015. A Formal Specification Method for PLC-based Applications. *15th International Conference on Accelerator and Large Experimental Physics Control Systems.* Melbourne, Australia, 91–94.

International Electrotechnical Commission, IEC 61131-6, 2013. *Programmable Controllers. Part 3: Programming Languages.*

Popescu, C. and Martinez, J.L. 2010. Formal methods in factory automation. pp. 465–476. *In*: Silvestre Blanes, J. (ed.). *Factory Automation.* InTech.

Ranjeeta, S. and Verma, H. 2018. Development of PLC based controller for pneumatic pressing machine in engine-bearing manufacturing plant. *Procedia Computer Science* 125: 449–458.

Vieira, A.D., Portela, E.A. and de Queiroz, M.H. 2017. A method for PLC implementation of supervisory control of discrete event systems. *IEEE Transactions on Control Systems Technology* 25(1): 175–191.

Wagner, F., Schmuki, R.,Wagner, T. and Wolstenholme, P. 2006. *Modeling Software with Finite State Machines: A Practical Approach*. CRC Press.

Wcislik, M., Suchenia, K. and Laskawski, M. 2015. Programming of sequential control systems using functional block diagram language. *IFAC–PapersOnLine* 48(4): 330–335.

Wcislik, M., Suchenia, K. and Laskawski, M. 2016. Method of programming of sequential control systems using LabVIEW environment. *IFAC–PapersOnLine* 49(25): 476–481.

Yue, J. and Yan, Y. 2022. Update law of simplifying Finite State Machines (FSMs): An answer to the open question of the unmanned optimization of FSMs. *IEEE Transactions on Circuits and Systems II: Express Briefs* 69(3): 1164–1167.

Chapter 13

Multilevel Automation

1. Introduction

The application of finite state machines (FSM) can be extended to the supervisory control level. A practical case of automation of a manufacturing robotic cell is presented, which is solved by designing finite state machines that implement the automated operation sequence of every unit of the cell and the integration of all the units. In many cases, for supervisory control, no methodology is used in automation for the design of control logic and even less for programming. For this reason, methods based on finite state machines help design structured PLC programming applied to the workstations and the supervisory controller. Code generation is demonstrated from the proposed finite state machines (Sanchez and Martell, 2019).

Multilevel automation achieves the automated execution of the total functioning of the cell. Also, it integrates the manual operation mode to allow the direct manipulation of the actuators and a semi-automated operation mode to test or execute the processing of the units individually and give the flexibility to handle interruptions of the cell work cycle. The resulting design satisfies the functional specifications and facilitates the comprehension, maintenance, and modification of the automated system. The strategy of this automation solution is modularity: a state machine is proposed for each subsystem, and the coordination of the individual machines is designed using another higher level state machine at the supervisory control level. In this way, several levels can be created. At the highest level, coordination and supervision of all subsystems is achieved. The advantage of this strategy is the possibility of maintaining relatively small state machines that facilitate the visualization of the process.

In this chapter, Section 2 describes multilevel automation, including the automation of level 1 workstations and the level 2 supervisory controller. Section 3 presents an FSM-based methodology for automating a manufacturing cell, including the supervisory control level. Section 4 reviews the application of FSM to improve the traceability of process data in the manufacturing cell and how FSM can be applied for data logging in SCADA systems. Section 5 concludes.

2. Supervisory Control

In a discrete manufacturing industry, parts go through a process line that consists of several stations interconnected by a transport system. The series of workstations describe the main production stages and their order. The processing in each workstation is a sequence of detailed and low level operations using field instrumentation, that is, sensors and actuators. The process line or manufacturing cell may have different purposes: physical and chemical transformation, assembly, quality inspection and testing, labeling, packaging, etc.

2.1 Automation Levels

In the manufacturing and process industries, the automation concept can be applied at different levels that implement progressive groupings (Groover, 2015). Level 0 differs for both industry types since it refers to the instruments in contact with parts or fluids that show and affect their conditions. However, at subsequent levels, the differences are smaller: at level 1, the nature of processes and equipment determine the type of basic control systems; at level 2, the groups of process units match the groups of machines in a manufacturing cell where supervision for proper integration is needed. In the processes industry, SCADA systems are located at level 2. At levels 3 and 4, planning and business functions based on more general production and economic principles, for example, but with different focuses, are required for both industries.

Automation in the manufacturing industries has been based mainly on using PLCs. Other automation technologies, such as industrial robots and computerized numerical control (CNC) have also contributed to the automation of manufacturing cells. Computer integrated manufacturing (CIM) relies on industrial communication networks to automate manufacturing processes, making them more robust and flexible. Materials transport systems, automated warehouses, and inspection

Table 13.1. Levels of automation in manufacturing industry.

Level	Description
Enterprise level (4)	Corporate information system. Enterprise level handles corporate information related to marketing, sales, accounting, research and design, production scheduling, etc.
Plant level (3)	Factory or production systems level produces operational plans such as processing, inventory, purchasing, and quality control.
Level	Description
Cell or system level (2)	Manufacturing cell or system level; a manufacturing cell is defined and operated as a group of machines or workstations interconnected by a material handling system; this level includes production and assembly lines.
Machine level (1)	Integrated operation of various devices, as in the case of CNC machines and similar production equipment, industrial robots, and material handling equipment.
Device level (0)	Actuators, sensors, and other hardware components that form individual control loops for the next level.

systems using video cameras are added to form manufacturing cells. Computational technologies, on the other hand, allow the development of manufacturing support systems (Sanchez and Martell, 2019).

For complex automation tasks, such as the automation of a manufacturing cell, formal programming methodologies are required (Popescu and Martinez, 2010). State machines provide a tool for problem analysis and logic control synthesis and lead to direct and structured programming at the same time. The concept of supervisory control has traditionally been associated with controlling discrete events through state machines to give robustness to control systems and maintain acceptable behavior or performance (Ramadge and Wonham, 1987). Under this supervision approach and with a correct logic design and coding methodology, a simulated case of a manipulator in a production system is reported (Music and Matko, 2002). Supervisory control involves monitoring the plant operation, including the controller, sensors, actuators, and process behavior. It should distinguish between normal operation, unstable behavior, and faulty conditions in the system's various components (Linkens and Abbod, 1992).

2.2 SCADA Functions

Modern SCADA systems allow collection of real-time data from the plant floor. SCADA systems are essential for improving the operation of production processes. Introducing modern IT standards and practices, such as using relational databases like SQL and web-based applications, into SCADA software has greatly improved the productivity, efficiency, security, and reliability of SCADA systems.

The SCADA systems perform advanced automation functions such as supervisory control, data acquisition and logging, network data communication, data visualization, and trending. SCADA systems can be programmed to perform process control decisions based on data collected from sensors and lower level controllers. Control functions may include powering on/off and adjusting set points and reference values.

Data Acquisition and Logging

SCADA systems acquire data devices through control networks connecting PLCs, sensors, and other devices. They measure process and control variables and parameters such as temperature, pressure, flow, speed, weight, etc. The collected and processed data are then sent to an HMI for a machine operator to analyze and make decisions.

Network Data Communication

Hardwired industrial networks are required to connect the SCADA system and transmit data between machines and operators. These industrial data communication networks allow the connected systems to be controlled from a central control room.

SCADA systems display data in HMI applications. The server workstations continuously monitor all sensors, alert the operator when there is a fault condition and alarm from deviations of process variables from their normal operational range.

3. Automation of a Manufacturing Cell

The case study is a manufacturing cell composed of three elements: (1) an automatic storage and automatic retrieval system (AS/RS); (2) a transportation conveyor; and (3) a packaging workstation with an industrial robot, also referred to as a manipulator arm.

In this multilevel automation example, state machines are used to automate individual devices, as well as complete machines, and to implement supervisory control for the coordination of the different components of the cell. The physical arrangement of the manufacturing cell is presented in Fig. 13.1.

A system configuration of the manufacturing cell is presented in Fig. 13.2.

The application example demonstrates the potential of state machines for generating control logic, as well as code. In addition, not only the programming of an automatic sequence is structured, but also the logic for manual operation, necessary to intervene in emergency situations and preparation or revision of the system. The list of input and output signals of the PLC equipment for the three work components, with their symbols for each unit, are shown in Tables 13.2, 13.3, and 13.4.

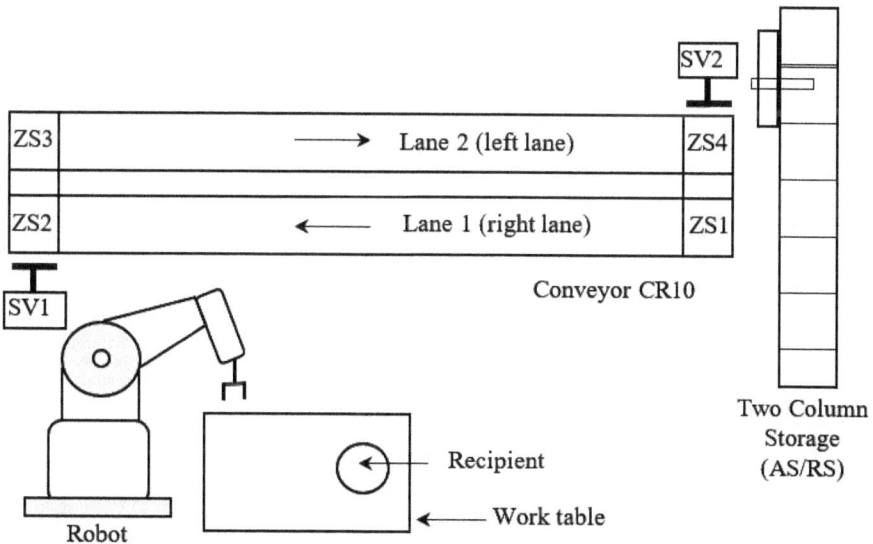

Fig. 13.1. Layout of the manufacturing cell.

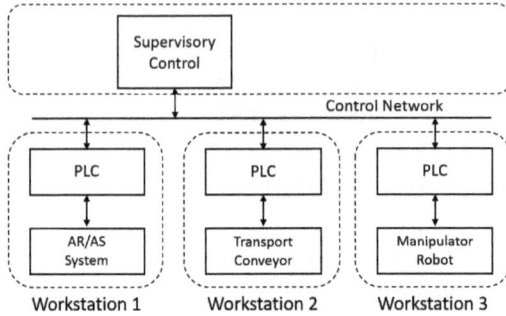

Fig. 13.2. System configuration of the manufacturing cell.

Table 13.2. Automatic storage and retrieval system (AS/RS).

PLC Inputs	Symbol	PLC Outputs	Symbol
X-axis left position sensor (actuator on the left side)	X_LT_ZS	Solenoid valve to retract actuator in X (to the right)	X_RET_SV
X-axis right position sensor (actuator on the right side)	X_RT_ZS	Solenoid valve to advance actuator in X (to the left)	X_EXT_SV
Y-axis actuator retracted position sensor (out of stock)	Y_RET_ZS	Solenoid valve for single acting piston in Y (in)	Y_SV
Y axis actuator extended position sensor (in stock)	Y_EXT_ZS	Solenoid valve to rotate 180° (towards band)	ROT_SV
Position the sensor at a 0° angle (towards storage)	ROT_STO_ZS	Solenoid valve to open gripper	GRIP_OPN_SV
Position the sensor at a 180° angle (toward the conveyor band)	ROT_BAND_ZS	Solenoid valve to close the gripper	GRIP_CL_SV
Gripper sensor in the open position	GRIP_OPN_ZS	Elevator - release brake	CR1
Gripper sensor in the closed position	GRIP_CLD_ZS	Elevator – slow velocity	CR2
Emergency stop push button	PB_STO	Elevator – go up	CR3
Encoder pulse	ZT1	Elevator – stop	CR4
		Elevator – go down	CR5

In the process of solving this automation, the graphic representation of the state machine is used: each stage or state is represented by a numbered circle; next to the circle or stage, the corresponding action is specified; the arrow represents the transition to the next stage. A transition includes conditions that are consequences

Table 13.3. Transport system.

PLC Inputs	Symbol	PLC Outputs	Symbol
Position sensor in lane 1, storage side	ZS1	Solenoid valve for single acting cylinder on robot side	SV1
Position sensor in lane 1, robot side	ZS2	Solenoid valve for single acting cylinder on storage side	SV2
Position sensor in lane 2, robot side	ZS3	Conveyor motor	CR10
Position sensor in lane 2, storage side	ZS4		
Emergency stop push button	PB_BAND		

Table 13.4. Robot interface signals.

Commands to Robot	Symbol	Robot Outputs	Symbol
Position sensor to detect container	di1	Robot gripper	c1
Task start signal	di5	Task end signal	c5

of the actions taken in the upstream stage and, at the same time, requisites for the actions in the downstream stage.

3.1 Automation of the Conveyor-Belt System

Part Placement in Front of Robot

Figure 13.3 represents the part transport sequence to a position within the reach of the robot. It is verified that no emergency 'stop' (SYS_OK) has been requested and that the start order has been given (either in automatic or semi-automatic mode). The conveyor motor is turned on to move the part in front of the robot. The band stops as soon as the part is detected in front of the robot, and the sequence is finished. The Band_Start1 signal is produced when the system is in automatic mode, and a part is detected in lane 1 (ZS1) or when the system is in semi-automatic mode (and not in automatic mode):

Band_Start1 =
Band_Auto_Start1 & ZS1 + Band_SemiAuto_Start1 & (Auto_Mode)' $\qquad$ (13.1)

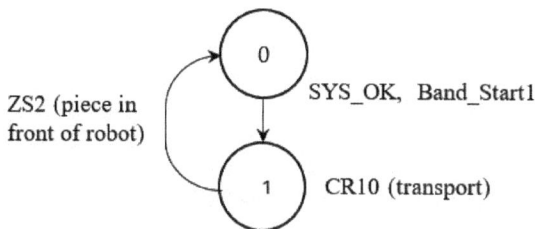

Fig. 13.3. FSM for part transport from the storage side to the robot side.

Remove Part from the Robot Side Back to the Storage Side

Figure 13.4 details the transport sequence of the packaged piece to the storage. The 'ready' status (SYS_OK), that is, the absence of emergency stops, and the start signal (in automatic or semi-automatic mode) are verified to activate the cylinder on the robot side that pushes the part towards lane 2. Then the conveyor motor is started to take the piece to the storage side. When the piece is detected on the storage side in lane 2, the motor stops, and the piece is pushed toward lane 1, where it can be taken to the storage structure.

The *Band_Start2* signal is produced in automatic mode when a part is detected in lane 1 on the robot side (ZS2) or when the automatic mode is not active and semi-automatic mode has been started:

$$Band_{Start2} =$$
$$Band\_Auto\_Start2 \ \& \ ZS2 + Band\_SemiAuto\_Start2 \ \& \ (Auto\_Mode)' \qquad (13.2)$$

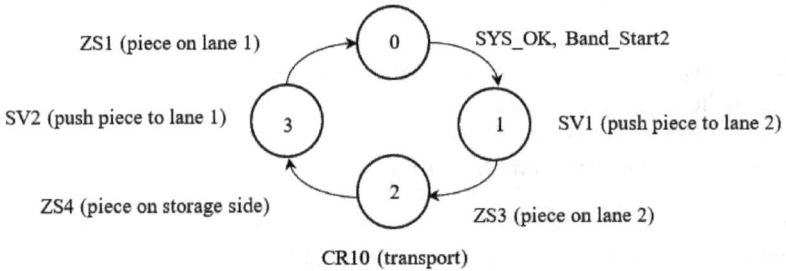

Fig. 13.4. FSM for part transport from the robot side to the storage side.

3.2 Control of the Robot System

When a start signal (automatic, semi-automatic) is received, the robot moves from its starting position to reach the part in conveyor lane 1, picks up the part, moves it to the worktable and packs it, takes the packed part, places it back on the belt in lane 1, and returns to its initial position. Once this sequence is ended, the manipulator task completion signal is sent. Figure 13.5 contains the finite state machine that controls the execution of the packing operation.

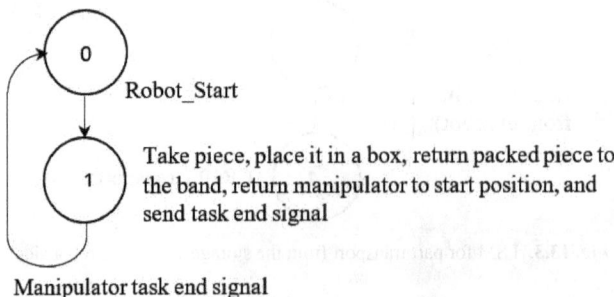

Fig. 13.5. FSM for robot task.

The *Robot_Start* signal is produced in automatic mode when a piece is detected on the belt next to the robot location (ZS2), and the position sensor on the work table (di1) is activated or in semi-automatic mode when automatic mode is not active.

Robot_start = *Robot_Auto_Start* & *ZS2* & *di*1+
+*Robot_SemiAuto_Start* & (*Auto_Mode*)'   (13.3)

The program to be executed in state 1 of the above FSM is exemplified with a high-level language for a 6-degree-of-freedom jointed commercial robot:

```
PROC main()
    MoveJ (start_pos, v1000, z100, tool1)
    SetDO do5, 0 ! Turn off the interlock signal to the PLC or robot task end
    signal.
    SetDO do1, 1! Open gripper
    IF (di5=1)! If start signal is received
        MoveJ(near_pce_band, v1000, z100, tool1) ! Approach piece on the band
        MoveL(pce_band, v100, fine, tool1) ! Reach piece position
        SetDO do1, 0! Close the gripper to hold a piece
        WaitTime 0.5
        MoveL(near_pce_band, v100, fine, tool1) ! Get apart from the band
        MoveJ(intermediate_pos, v1000, z100, tool1) ! Go to the intermediate
        position for a safe trajectory
        MoveJ(near_box, v1000, z100, tool1) ! Approach box on the work table
        MoveL(inside_box, v100, fine, tool1) ! Introduce a piece inside a box
        SetDO do1, 1! Open gripper
        WaitTime 0.5
        MoveL(near_box, v100, fine, tool1) ! Move away from the box
        MoveL(box, v100, fine, tool1) ! Reach the position to hold the box
        SetDO do1, 0! Close the gripper to take the box with the piece inside
        MoveL(near_box, v100, fine, tool1)
        MoveJ(intermediate_pos, v1000, z100, tool1)
        MoveJ(near_box_band, v1000, z100, tool1) ! Approach band
        MoveL(box_band, v100, fine, tool1)! Reach the position to leave the
        packed piece on the band
        SetDO do1, 1! Open gripper
        WaitTime 0.5
        MoveL(near_box_band, v100, fine, tool1) ! Get apart from the band
        MoveJ (start_pos, v1000, z100, tool1)
        SetDO(do5, 1)
    ENDIF
ENDPROC
```

MoveJ moves the robot's joints toward a target end-effector position with a given speed (v####), tolerance (z####), and tool (tool#). *MoveL* allows the robot to move along a linear path toward the target position, generally with a lower speed and tolerance to properly interact with the object to be manipulated. *WaitTime* produces a waiting time to ensure stable placement of the manipulated object when leaving it or correct gripping when picking it up. *SetDO* assigns a value to the indicated variable. The di5 signal is the *Robot_Start* signal at the transition from stage 0 to 1 or the signal from the PLC to the robot controller to start its routine. Signal do5 is the manipulator task completion signal sent by the robot controller to the PLC or transition from stage 1 to 0. Signal do1 is the output signal from the robot controller to the end effector or gripper.

3.3 Automation of AS/RS

Automatic Retrieval: Taking Part from Storage to Conveyor Band

Figure 13.6 shows the FSM that implements the sequence to take a part from the storage. The single acting cylinder of the Y direction is retracted by default for safety, and the robot or elevator is turned towards the storage also by the default position of the single acting rotation actuator. The start signal, either in automatic or semi-automatic mode, is given by the following equation:

$AS/RS\_Start1 =$
$AS/RS\_Auto\_Start1 + AS/RS\_SemiAuto\_Start1 \ \& \ (Auto\_Mode)'$ \hfill (13.4)

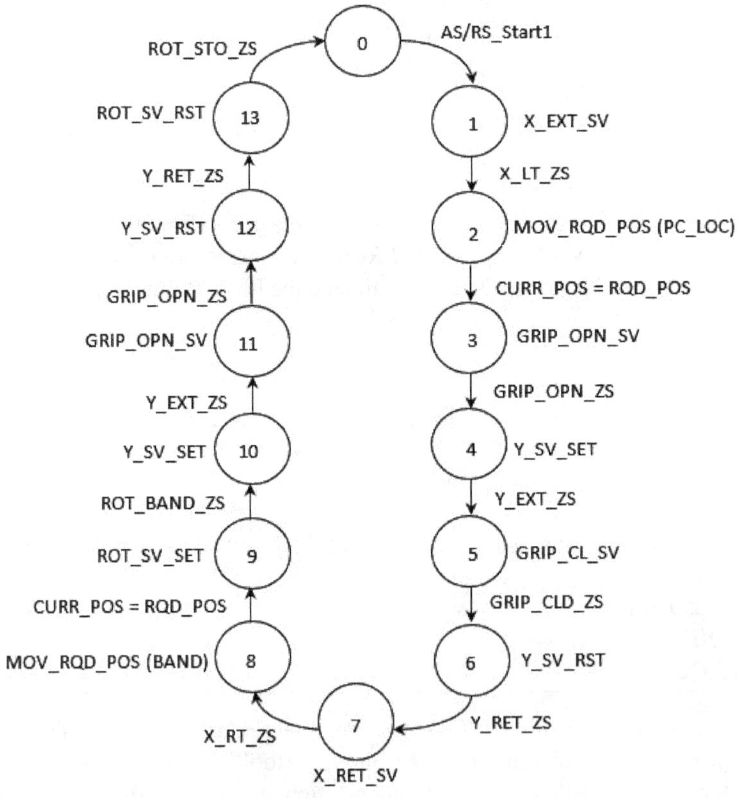

Fig. 13.6. FSM for taking part from the storage to the band.

The retrieval sequence from the mentioned initial conditions involves the following activation of the different actuators. The elevator moves to the left side with the cylinder for the X direction and then moves vertically with the Z direction cylinder to the desired position (PC_LOC). The gripper opens, the Y actuator extends to enter and reach the pallet on which the piece is placed, the gripper closes to seize the pallet, the Y cylinder retracts to extract the part from the storage structure, the elevator moves to the right and then vertically to the height of the band (BAND). The elevator turns towards the band, the Y actuator extends to place the pallet with the part, the gripper opens, the Y cylinder retracts, and the elevator is rotated towards the storage returning to the initial position.

Automatic Storage: Taking Part from Conveyor Band to Storage

The sequence to take the packed piece from the band and place it in the storage structure is represented in Fig. 13.7. The AS/RS_Start2 signal is produced in automatic mode when a part placed on lane 1 (ZS1) is detected or when the automatic mode is not active, and the start has been given in semi-automatic mode:

AS/RS_Start2 =
AS/RS_Auto_Start2 &ZS1 + AS/RS_SemiAuto_Start2 & (Auto_Mode)' $\qquad$ (13.5)

Because of the default position of the single acting actuators, the robot is retracted in the Y axis and turned to the AS/RS unit. Once the start signal is received,

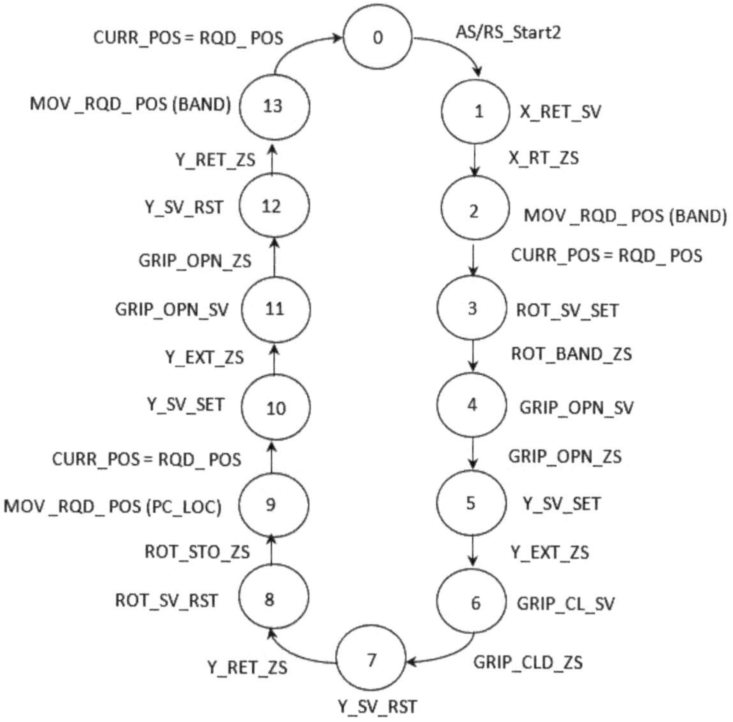

Fig. 13.7. FSM for taking part from the band to the storage.

the Cartesian robot moves towards the right lane (X direction), the gripper is placed on the band level (BAND position in Z direction) and is turned towards the conveyor band. The gripper opens, and the cylinder in Y direction extends in to take the pallet with the processed part; the gripper closes, retracts in Y, and turns towards the storage structure. The elevator moves vertically to the desired height (PC_LOC). The cylinder in the Y direction extends to place a pallet with the piece inside the structure, the gripper opens, the Y cylinder retracts, and the elevator returns to the band's height, returning to the starting position.

3.4 Control of the Elevator of the AS/RS

The vertical movement of the Cartesian robot of the AS/RS is carried out at normal speed until approaching the required position within a certain margin or tolerance. Once inside that margin, the movement must continue at a low speed until reaching the desired height. This controlled operation is implemented with the FSM of Fig. 13.8. In the event of a motor overload; stage 0 must be activated to stop the motor. Additionally, it should be considered that when starting the motor (in stages 1 or 3), the brake must be held for a short time (it can be from 100 to 200 milliseconds) and then released so that the motor continues its operation and produces the desired movement to ensure correct energization of the motor windings.

The FSM of Fig. 13.8 is activated or called from the FSM of Figs. 13.6 and 13.7 when the action of moving the elevator to the required position (MOV_RQD_POS) is taken during the operation of the AS/RS.

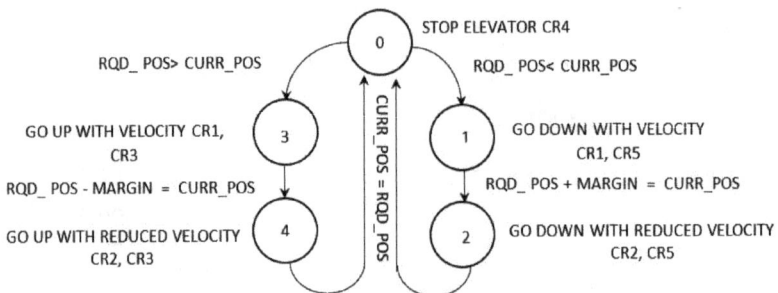

Fig. 13.8. FSM for vertical movement of AS/RS elevator.

3.5 Manufacturing Cell Integration

The coordination of the operation of the individual units of the cell constitutes the automation at a supervisory level. The supervisory finite state machine (Fig. 13.9) activates each cell element and waits for the corresponding termination signal to proceed with the proper order. The cell integration is carried out from the human-machine interface developed in a computer workstation with communication to the PLC of the units via Ethernet.

The states of the supervisory or master FSM of Fig. 13.9 produce the start signal of the lower level FSM (Figs. 13.3 through 13.7) and receive their termination

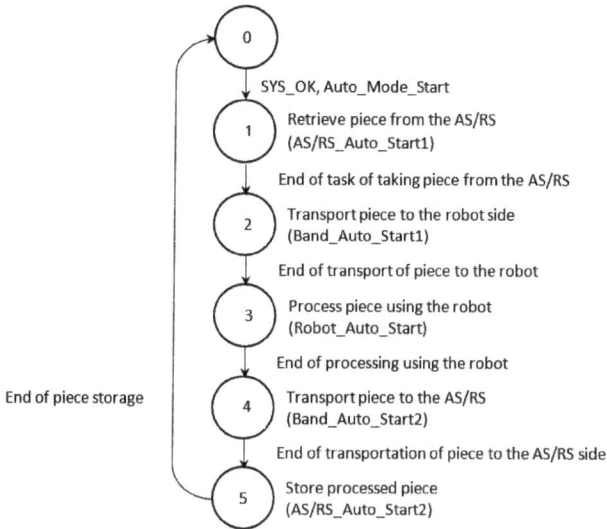

Fig. 13.9. FSM for supervisory control.

signals as transitions. The execution of the supervisory FSM is the operation of the manufacturing cell in automatic mode.

State 0. The system is ready, awaiting an automatic start signal that can be triggered from an automatic selector in the SCADA/HMI system.

State 1. The supervisory system generates the start signal for retrieving the part from the storage structure. The AS/RS FSM of Fig. 13.6 is executed and sends a task completion signal that allows the supervisory state machine to proceed with the next state.

State 2. The start signal for the FSM of Fig. 13.3 is produced. The conveyor transports the pallet with the part retrieved in the previous state to the robot station, and the supervisory controller waits for the completed task signal from the PLC of the conveyor to continue.

State 3. This state sends the start signal to the workstation, where the articulated robot puts the part inside a box (FSM of Fig. 13.5). When the robot leaves the box with the piece to the conveyor and returns to its home position away from the conveyor, the task end signal is sent to the supervisory FSM.

State 4. The conveyor is activated to return the part in the box on the pallet to the position where it will be picked up by the AS/AR system (FSM of Fig. 13.4 is started), and the signal of the completion of this process is awaited.

State 5. The AS/RS is activated to pick and store the processed part (by the start of FSM of Fig. 13.7) and wait for the finish signal to start the master or integration sequence again.

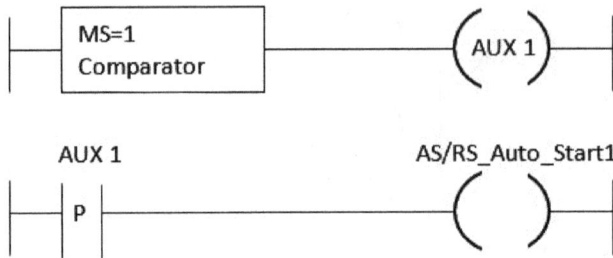

Fig. 13.10. Startup signal generated by the supervisory FSM.

A button gives the start of the automatic mode of the cell (Auto_Mode_Start), and this operation mode (*Auto_Mode*) is ended by another button, both virtual instruments of the HMI.

The start signal of the state machine of each unit in automatic mode is produced by the rising edge (or positive edge, P) of the states of the supervisory machine. For example, if the integer register assigned to the supervisory state machine is labeled MS, the automatic start of the sequence to fetch a part from the storage occurs according to the ladder logic of Fig. 13.10.

The task completion signal of each unit in the cell occurs with the same transition to the zero or initial state of the corresponding machine. The change of the state in the supervisory machine must be programmed before the change to the zero state of the FSM of the unit. For example, if the integer register for controlling the FSM for the automatic retrieval of a part from the storage (Fig. 13.6) is labeled as AR, at the end of this task, the value of the MS register of the supervisory machine is incremented, and the value of AR is returned to zero, in the sequence shown in Fig. 13.11.

The semiautomatic mode allows the execution of a particular FSM separately from the rest of the units. The manual operation of the actuators is performed by virtual buttons when the automatic and semiautomatic modes are not activated. For example, if numeric registers AR and AS are used to represent the current state in the automatic retrieval (Fig. 13.6) and storage (Fig. 13.7) FSM, respectively, the program shown in Fig. 13.12 could be used to activate the solenoid that produces the rotation of the Cartesian robot towards the band, using virtual buttons in manual mode.

Structured programming of the desired operation is an important goal of using an FSM in the development of automation and integration of the manufacturing

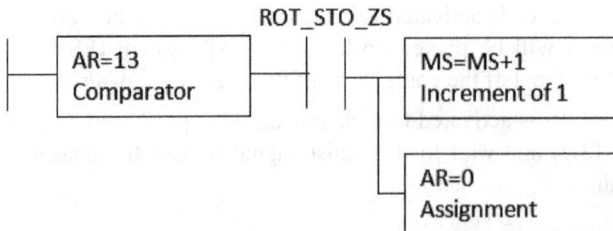

Fig. 13.11. Unit task completion signal as transition in the supervisory FSM.

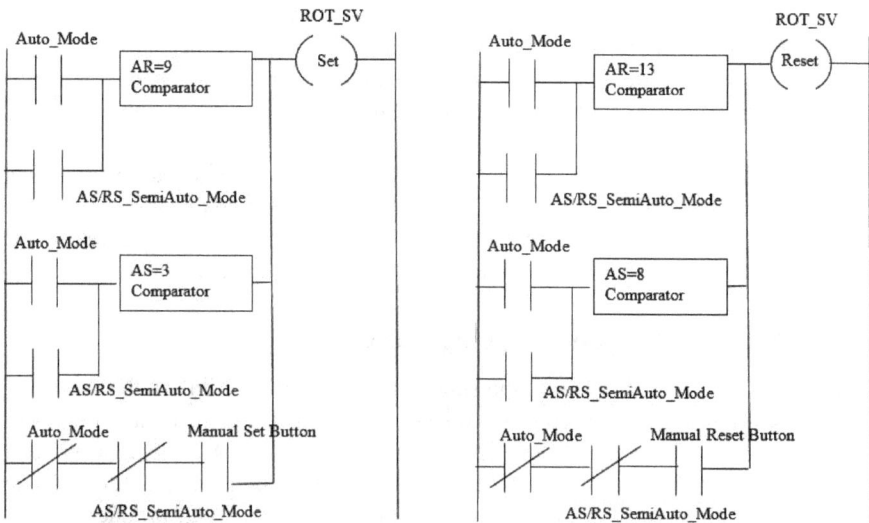

Fig. 13.12. Operation of a single acting actuator in automatic, semiautomatic, and manual modes.

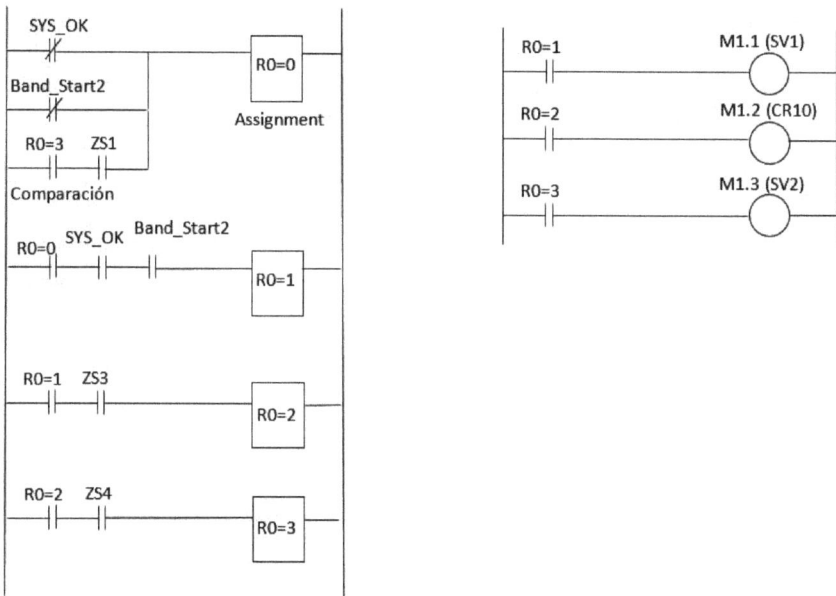

Fig. 13.13. Example of code generation from the FSM of Fig. 13.4.

cell. The program of the FSM of the conveyor band to transport the part in the box from the robot side to the pick-and-place position on the storage side (Fig. 13.4) is presented as an example of the use of the FSM for the implementation of a sequence. Figure 13.13 shows the program in a ladder diagram. The numeric register R0 is used to control the active state, and the elements of the register M1 are used to activate the outputs or actuators. The value of the current stage and the satisfaction of the exit transition conditions, that is, the detection of completion of execution of the actions corresponding to the current stage, allow the stage number to be increased or advanced. Finally, the loads are activated with the corresponding state number or register value. The PLC code for the rest of the units and their integration is obtained in a similar way.

The presented automation system includes the activation of actuators in manual mode, the individual operation of each unit in semi-automatic mode, and the integration of units in an automatic mode. An FSM-based analysis and development allow structured and modular programming for coordinating different process units.

4. Data Acquisition and Logging Functions

In industrial production systems, the supervisory control coordinates the operation of the workstations but also serves to collect production data and pass it to the upper layer MES and ERP systems (Fig. 13.14) or 'level 2' SCADA systems in process industries. SCADA software utilizes the power of relational databases to exchange data with existing MES and ERP systems, allowing process and production data access through an entire organization. Historical data from a SCADA system can be logged in a database, which allows easier analysis through data trending.

Production information can be generated with basic sensors installed in the production processes. Simple presence binary sensors can be used by applying basic timing and counting functions to detect and generate production data such as the number of production pieces, duration of work cycles, lead times, production rates, production delays, etc. In material handling systems and other industrial processes, traceability makes it possible to determine at any time when, where, and how a product was processed, produced, stored, transported, consumed, or disposed. Process tracking or tracing responds to the need and convenience of manufacturers and consumers to record and document changes in the condition and location of a product digitally in order to generate historical information. This targeted data acquisition makes it possible to take actions to optimize or correct production processes.

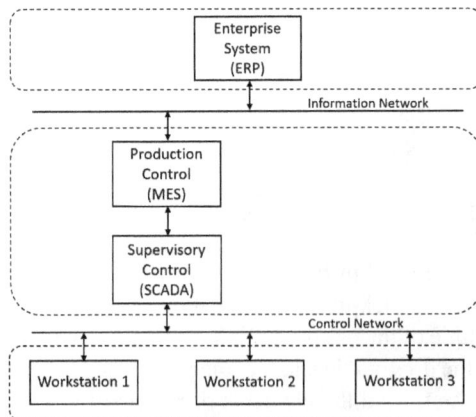

Fig. 13.14. Example manufacturing cell connected to MES and ERP systems.

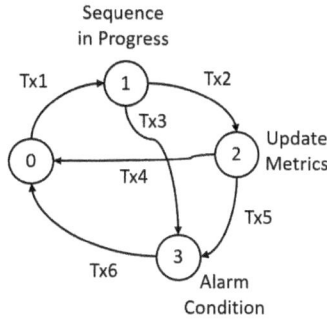

Fig. 13.15. FSM for process and production data generation and logging.

An FSM-based programming is a robust method to implement sequential logic control and is also helpful for generating production data. Active states and transitions of FSM supervisory controllers can be used for the traceability of manufacturing production systems. Moreover, taking advantage of the modularity of the FSM, the supervisory control can execute state machines such as the one shown in Fig. 13.15 for diagnostics and traceability of each workstation.

This supervisory FSM may be applied over the entire manufacturing cell, that is, over groups of workstations, as an alternative to other data processing means to generate cell metrics. The states, transitions, and actions for the proposed supervisory FSM for the generation and collection of process and production data of a workstation in a manufacturing cell are described in Table 13.5.

An FSM is essentially related to product traceability because different transitions of the state machine can be used for the generating and collecting process and

Table 13.5. Supervisory FSM for workstation.

Current State	Transitions	Next State/State Actions
State 0 System Ready	Tx1: Activate Sequence	State 1: Sequence in Progress Action 1: Increment Counter Action 2: Start Sequence Timer
State 1 Sequence in progress	Tx2: End of Sequence	State 2: Update Metrics Action 1: Wait for Workstation data and update on data received Action 2: Start Data Wait Timer
State 1 Sequence in progress	Tx3: Sequence Timer Done	State 3: Alarm Condition Action 1: Activate Delay Timer Action 2: Set Alarm Code 1
State 2 Update Metrics	Tx4: WS Data Received	State 0: System Ready
State 2 Update Metrics	Tx5: Wait for Timer Done	State 3: Alarm Condition Action 1: Activate Delay Timer Action 2: Set Alarm Code 2
State 3 Alarm Condition	Tx6: Reset Alarm	State 0: System Ready Action 1: Set Alarm Code 2 Action 2: Reset Delay Timer

product data. A work cycle can be timed from the first state change when the machine starts until the last state finishes. Processed parts can be counted whenever the last work-cycle state is successfully completed. Alarms or fault conditions can be detected at any active state.

5. Conclusions

Modern SCADA systems have four primary functions: supervisory control, data acquisition, network communication, and data visualization. The software and hardware elements of a SCADA system and their connectivity to lower level controllers can enhance the monitoring and control of industrial processes. The SCADA system components work together to perform the functions such as collecting, analyzing, and displaying real-time data from factory operations.

This chapter illustrates the use of a methodology based on state machines for the multilevel automation of a manufacturing cell that can be implemented as a supervisory control function of a SCADA system. Finite state machines, as demonstrated, are not only an analytical tool but also a modeling tool for structured programming and control solutions; they can not only be applied for each workstation but also for supervisory control as master FSM to coordinate the operation of the different workstations. The methodology is complemented by the definition of operating modes, which implies the development of a human-machine interface that provides the appropriate virtual instrumentation.

With a programming methodology, the integration of the different units and cell processes is facilitated, as well as the revision and correction of the programs. The methodology, by including the design of a supervisory state machine, allows planning of the necessary communications between different PLCs in charge of separate cell units. The FSM-based modeling for the automation and control of multiple workstations in a manufacturing cell or manufacturing line is also suitable for incorporating production metrics in each cell unit. Supervisory control for metrics generation may be implemented by using additional state machines particularly designed for data acquisition and logging functions. The FSM active states of each of the supervised workstations can be used for tracking and traceability of the manufacturing cell or a higher level traceability FSM may be proposed with the scope of the whole cell.

The practical experience in the application of state machines develops a competence in analysis and synthesis, as well as in the design and implementation of advanced automation solutions, contributing to the engineer's training to support manufacturing systems and incorporating new technological trends of Industry 4.0.

References

Groover, M.P. 2015. *Automation, Production Systems, and Computer-integrated Manufacturing.* Fourth Edition. Boston: Pearson.

Linkens, D.A. and Abbod, M.F. 1992. Real Time Supervisory Control for Industrial Processes. *IFAC Artificial Intelligence in Real-Time Control.* Delft, The Netherlands.

Music, G. and Matko, D. 2002. Discrete event control theory applied to PLC programming. *Automatika* 43(1–2): 21–28.

Popescu, C. and Martinez, J.L. 2010. Formal methods in factory automation. pp. 465–476. *In*: Silvestre Blanes, J. (ed.). *Factory Automation*. InTech.

Ramadge, P.J. and Wonham, W.M. 1987. Supervisory control of a class of discrete event processes. SIAM J. *Control and Optimization* 25(1): 206–230.

Sanchez, I.Y. and Martell, F. 2019. Multilevel automation of a manufacturing robotic cell using finite state machines. *Ing. Invest. y Tecnol.* 20(4): 1–12.

Chapter 14
Hybrid Dynamical Control

1. Introduction

Hybrid Dynamical Systems (HDS) are complex systems involving interacting discrete-events systems and continuous-time dynamical systems. They are important in system modeling and control design of embedded systems, power converters, cyber-physical systems, robotics, and manufacturing systems. Still, their use is increasing in many other application areas. Examples of such systems include computer and communication networks, industrial production systems, air and highway traffic control systems, smart grids, and interconnected power systems (Goebel et al., 2009). In mechatronics and automation, HDS have been used in designing advanced control strategies and for verification of control systems.

HDS involve discrete and continuous variables. Their modeling equations contain mixtures of logic, discrete or digital dynamics, and continuous or analog dynamics. The continuous dynamics of such systems can be continuous-time, discrete-time, or mixed (sampled data) and are usually given by differential equations. The discrete variable dynamics of hybrid systems are generally represented by a digital automaton or input-output transition system with a finite number of states. Continuous and discrete dynamics interact at 'event' or 'activation' moments when the continuous state reaches certain prescribed sets in the continuous state space (Branicky, 1995).

The hybrid automata are one of the mathematical foundations of HDS. Therefore, they can serve as computational models of a control system. Many automation and control applications can be better designed if the systems are modeled and controlled as an HDS. Model-based control requires the base of knowledge incorporated by the process modeling, and the designed control can be regarded as an intelligent and autonomous control. HDS are also applied in supervisory control systems; they are particularly useful in modeling applications where high level decision-making is used to implement advanced supervisory control functions. This occurs in distributed control systems used to control the physical plant in a decentralized manner, like in chemical process control or flexible manufacturing facilities (Lemmon and Markovsky, 1998).

This chapter provides an introduction to Hybrid Dynamical Control (HDC). In Section 2, hybrid dynamical systems are described, and their modeling with hybrid automata is explained. Section 3 briefly reviews the use of an HDS to design control systems. Section 4 presents an HDC approach for the position and speed control of a DC motor. Section 5 gives the example of the automation of a four-story elevator using finite state machines. Section 6 describes the application of finite states model predictive control in power electronics converters and conclusions are in Section 7.

2. Hybrid Dynamical Systems

In real-world applications, many processes and engineered systems can be considered hybrid dynamical systems (HDS) which are typically composed of continuous and discrete-valued (digitally quantified) signals. Continuous or digital signals depend on independent variables such as time, which may also be continuous-time or discrete-time. Another distinction in discrete-event systems is that some signals can be time-driven while others are event-driven.

A typical example of a simple HDS is the air-conditioning system to control the room temperature; this system is an on-off control of a continuous variable. The ambient temperature is controlled by a compressor and an evaporator connected in a thermodynamic cycle to produce cooling. The room temperature generates event conditions; when the temperature is higher than a threshold, the compressor is turned on, heat is extracted from the room until the temperatures drop, and then the compressor is turned off, as indicated in Fig. 14.1.

The room temperature system operates in 'on' and 'off.' Each mode of operation can be described by a discrete state, $S = \{s_1 = OFF, s_2 = ON\}$. The continuous-time temperature variable, $T(t)$, can be described by two differential equations, one for heating and another for cooling.

$$\dot{T}(t) = f(T(t), s_1) \tag{14.1}$$

$$\dot{T}(t) = f(T(t), s_2) \tag{14.2}$$

This system has a hybrid state composed of a discrete state and a continuous state $(S, T(t))$. The change of the discrete state is triggered by reaching some temperature values in the continuous state. One transition for changing from 'off' to 'on' and another from 'on' to 'off'.

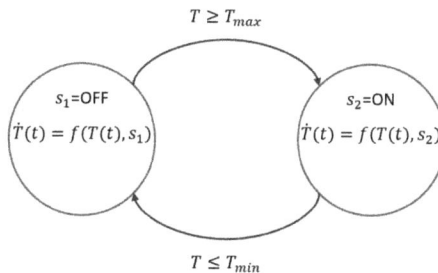

$T \geq T_{max}$

$s_1 = OFF$

$\dot{T}(t) = f(T(t), s_1)$

$s_2 = ON$

$\dot{T}(t) = f(T(t), s_2)$

$T \leq T_{min}$

Fig. 14.1. Room temperature control.

$\delta_1: T \geq T_{max}$ (14.3)

$\delta_2: T \leq T_{min}$ (14.4)

2.1 Modeling Hybrid Systems Behavior

Four different phenomena can occur in hybrid systems that are related to either autonomous conditions or affected by controlled actions (Heemels et al., 2011: (1) autonomous switching of the dynamics; (2) autonomous discrete state jumps; (3) controlled switching of the dynamics, and (4) controlled discrete state jumps.

Autonomous switching of the dynamics is a behavior when the state dynamics changes triggered by a time-driven event or by an asynchronous event. In autonomous state jumps, at some point, the state dynamics may jump from one dynamic behavior to another, implying discontinuities. Controlled switching occurs if the system has a discrete input that invokes the switching among different continuous dynamics. Controlled state jumps are discontinuities in the state trajectory in response to a control command.

To model the previously described behaviors, there are a minimum set of discrete states, space state, and functions that need to be considered:

- A set of discrete states: $S = \{s_0, s_1, s_2, \ldots s_n\}$
- A continuous state space: $X = \{x_1, x_2, \ldots x_m\}$
- A set of vector fields, f, or flow equations describing the state space dynamics for all discrete states of S.
- A set of discrete state transition functions, δ.

With the previous definitions, the hybrid state of a system, H, is given by $(s, x) \in S \times X$. Therefore, a hybrid system can be defined by combining a State Space system with a finite state automaton but requires some additional elements to include state jumps and complete the representation of the interaction between the continuous state dynamics and the discrete states.

2.2 Hybrid Automata

Hybrid automata are probably the most famous and expressive generalization of timed automata that can model hybrid systems (Waez et al., 2013). Hybrid automata extend finite state machines that combine discrete states with continuous variables. The continuous behavior of the system is related to discrete state transitions.

A hybrid automaton, H, is a collection:

$$H = \{S, X, f, Init, Inv, \delta, \mathcal{G}, \mathcal{R}\}$$ (14.5)

where:

$S = \{s_0, s_1, s_2, \ldots s_n\}$ is the set of discrete states;
$X = \{x_1, x_2, \ldots x_m\}$ is the continuous state space;
f is the set of flow equations describing the state space dynamics for all discrete states;

Init is the set of initial values (s_0, x_0) for the hybrid state;

Inv is the set of invariants for each discrete state;

δ is the set of discrete state transitions;

$\mathcal{G}$ is the set of guard conditions prescribing when a transition is executed;

$\mathcal{R}$ is the set of reset maps prescribing when a transition is executed.

A guard, $\mathcal{G}$, describes a region in the state space, X, where a discrete state transition may occur. Each discrete state has an invariant, *Inv*, associated with it, which describes the conditions that the continuous state has to satisfy at this discrete state. Invariants and guards are complementary: whereas invariants describe "bounded values" or "region limits" for the continuous variables, the guards serve as "enabling conditions" for triggering discrete state transitions. The reset map, $\mathcal{R}$, is a set-valued function that specifies how new continuous states are related to previous continuous states for a particular transition.

Formal definitions of hybrid automata are not unique since they are computational models for many diverse types of hybrid systems. In engineering practice, as for finite state automata, the graphical representation of hybrid automata represents a more practical design approach. Circles represent the discrete states, and the arrows among the discrete states graphically represent the transition functions.

3. Hybrid Dynamical Control

A hybrid control system is a system in which the behavior of interest is determined by interacting processes of distinct characteristics, in particular, interacting continuous and discrete dynamics (Antsaklis and Koutsoukos, 2003). If the quantization of the continuous variables or signals is considered, then the hybrid control system contains digital signals; this is discrete-valued digital signals in discrete time. Digital computer-based control systems, formally modeled or not by the theory of HDS, can be regarded as HDS control systems, see Fig. 14.2.

A classic example of an HDS is the internal combustion engine, see Fig. 14.3. In modern motors, the fuel injection is performed by digital valves that are operated depending on the operational modes of the combustion engine. The vehicle

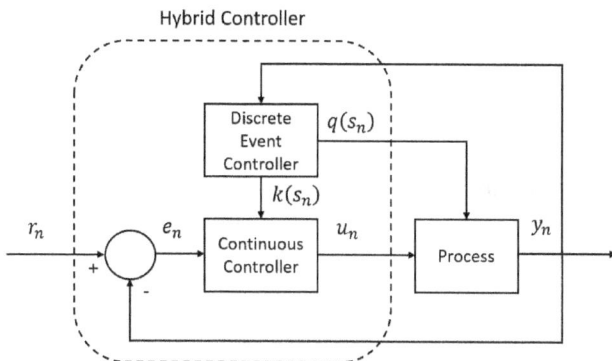

Fig. 14.2. Hybrid dynamical control systems.

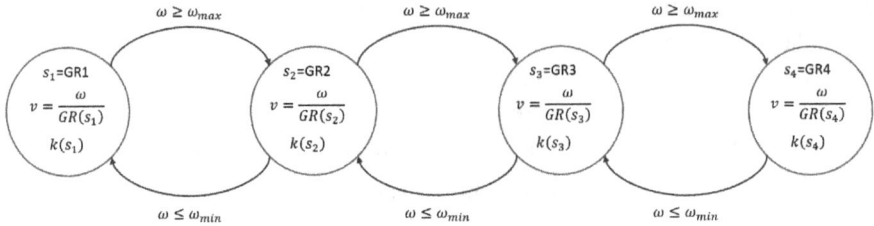

Fig. 14.3. Automatic transmission control.

speed is developed by the power train that depends on the motor speed and torque produced by the pistons, and that is transferred by the mechanical transmission to the wheels at different gear ratios. The mechanical transmission is a switching system with controlled variables such as motor speed and several operational modes that correspond to the different gear ratios. Based on the motor angular speed and the current gear ratio, the discrete event controller triggers an automatic change of the gear ratio to develop the vehicle speed within the operational range for that gear ratio. For each operational mode given by the gear ratio, the gains of the combustion engine controller can be switched to control the motor speed efficiently.

3.1 Finite State Model Predictive Control

Power converters are other devices that can be modeled and controlled as HDS; such applications are found in aerospace and power systems (Rivera et al., 2017). Power converters are switched systems where the dynamic behavior of interest can be adequately described by a finite number of dynamical models, typically sets of differential or difference equations, together with a set of rules for switching among these models. These switching rules are described by logic expressions or a discrete-event system with a finite state automaton.

Finite State Model Predictive Control, FS-MPC, has received attention from researchers and engineers because the predicted behavior of the variables in a power converter can be mathematically modeled (Rodriguez and Cortes, 2012; Wang et al., 2017). FS-MPC is an alternative to the use of modulators and linear controllers. Since power converters and drives combine discrete states and nonlinear dynamics, FS-MPC is suitable for their control and promises to strongly impact control performance in the power electronics field.

In the case of power converters, a finite number of inputs corresponding to the power electronic switches produce changes in the values of the controlled variables. Thus, the control system is a hybrid system in which the inputs are discrete, and the voltage and current controlled variables are continuous. Figure 14.4 shows a basic application of FS-MPC for the current control of a three-phase motor; there is a finite number of combinations of the power switches that can be optimally selected by testing the model online and then applying the optimal selection to the power converter to simultaneously control the motor current and the developed torque and speed.

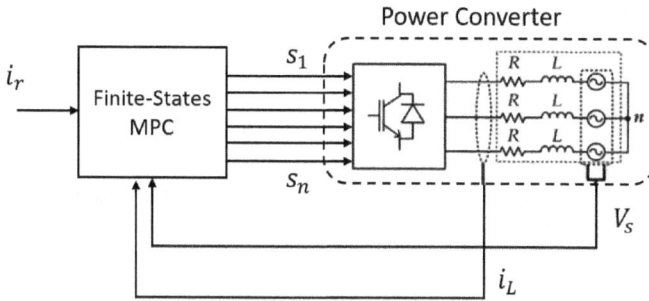

Fig. 14.4. Model predictive control in power converters.

3.2 Hybrid Control in Industrial Processes

HDS requiring hybrid control can also be found in many industrial applications. Generally, the plant or process is a continuous-time system controlled by algorithms employing discrete states. This type of control can be implemented in programmable logic controllers and other electronic control units used in industrial automation. In these applications, discrete states are used to implement combinational or sequential control logic that incorporates decision-making capabilities into the process control. Hybrid control strategies can optimize material handling systems and other type of batch-control applications.

In a manufacturing line, parts may be processed through different stations, and the arrival of a part triggers the process of a particular machine. In this case, the automation is composed of the event-driven control system of the parts moving among different machines and the time-driven control of the processes within particular machines. The event-driven dynamics are implemented via finite states automata, Petri net models, or logic expressions (rule-based control). The time-driven control of process variables is implemented using linear control system schemes, mostly PID control.

3.3 Supervisory Control and Cyber Physical Systems

Supervisory control algorithms implemented in digital computing platforms provide two kinds of behavior, either close loop regulatory control or logic and decision-making functions. These examples also fit into the class of hybrid dynamical systems (Gobel et al., 2009). HDS modeled by hybrid automata are suitable to be implemented computationally. Therefore, they serve as computational models of supervisory control systems (SCS). Combining automata theory and regulatory control can help in the practical modeling and control of SCS (Possan and Bittencourt, 2012). An HDS can also be the logical and mathematical foundation for Cyber-Physical Systems (CPS). The HDS theory is fundamental for the modeling and control of CPS until more advanced computational models are well established for their use.

4. FSM and PID Control of a DC Motor

The DC motor is a typical actuator that can be controlled with diverse control strategies, including HDC. This section presents the design of a cascade position-speed control scheme of a DC motor combining a finite state machine for the position control with a conventional IMC-based PID control of the motor speed. The HDC strategy is compared with a conventional proportional controller for position control combined with the same IMC-PID for speed control. The FSM position controller requires some parameters, such as maximum and minimum speeds and deceleration windows, that can be adjusted for tuning the response of the state machine based position controller.

The HDC scheme where an FSM is used as the motor position controller and the PID for the motor speed control is depicted in Fig. 14.5. The block diagram represents the DC motor model and the inner and outer control loops. The position error signal, e_1, enters the FSM controller block. The FSM controller output generates a speed reference, ω_{SP}. The speed error e_2 is calculated and sent as the input to the PID controller. The PID controller output is the armature voltage for the DC motor. The DC motor develops a rotational speed, ω, which is integrated to produce the angular displacement, θ, of the motor shaft.

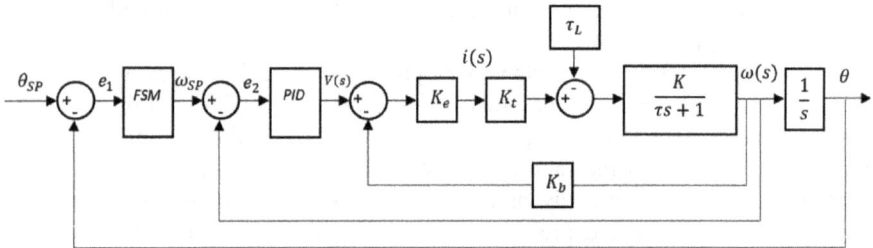

Fig. 14.5. Cascade position (FSM) and speed (PID) control.

4.1 Finite State Machine Based Position Control

For the DC motor position control, the automaton is defined with the following elements: the initial state $x_0 = 0$ corresponds to zero speed output: $S_m = \{x_0\} = \{0\}$, that is, the initial state ensures that when the position controller is activated, the system can start a positioning sequence. The states are defined according to the control sequence as $S = \{x_0, x_1, x_2, x_3, x_4, x_5\} = \{0, 1, 2, 3, 4, 5\}$, and the transition functions are $\delta = \{\delta_1, \delta_2, \delta_3, \delta_4, \delta_5, \delta_6, \delta_7\}$.

Figure 14.6 shows the state diagram for the FSM-based position control. The diagram has six states that control the motor's position; the initial state is 0 and applies zero angular speed, $\omega_{SP} = 0$. If the position set point is changed, an error is computed. If the error is positive, $e_1 > S_p$, where S_p is a small positive value, the transition δ_1: $\theta_{SP} > \theta$ triggers a change to state 1, which applies a constant maximum speed, ω_{Max}, until it reaches a deceleration window, $e_1 < D_{p1}$, and then transition δ_2: $\theta < \theta_{SP} - \theta_{w1}$ ($\theta_{w1} > 0$) triggers a change to state 2, that applies a minimum speed, ω_{Min}.

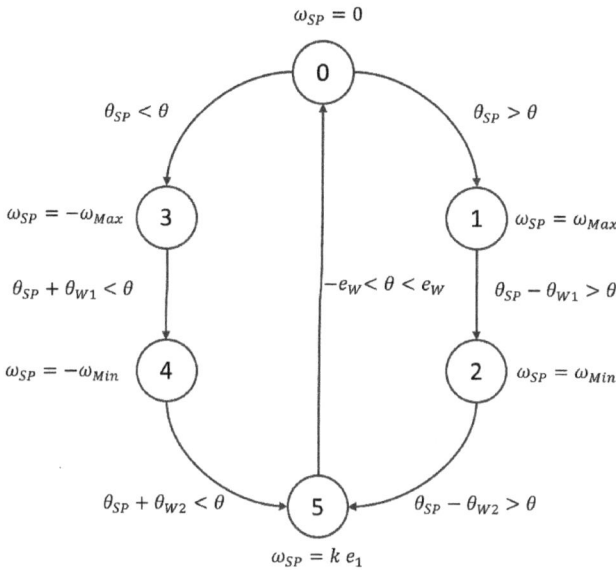

Fig. 14.6. State diagram of the position controller.

Once a tighter deceleration window is reached, $e_1 < D_{p2}$ with $D_{p2} < D_{p1}$, transition δ_2: $\theta < \theta_{SP} - \theta_{w2}$, with $0 < \theta_{w2} < \theta_{w1}$, triggers a change to state 5 that enables a proportional controller for final positioning, until the error has a magnitude less than e_w (δ_5: $-e_w < e < e_w$, $e_w \rightarrow 0$). When the position feedback approaches the desired position, transition δ_5 is generated to change to state zero, which applies zero speed and stops the motor.

In the reverse direction, if a negative error is computed, $e_1 < S_n$ (where S_n is negative with a small magnitude), the transition δ_0: $\theta > \theta_{SP}$ triggers a change to state 3 that generates a constant maximum negative speed, $-\omega_{Max}$, until it reaches a deceleration window, $e_1 < D_{n1}$ (with $D_{n1} < 0$ and $|D_{n1}| < |S_n|$), and then transition δ_3: $\theta > \theta_{SP} + \theta_{w1}$ (where $\theta_{w1} > 0$) triggers a change to state 4, which applies a minimum speed, $-\omega_{Min}$. Once the tighter deceleration window is reached, $e_1 > D_{n2}$, with $D_{n2} < 0$ and $|D_{n2}| < |D_{n1}|$, transition δ_4: $\theta > \theta_{SP} + \theta_{w2}$, with $0 < \theta_{w2} < \theta_{w1}$, triggers a change to state 5 that enables a proportional for final positioning to achieve δ_5: $-e_w < e < e_w$ ($e_w \rightarrow 0$). When the position feedback approaches the desired position, a transition δ_5 is generated to change to state zero that applies zero speed, $\omega = 0$.

4.2 FMS-PID Dynamical Response

The dynamical response of the FSM-PID controller is shown in Fig. 14.7. The FSM controlled the positioning of the DC motor shaft from a zero reference to 360 degrees. In this test, the FSM maintains the maximum speed reference as long as possible, allowing a faster response in the position control. Then the FSM abruptly reduces its output and applies the minimum speed to reach the desired position without overshooting. Speed values are read in the secondary axis of the graph.

The FSM position controller (combined with a continuous proportional controller within a specific error range) was compared against a conventional PD controller

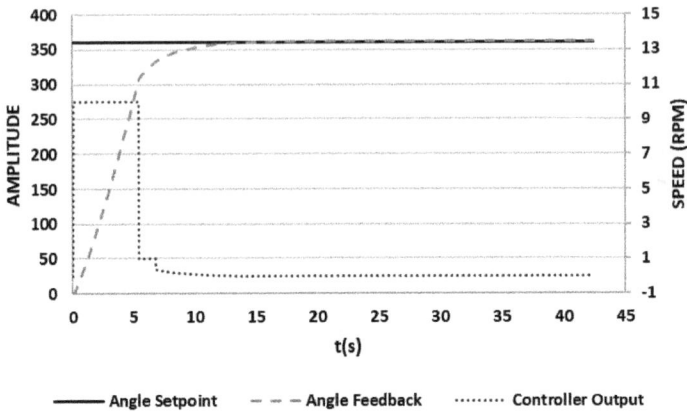

Fig. 14.7. Angular position control with FSM controller.

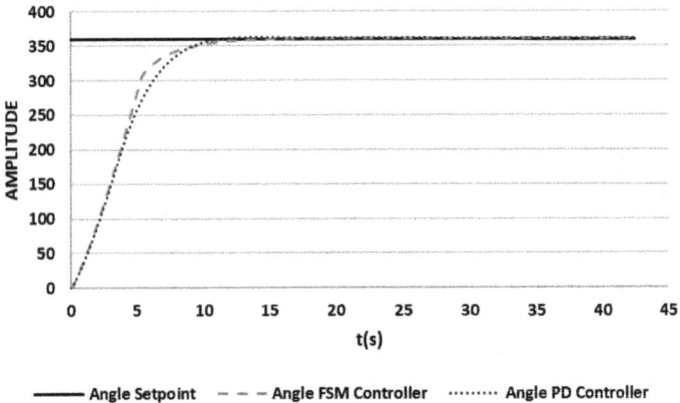

Fig. 14.8. Position control comparison: FSM versus PD controller.

for the position, using an inner IMC-based PID speed controller. Maximum speed references are selected and applied to both controllers. Figure 14.8 compares the two position controllers (FSM versus PD) responding to a position reference change. The dynamical response of the FSM controller is nonlinear; it approaches the reference signal faster than the linear PD position controller. The stability of the FSM controller is assured by the event controlled transition from one state to another.

In this comparison, the FSM-PID controller performs better than the conventional PD-PID controller; the PD even adjusted for its best performance, this is, with the highest gain without oscillation. A cascaded FSM-PID control strategy can accomplish the position and speed control of a DC motor. The FSM assures stable motor operation if it is correctly programmed and tuned.

The speed references generated as outputs by both FSM and PD position controllers are compared in Fig. 14.9. The speed reference signals applied to the inner speed controllers are initially limited by the maximum speed, as expected, and later computed to be directly proportional to the position error by the PD controller.

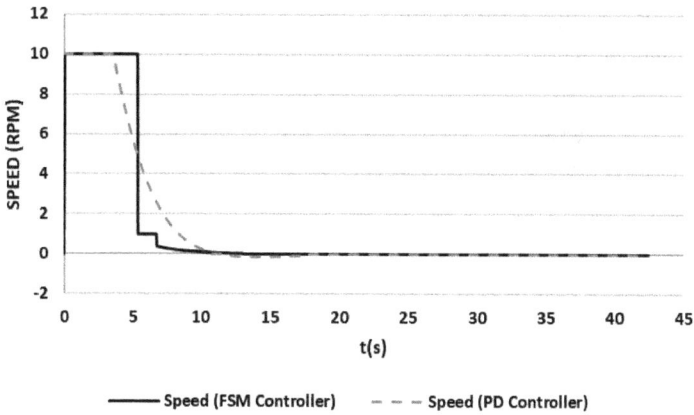

Fig. 14.9. Speed reference comparison: FSM versus PD controller.

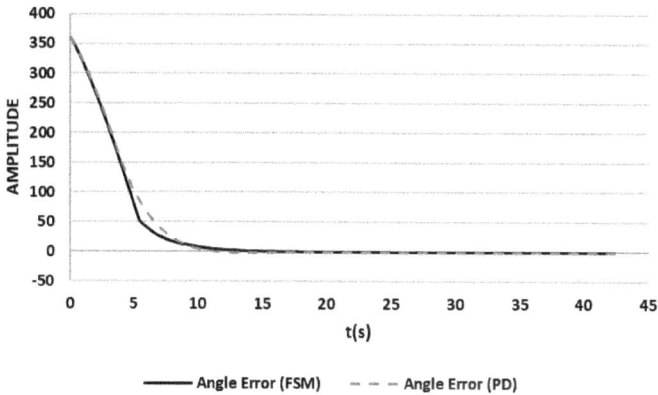

Fig. 14.10. Error comparison: FSM versus PD controller.

The FSM position controller generates two constant speed references: first, it applies the maximum speed and then the minimum speed for constant deceleration. The final approaching is with a speed directly proportional to the small error, and this is incorporated with a final deceleration window. The FSM controller applies predefined maximum and minimum speed parameters that must be tuned for the selected motor and desired application.

Figure 14.10 compares the position errors computed using the FSM and PD position controllers. The error curves are pretty similar except for the final approaching phase. The FSM leads to minor errors due to an early deceleration to reach the final desired value without oscillation. The time integral of the absolute position error during the simulation test results is 6% less for the FSM than for the PD controller. The complete control scheme, FSM, for positioning and servo-controlled speed, can be regarded as a hybrid dynamical control system with a nonlinear positioning performance.

5. HDC of 4-Story Elevator

The automation of an elevator is another example to show the application of HDC for positioning control. Consider an elevator with four floors that is driven by an electric motor. A speed servo drive controls the high and low speeds in both directions to lift and lower the elevator. The controller sets binary outputs to activate predefined high and low speeds for rapid displacement and deceleration. The controller applies the high speed for as much time as possible, while the low speed is applied only for deceleration and the final approach to the commanded floor position. The elevator is depicted in Fig. 14.11.

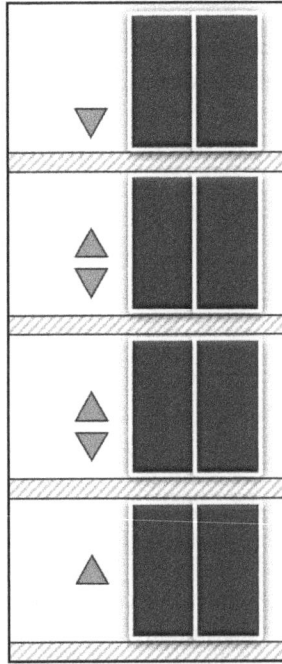

Fig. 14.11. 4-Story elevator

5.1 Elevators Sensors and Actuators

The elevator is instrumented with additional binary sensors, actuators, and user interface devices for commands and indications, connected to the PLC digital inputs and outputs. There are push buttons inside the elevator cabin and call buttons on each one of the floors (ground, first, second, and third floors). Proximity position sensors are installed on each floor for exact and repetitive positioning. There are additional proximity sensors slightly above and below each floor to generate deceleration events so that the controller can switch from high to low travel speeds. The input/output list corresponding to sensors and actuators to control the 4-story elevator is shown in Table 14.1.

Table 14.1. PLC inputs and outputs.

PLC Inputs	Symbol	PLC Outputs	Symbol
Elevator quadrature encoder count	ZT1	Elevator lift	CR4
Elevator door closed	ZS4	Elevator lower	CR5
Elevator door open	ZS5	Elevator high speed	CR6
The ground floor door closed	ZS0a	Ground floor close door command	CR0a
Ground floor door open	ZS0b	Ground floor open door command	CR0b
Floor 1 door closed	ZS1a	Floor 1 close door command	CR1a
Floor 1 door open	ZS1b	Floor 1 open door command	CR1b
Floor 2 door closed	ZS2a	Floor 2 close door command	CR2a
Floor 2 door open	ZS2b	Floor 3 open door command	CR2b
Floor 3 door closed	ZS3a	Floor 3 close door command	CR3a
Floor 3 door open	ZS3b	Floor 3 open door command	CR3b

Table 14.2. Control system push-buttons and pilot lamps.

PLC Inputs	Symbol	PLC Outputs	Symbol
Call button on to ground floor.	PB0	Elevator at the ground floor pilot light	PL0
Call button to floor 1	PB1	Elevator at floor 1 pilot light	PL1
Call button to floor 2	PB2	Elevator at floor 2 pilot light	PL2
Call button to floor 3	PB3	Elevator at floor 3 pilot light	PL3
Call button at ground floor to go up	PB0a		
Call button at floor 1 to go up	PB1a		
Call button at floor 1 to go down	PB1b		
Call button at floor 2 to go up	PB2a		
Call button at floor 2 to go down	PB2b		
Call button at floor 3 to go down	PB3b		
Emergency-stop	PB4		

The input/output list corresponding to push buttons and pilot lights to operate the 4-story elevator is shown in Table 14.2.

5.2 *Description of Operation*

At the initial state, the elevator controller must detect the current position and stay at that floor position. Then the corresponding state to each floor is activated, and the controller waits for a call button to be pressed.

Several possible operation commands can be activated with the call buttons to start the elevator motion to either lift or lower at high speed. Once the elevator is displacing at high speed, several deceleration conditions need to be detected to reduce the speed when lifting or lowering. The detection of the deceleration conditions depends on the target destination of the elevator is. Therefore, it is important to keep the memory of the target floor activated initially with either the call button at the floors or the button inside the elevator cabin.

A functional description can be specified with steps for the elevator motion and the opening and closing of the floor and elevator doors, as follows:

Step 1. The elevator is idle, waiting for a command.

Step 2: If a call button is pressed at the current floor of the elevator, go to step 9 to open the doors; if a call button is pressed from a lower floor, go to step 6; if a call button is pressed from an upper floor, follow step 3.

Step 3. If a call button is pressed from an upper level, lift the elevator at high speed to reach the deceleration condition.

Step 4. While lifting at high speed and the destination floor's deceleration condition is reached, switch to low speed.

Step 5. While lifting at low speed and when the destination floor is reached, stop the elevator and go to step 9 to open the doors.

Step 6. If a call button is pressed from a lower level, lower the elevator at high speed to reach the deceleration condition.

Step 7. While lowering at high speed and the destination floor's deceleration condition is reached, switch to low speed.

Step 8. While lowering at low speed and when the destination floor is reached, stop the elevator and go to step 9 to open the doors.

Step 9. Open both elevator cabin and destination floor doors.

Step 10. Wait for a time delay and the deactivation of the presence sensor to close the cabin and destination floor doors. Go to Step 1.

5.3 Elevator Control with Hybrid Automata

An alternative hybrid automata based control solution uses an incremental encoder to detect the elevator position (height) at any time by a digital counter and to define precise position values in millimeters to detect the floor heights at which the ground, first, second and third floors are located. This implementation requires a quadrature pulse encoder connected to a high-speed counter that is conventionally a PLC module or an electronic card connected to a PLC.

In this configuration with a position encoder sensor, the elevator can also be controlled by a hybrid automaton or finite state machine (see state diagram of Fig. 14.12). Its operation is described as follows: the elevator is initially idle (state 0). When a call button is pressed, the corresponding desired destination position can be updated as the controller's target position or position set-point command. The comparison of the required position with the actual elevator position triggers the transition to either lift (state 1) or lower the elevator (state 3). Once the elevator displaces at high speed, several deceleration positions need to be detected to reduce the speed when lifting (change to state 2) and lowering (change to state 4). Once the elevator reaches the target position count, it changes back to the idle state to stop the elevator and then activates the sequence to open and close the elevator and floor

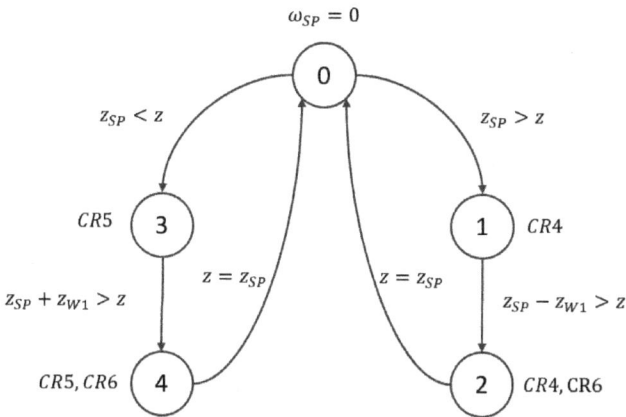

Fig. 14.12. Hybrid automaton for the elevator control.

doors. The opening and closing of the doors can be implemented with additional states diagram. The deceleration positions depend on the target floor and are given in position counts o may be activated using proximity sensors.

5.4 Elevator Control Implementation

A hybrid control is designed for the operation of the elevator. The FSM elevator controller is tested with the 4-story virtual reality elevator shown in Fig. 14.9. The control program is developed using the Siemens "Totally Integrated Automation" (TIA) portal software, which is downloaded and runs on a programmable automation controller S7-1200. The complete control system incorporates a Human Machine Interface (HMI) application, which runs on a touchscreen operation panel.

The elevator is successfully controlled by considering it a hybrid dynamical system. Since FSM and hybrid automata are computational models, some advantages can be obtained when used in HDCS schemes:

1) The states serve to synchronize the computational state in the controller with the system's physical state; this is, without the need to latch memory bits.

2) There are clearly defined states that energize the actuators. Therefore, state duration can be used to determine metrics of its performance or to measure operation hours.

3) The state-based programming presents some advantages for fault detection since states wait for event conditions. Alarms or faults can be detected if the sensors are not activated in the corresponding state.

In summary, including an FSM represents a convenient approach to automating and controlling mechatronic systems and processes. The positioning with encoders is common for many applications, including CNCs, Routers, Cartesian or articulated robots, automated storage systems, etc. The implementation with FSMs and their enhancement as an HDC represent several computational advantages. Supervisory control and data logging functions are easier to implement with state machines. The states and their duration provide production or operation information like counting execution cycles or timing work cycles.

6. Conclusions

Hybrid Dynamical Systems (HDS) exhibit continuous and discrete variables and continuous and discrete evolution. These systems have been studied from diverse theoretical perspectives to model various physical processes and engineering systems. Modeling of HDS is particularly relevant in mechatronics and control. Different models are used, from detailed models that may include equations useful for computer simulation to models that serve for analysis but are not easily implemented for control or control system verification. Hybrid automata are a tool for modeling and control of an HDS. The study of hybrid systems is essential in designing sequential supervisory controllers, and it is central to designing intelligent control systems with a high degree of autonomy.

Two application examples were presented in this chapter. In the first example, the FSM controller was combined with a conventional IMC based PID for speed control to control the position and speed of a conventional DC motor. An FSM controller is defined and proposed for the position control of a DC motor. The FSM-PID with a PD-PID for the motor position control was compared using the same IMC-based PID for speed. The FSM applies the maximum speed for as much time as possible, while the minimum speed is applied only for the time needed for the deceleration and final approach to the required position. The complete FSM-IMC-PID control scheme can be regarded as a hybrid control system with a nonlinear fast positioning control.

A second example was the control of a 4-story elevator modeled and simulated with a virtual system; two control schemes were presented, one using only binary sensors and another solution considering an incremental encoder. The position can be detected alternatively using a quadrature encoder; the 4-floor positions and the deceleration positions are defined in counts, and one proximity sensor is installed at home position for position reset and synchronization. Both presented examples can be considered simplified applications of hybrid automata and are adequate control schemes for positioning control in many diverse motion control applications. Advanced monitoring functions can easily be incorporated if control is implemented with hybrid modeling and control schemes. HDC benefits from collecting additional data that may be required to measure the system's performance.

References

Antsaklis, P.J. and Koutsoukos, X.D. 2003. Hybrid system control. pp. 445–458. *In*: Meyers, R.A. (ed.). *Encyclopedia of Physical Science and Technology*. Third Edition. Academic Press.

Branicky, M.S. 1995. *Studies in Hybrid Systems: Modeling, Analysis, and Control*. Sc.D. Thesis, MIT, Cambridge, MA.

Goebel, R., Sanfelice, R.G. and Teel, A.R. 2009. Hybrid dynamical systems. *IEEE Control Systems Magazine* 29(2): 28–93.

Heemels, W., Lehmann, D., Lunze, J. and De Schutter, B. 2009. Introduction to hybrid systems. pp. 3–30. *In*: Lunze, J. and Lamnabhi-Lagarrigue, F. (eds.). *Handbook of Hybrid Systems Control: Theory, Tools, Applications*. Cambridge University Press.

Lemmon, M.D., He, K.X. and Markovsky, I. 1988. *A Tutorial Introduction to Supervisory Hybrid Systems.* Technical Report of the ISIS Group at the University of Notre Dame. ISIS-98-004.

Possan Junior, M.C. and Bittencourt, A. 2012. Modelling and implementation of supervisory control systems using state machines with outputs. pp. 285–306. *In*: Aziz, F.A. (ed.). *Manufacturing System.* First Edition. IntechOpen.

Rivera, M., Wheeler, P., Rodriguez, J. and Wu, B. 2017. A Review of Predictive Control Techniques for Matrix Converter Applications. *IECON 2017–43rd Annual Conference of the IEEE Industrial Electronics Society*, Beijing, 7360–7365.

Rodriguez, J. and Cortes, P. 2012. *Predictive Control of Power Converters and Electrical Drives.* Wiley-IEEE Press.

Waez, M.T.B., Dingel, J. and Rudie, K. 2013. A survey of timed automata for the development of real-time systems. *Computer Science Review* 9: 1–26.

Wang, F., Mei, X., Rodriguez, J. and Kennel, R. 2017. Model predictive control for electrical drive systems: an overview. *CES Transactions on Electrical Machines and Systems* 1(3): 219–230.

Chapter 15
Cyber Physical Systems

1. Introduction

Industry 4.0 (I4.0) is a technological trend promoted by German technology producers aiming to digitalize production processes to make factories more productive and efficient. The ultimate objective is to incorporate data-driven models and artificial intelligence to optimize the production systems. In the United States of America, this technological trend is named Smart Factories or Smart Manufacturing. Greater digitalization implies a higher degree of automation and data connectivity, allowing manufacturing processes and supply chains to focus on a variable and increasingly personalized demand for products.

Cyber-Physical Systems (CPS) constitute the main enabling technology of Smart Manufacturing (Lu et al., 2020). A CPS is any device that integrates computing and communication capabilities to control and interact with a physical process. In a CPS, the machine or process has its virtual model, called digital twin, that can simulate the dynamical response and behavior of the physical system. The mathematical and data models allow better control and optimization of the physical systems. Model-based analysis methods are tools for verifying the correct functioning of the automated systems. The evolution toward CPS has led to new possibilities for better devices and systems integration, from the low level control of the physical devices and machines, through the intermediate level of process Supervisory Control and Data Acquisition Systems (SCADA) and Manufacturing Execution Systems (MESs), up to the higher level of Enterprise Resource Planning (ERP) systems.

This chapter reviews the CPS and Digital Twin concepts and functions. This chapter aims to show how a virtual mechatronic system can be developed and how this can be used as a digital twin for implementing a basic CPS configuration. Section 2 reviews the technological trends of Industry 4.0. Section 3 evaluates the functional characteristic of the CPS and the Digital Twin. Section 4 presents a design approach for the implementation of a CPS. In Section 5, the design of a virtual router CNC router is presented as a case study. Conclusions are provided in Section 6.

2. Industry 4.0 and Smart Manufacturing

Over the years, the industry has evolved in terms of the technology implemented to produce more with fewer resources, as we can see through the industrial revolutions (see Fig. 15.1). The economic benefits of having smart factories and implementing the emerging technologies of Industry 4.0 include, among others, reduced set-up times, shorter lead times, reduced labor and material costs, increased production flexibility, higher productivity, and enhanced product customization (Qi et al., 2019). A manufacturing system involves numerous types of decision-making at all levels and has to optimize a wide range of operations. However, the traditional hierarchical control architecture provided by the automation pyramid and its rigid nature do not offer the agility, flexibility, and scalability that companies require to adopt the new technologies.

2.1 Industry 4.0 Fundamental Technologies

Industry 4.0 has particular characteristics or design specifications: (1) interoperability; (2) transparency and availability of information; (3) technical assistance; and (4) decentralization of tasks and decision-making. A series of technologies support I4.0, some of them still emerging. The Internet of Things (IoT) and CPS are its main enabling technologies. The fact that devices such as sensors, actuators, and controllers are interconnected through networks such that they can be monitored from the Internet confers the attributes of interoperability and transparency of information to industrial processes.

Cyber-Physical Systems (CPS) and the Industrial Internet of Things (IIoT) constitute key enabling technologies of Smart Manufacturing. However, they are complemented by other technologies such as artificial vision, virtual and augmented reality, additive manufacturing, cloud computing, and data analytics (Big Data). Topics such as cybersecurity, vertical and horizontal integration, simulation, and prototyping are also relevant. For instance, in cloud computing, tasks and decision-making are decentralized, and users are enabled to configure products from the Internet and generate a production order online; all these functions require a highly integrated and secured platform.

The IIoT is possible thanks to the extensive use of industrial Ethernet protocols (ProfiNET, Ethernet/IP, EtherCAT, etc.) that allow adding connectivity capacity to sensors and actuators and real-time execution of control loops. Currently, industrial

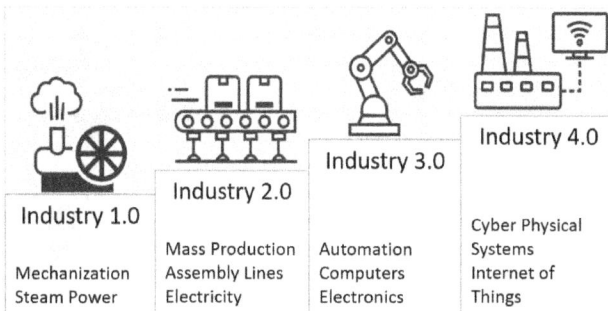

Fig. 15.1. Industrial revolutions.

Ethernet allows horizontal integration (at the same automation level) and works with other technologies such as OPC (Ole for Process Control) with client/server architectures. They allow vertical integration, that is, from the plant floor level to the servers of the MES and ERP systems.

2.2 From Industry 3.0 to Industry 4.0

What is now called Industry 3.0 (I3.0) began with the development of electronics and computing. The automation and computerization of manufacturing systems are considered to have started with the invention of the first PLC in 1969. The Programmable Logic Controllers (PLC), in conjunction with sensors and actuators equipped with electronic conditioning, allowed to implement more precise and repetitive control systems, both in the manufacturing industry (metalworking, automotive, electronics, etc.) and in the process industries (iron and steel, chemical, petrochemical, cement, etc.). In the process industries, Distributed Control Systems (DCSs) were also developed to implement process digital control, and they required the application of industrial data communications protocols (Martell et al., 2023).

Industry 4.0 is regarded as an industrial revolution, but the required steps for its implementation in industrial plants depend on implementing more conventional operational (OT) and information technology (IT). In this sense, the automation and computerization technologies and systems developed and implemented during the I3.0 can be considered precursory for I4.0 (OT and IT Technologies).

Computerized Numerical Control (CNC), industrial robots, Automatic Storage and Automatic Retrieval Systems (AS/AR), and Material Handling Systems (MHS) have been developed over time and, eventually, were incorporated into computer integrated manufacturing (CIM) and later, into the vertical and horizontals integration, conceptualized in the industrial automation pyramid (SCADA-MES-ERP systems). Figure 15.2 shows the evolution from the technologies used in the so-called automation pyramid to the more networked and distributed systems under the conceptualization of CPS with IIoT.

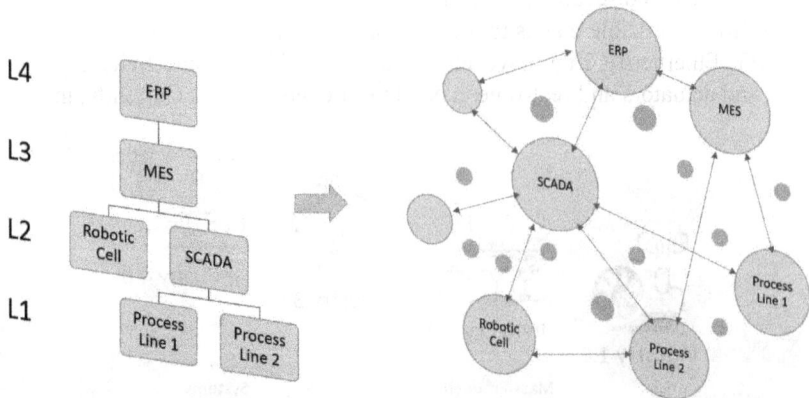

Fig. 15.2. Evolution from automation pyramid to CPS.

3. Cyber Physical Systems and Digital Twins

Cyber physical systems (CPS) are a new generation of systems with integrated computation, communication, and control capabilities that enhance the physical world and can interact with humans through many new modalities. The conceptualization of CPS is not unique due to its vast areas of applications and the multiple technologies they use. One of the most recurrent definitions of CPS in scientific literature comprises five cognitive levels (Lee et al., 2015; Pivoto et al., 2021). The 5-level CPS is called 5C architecture and is shown in Fig. 15.3. It consists of five levels: connection, conversion, cyber, cognition, and configuration.

In the 5C architecture, the smart connection level (level 1) acquires accurate and reliable data from machines and devices. Level 2 ensures that the data is converted to helpful information. The cyber level (level 3) runs a virtual twin of the physical system to add modern automation functions. Level 4 is the cognition level to add Artificial Intelligence (AI)-based decision-making, interaction, and visualization. Configuration level 5 adds supervisory functions to make machines self-configurable and self-adaptive. Table 15.1 describes additional functions of the 5C abstraction levels.

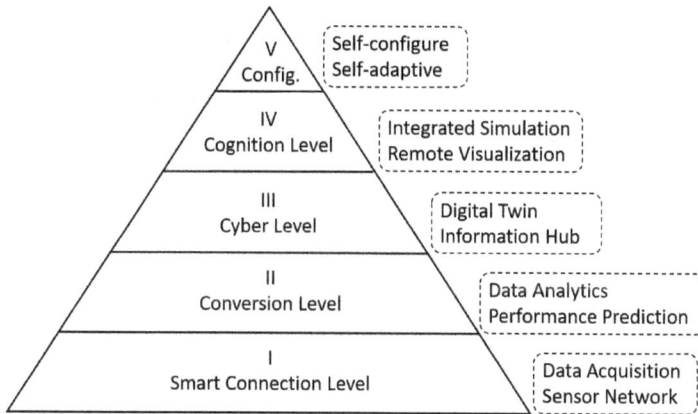

Fig. 15.3. 5C architecture for a CPS.

Table 15.1. CPS levels of the 5C Architecture.

CPS Levels	Description
Level 1. Plug and Play Devices	Plug and play communications, machine to machine communications
Level 2. Sensory Data	Sensory data conversion, the health of the sensor
Level 3. Cyber copies	Digital twin, control of the real machine
Level 4. Cognitive Modules	Data analysis, smart decisions based on artificial intelligence
Level 5. Configuration	Self-optimized, self-predicted and self-adjustable

3.1 Cyber Physical Production Systems

From an industry perspective, a CPS is an internet-enabled physical system or device embedded with computers and control components consisting of sensors and actuators. Such a device is capable of self-monitoring, self-regulation, and autonomous operation, generating information about its functioning, and communicating with other connected devices or systems.

A Cyber Physical Production System (CPPS) consists of a machine or process with its virtual model, called a Digital Twin (DT), consisting of a 3D computer simulation that can model the characteristics and appearance of the physical production system. DT replicates the visualization and behavior of physical systems in a digitized environment, incorporating more elaborated automation, control, and monitoring functions into the industrial processes. The CPPS has a digital environment characterized by having a knowledge base of the ideal functioning of the physical system. Thus, CPPS is a physical system extended by information and communication technologies. For example, a robot enabled for I4.0 has its virtual twin, in which operating sequences can be modified and tested, or failure scenarios can be analyzed. The DT covers different areas necessary for proper operation, such as design, simulation, programming, communication networks, and actuator control. Model-in-the-loop (MiL) and hardware-in-the-loop (HiL) practices for verifying mechatronic systems can be extended to twin-in-the loop architectures for the real-time operation of digital twins (Samir et al., 2019). Figure 15.4 shows the concept of a DT incorporated into a CPS.

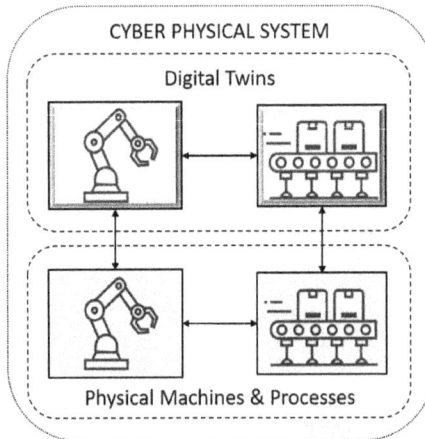

Fig. 15.4. CPS with digital twin (Adapted from Lu et al., 2020).

3.2 Digital Twins

The available computational resources for modeling physical systems have increased in quality and quantity, allowing the manufacturers to work with virtual twins of machines and processes. Digital twins are not only 3D representations of physical systems; they enable a model-based process operation for a more efficient response.

A digital twin aims to replicate the real system with high precision. The digital twin is considered an evolved monitoring tool. Computer simulations of the real system bring various advantages to monitoring the process, preventing failures, and optimizing the operation of the real system (Havard, 2019).

The digital twin is a 3D computer simulation that performs a numerical simulation of the dynamics of the actuators, mechanisms, or any other mechanical, thermal, or fluidic process (Cimino et al., 2020). The definition of a digital twin shares similarities with model-based development (MBD) and implementation methods such as MiL simulation (MiLS) and HiL simulation (HiLS). MiLS is based on models of each system component, while HiLS performs a combination of model components and hardware to obtain more accurate results. A distinguished feature of strict simulation schemes is the high coupling and interaction between the discrete and continuous variables. This characteristic leads to the modeling of hybrid dynamical systems.

4. Cyber Physical Systems Implementation

Modeling techniques of CPS incorporate good design practices for mechatronics systems and industrial automation (Bradley, 2012). However, other advanced design and modeling tools have been suggested, since the implementation of CPS requires more advanced automation functions and deal with new aspects derived from integrating networked devices and systems such as time synchronization, cybersecurity, etc. Table 15.2 shows the diverse design and modeling tools suggested for the CPS design and implementation.

Table 15.2. A CPS design and modeling tools.

CPS Requirements	Design and Modeling Tool
Sequential modeling	Finite-state machines
Concurrency models	Petri net based specification
Modeling dynamic behaviors	Hybrid dynamical systems
Computational models	Unified modeling language
Design methodologies	Model based design, concurrent engineering
Hardware realization	PLC, FPGA, embedded processors
Analysis of controllers	Local and global time synchronization
Verification	Formal methods, model checking
Analysis of performance	Optimization techniques
Security aspects	Algorithms, protocols, e-services

4.1 CPS for Advanced Automation and Control

Even though the CPS is an innovative concept, its implementation needs to add automation and digitalization functions already provided by traditional SCADA and MES systems. Figure 15.5 shows the incorporation of a DT into a physical system

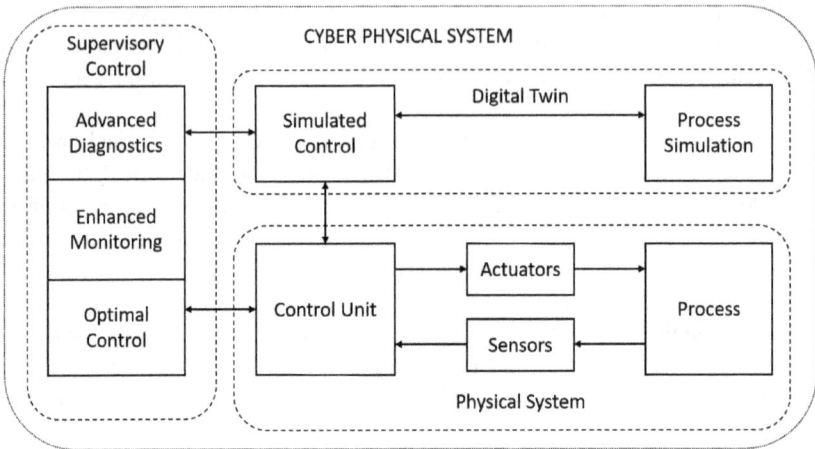

Fig. 15.5. Basic concept of a CPS showing a DT and advanced automation functions.

integrated with a supervisory controller to add some tailored automation functions for process monitoring, optimization, and diagnostics (Iannino et al., 2022).

Including a digital twin application implies and enhances monitoring systems that can register data from both the real process and the virtual twin. Modern computational algorithms can be implemented in real time to optimize the response of the control systems and reach higher productivity and quality. The simultaneous operation of the digital twin with the physical system allows for to detection of failures promptly, thus reducing delays in manufacturing production processes.

4.2 Basic Approach to a CPS Design

From the control and automation perspective, the goal is to develop new engineering methods to build robust and reliable CPS in which physical designs and cyber models are compatible, synergistic, and integrated (Platzer, 2018). A relevant aspect of the digital twin is the interconnection with its physical counterpart for the transfer of information that allows the user to see and interact in real time with the system, giving the advantage of observing, monitoring, and adjusting the performance of both systems. For example, the real time concurrent execution of the physical system with its virtual twin can help diagnose multiple failures in the sensors, actuators, and control unit.

The digital twin is often conceptualized as a 3D model of the physical system. However, numerical simulations in real time are required to model the dynamical and functional characteristics of the physical systems (Qi et al., 2021). The design process starts with the digital twin development, following the integration with the physical system and then implementing more CPS automation and monitoring functions. Figure 15.6 shows a design approach for implementing CPS.

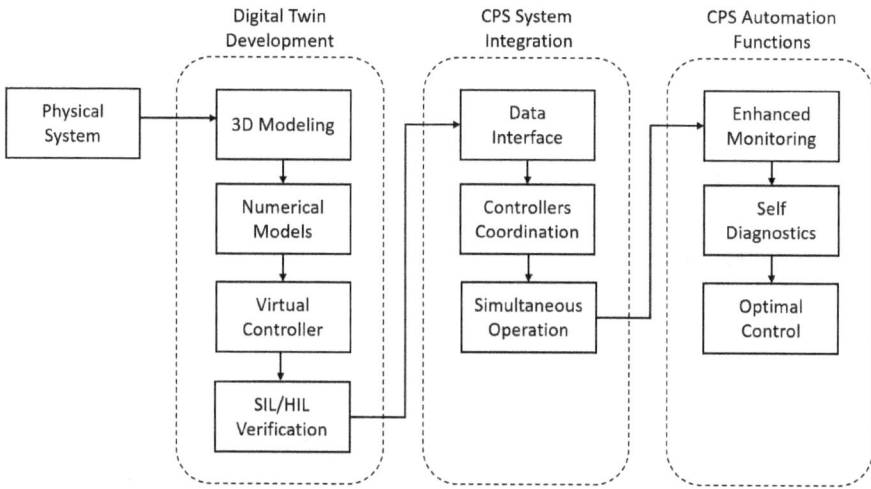

Fig. 15.6. Design approach to a digital twin and a CPS.

The first stage of the presented design process is the development of the virtual twin. The objective is to design a virtual machine that reproduces the physical system's dynamical response and 3D appearance. This stage is composed of four steps that involve good design practices found in mechatronic systems design:

- Development of the virtual machine 3D model;
- Development of mathematical and numerical models;
- Development of the virtual control system and the user interfaces;
- System verification in an HIL configuration.

In a CPS system, the sensors, actuators, control unit, and monitoring computer are networked systems (Lee et al., 2011). The second stage integrates the physical machine and the digital twin to compose the CPS. This stage deals with the synchronizations of the physical system with the virtual counterpart and the controllers and is composed of three steps:

- Implementation of an industrial data communication protocol;
- Coordination of the virtual controller and the physical system controller;
- Simultaneous operation of a physical system and the virtual system.

The real-time execution of the numerical model presents several advantages when implementing model based and model reference control strategies or even model predictive control to optimize the operation of the physical system. Another main objective of the CPS is to monitor and diagnose failures in the hardware components and the execution of the sequential logic control. The third stage is developing and implementing the CPS peculiar automation functions to be added to the supervisory control. This stage completes a CPS that later can incorporate other cognitive and configuration functions:

- Enhanced monitoring and logging of the physical system and the virtual system;
- Advanced diagnostics, including fault detection and location;
- Implementation of model based optimal control strategies.

This design and development approach for a CPS implementation encompasses the automation at the operational technologies level and some other special functions supported by data analytics and AI technologies (that can be implemented once a basic CPS is constructed).

5. Case Study: CPS of a CNC Router

Building a virtual CNC router (Model 3018 PRO) in Unity software, including the numerical simulation of mathematical models for the dynamics of the actuators, which are then interconnected and coordinated to the physical router, is presented as a case study. More details of this development process are in (Garcia-Martinez et al., 2021) and in (Hernandez-Vazquez et al., 2022).

The software used to generate the tridimensional router model was SolidWorks and Unity. SolidWorks is a computer aided design (CAD) software that allows to create 3D models, assemblies, and 2D drawings. The control of the Cartesian robot is designed and implemented in the CoDeSys platform, as well as the trajectory generator and the closed-loop control for the movement of each of the motion axes. The external controller configured in hardware in the loop simulation scheme uses the Modbus open source industrial protocol for data communication, see Fig. 15.7.

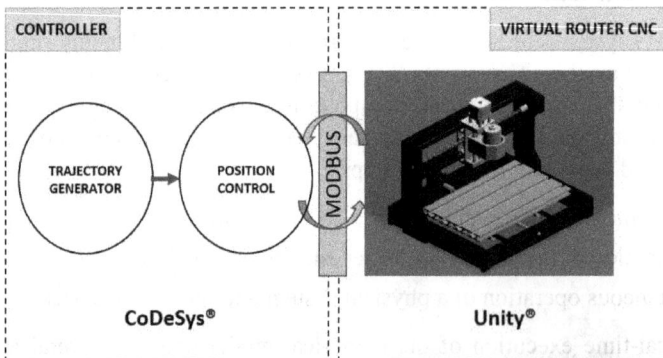

Fig. 15.7. Virtual router in an HiL scheme.

5.1 Virtual Twin Development

The development of the virtual router considers hardware components and software tools: the 3D model design, the mathematical models implemented to simulate and control the actuators, and the equations for the trajectory generation. Building a virtual CNC router in Unity software is presented to illustrate a digital twin development.

3D Model Design

One of the digital twin's important aspects is the robot's representation in the virtual environment. Therefore, the 3D digitized model should replicate the real dimensions of the physical model. However, it is convenient to rescue the important details of the model to have a lighter 3D object for faster rendering. The virtual 3D representation of the router CNC 3018 PRO was built in SolidWorks, adjusting the dimensions of each part according to the model datasheet. Each component, including the base and the three endless screws, was grouped depending on the axis of movement it is related to (X, Y, or Z). As an animation and video game development platform, Unity offers tools to create interactive graphical interfaces. Unity is used to add animation and replicate physical behavior and dynamics for a more realistic simulation.

Mathematical and Numerical Modeling

The motion dynamics depend on the actuator's transient response. The controller on each motion axis produces a percentage pulse frequency output, u_n. Stepper motors are considered to achieve an instantaneous response ideally. To avoid abrupt changes in the stepper motor coupled to the corresponding leadscrew, the control signal, m_n is ramped by a first-order differences equation (15.1), where the parameter α of the recursive equation can be adjusted to match the velocity of the physically installed motor. The parameter, α, is expressed in terms of a time constant, τ, and the sample time T_s of 20 milliseconds (same value configured in Unity), as shown in (15.2).

$$m_n = \alpha\, m_{n-1} + (1 - \alpha)\, u_n \tag{15.1}$$

$$\alpha = \frac{\tau}{\tau + T_s} \tag{15.2}$$

The signal m_n is considered proportional to the motor rotational velocity and is used for calculating the linear displacement d_n by an integration process. The signals are as follows: u_n is the controller output or percentage of maximum pulse frequency, m_n is the ramped output, and d_n is the linear displacement along the corresponding axis.

The displacement of each axis is given by a numerical integration represented in (15.3), whose parameters are associated with the internal characteristics of the stepper motor and the signal m_k, calculated by (15.1). F_{max} is the maximum pulse frequency, n_s is the number of steps, r_g is the gear ratio (load revolution per motor shaft revolution), and P is the ratio of advanced millimeters per revolution.

$$d_n = d_{n-1} + T_s m_n F_{max} \left(\frac{1}{n_s}\right) r_g P \tag{15.3}$$

The angular displacement, θ_n, of the endless screws correspond to the linear displacement, d_k, and is calculated similarly by substituting the parameter pitch (P) with the ratio of 360° per revolution, as shown in (15.4):

$$\theta_n = \theta_{n-1} + T_s m_n F_{max} \left(\frac{1}{n_s}\right) r_g (360°) \tag{15.4}$$

Previous equations (15.1) to (15.4) are implemented to incorporate a basic dynamical response to the 3D model. Defining the sample time within Unity, the simulation of the movement along the axes is important. In this case, the fixed sample time was defined as 20 ms.

Virtual Controller

The control of the Cartesian robot, the trajectory generator, and the closed control loops for the position at the motion axes are programmed in the CoDeSys® platform. The virtual router controller is connected through Modbus/TCP protocol to the virtual CNC in hardware in the loop simulation scheme. CoDeSys is an automation software for developing PLC programs in the standard IEC61131-3 (ladder diagrams, function blocks, instruction lists, and structured text).

Trajectory generators should be suitable for real-time execution and based on a simple description of the desired effector position and orientation provided by the user. Mathematical The trajectory generator is shown in (15.5), where, using a cubic polynomial, the set point position, SP, at the time, t, at a particular axis is calculated using the coefficients a_0, a_1, a_2, and a_3, which are functions of the initial position d_i, the final position d_f, and the estimated travel time t_f. Equations (15.5) through (15.9) must be stated for each movement axis (X, Y, and Z):

$$SP = a_0 + a_1\, t + a_2\, t^2 + a_3\, t^3 \tag{15.5}$$

with:

$$a_0 = d_i \tag{15.6}$$

$$a_1 = 0 \tag{15.7}$$

$$a_2 = \frac{3}{t_f^3}(d_f - d_i) \tag{15.8}$$

$$a_2 = -\frac{2}{t_f^3}(d_f - d_i) \tag{15.9}$$

The coefficients for the trajectory generator are calculated after the user enters the desired positions (d_f for each axis). With the current initial positions (d_i), the desired final positions (d_f), and the estimated time (t_f), the four necessary coefficients for the trajectory are calculated once. In contrast, the intermediate positions are obtained at each sample time (controller execution time).

A discrete PID controller is implemented to control the position set points given from the trajectory generator by (15.5). The error variable, evaluated by (15.10), is used in the recurrence equation for the discrete PID stated by (15.11). The controller parameters (k_p, k_i, and k_d) and sample time (T_s) are set up in the controller. The results of the control equations for each movement axis are the data sent to the Unity virtual router under the automated mode.

$$e_n = SP - d_i \tag{15.10}$$

$$u_n = u_{n-1} + k_p(e_n - e_{n-1}) + \frac{k_i T_s}{2}(e_n + e_{n-1}) \tag{15.11}$$

The implemented virtual controller with the trajectory generator shows a quick and straightforward solution to the motion control specified by desired Cartesian positions. The issue of relying only on data to design and use a digital twin is subject to discussion, emphasizing the weak definition of limitations and validity of data concerning the explicit range and conditions of a dynamic model.

Hardware in the Loop Verification

The integrated system is communicated in hardware in the loop simulation scheme. CoDeSys software emulates a SoftPLC in the same computer that runs the virtual router without an external device. The virtual CNC router uses Modbus/TCP as a data communication protocol; it can be connected to any controller that can be configured as a Modbus slave/server device. Modbus is an open-source communication protocol supported by many PLC, smart instruments, and IO modules. Modbus has a standard data exchange that consists of coils (read and write bits), discrete inputs (read only bits), input registers (read only), and holding registers (read and write); registers are 16 bit integers.

The virtual controller is responsible for controlling the virtual router. First, it is important to define the sample time to execute the control tasks. For this application, a sample time of 60 milliseconds is defined, three times the virtual model's sample time to allow the robot to move without a data glut. The dynamical response was adjusted in the HIL. The controller has the required algorithm for positioning control and trajectory generation. Since the trajectory generator equations depend on the current execution time, the control must produce overdamped or critically damped responses.

The SoftPLC provides means to implement user interfaces. The developed interface was used to move, control, and interact with the model in Unity. For the automatic operation following a trajectory, the user indicates the coordinates of the desired position and the time to reach it. To implement a linear displacement of the router tool from the initial position to the target position, it is necessary to calculate a sequence of intermediate points along the specified time.

Each point of the calculated trajectory is specified in terms of the position and orientation of the tool; however, in the case of the CNC router, since the tool orientation is fixed, only the XYZ Cartesian positions are required. For the communication of the current positions at each axis, three holding registers were configured within the Modbus slave device; for sending the calculated pulse frequency manipulations to the actuators, three input registers were configured in the same section; and an input register was added to use the bits for Boolean variables such as the selection of movement mode (auto/manual) and the activation of each button for manual movement (start/stop).

5.2 *Virtual Twin Coordination with the Physical Router*

Once the virtual twin is verified by reproducing a linear low order dynamical response of the real CNC, it can be integrated with the real machine to compose a CPS.

The simultaneous operation of the digital twin and the physical system is a critical task since coordinating the physical and virtual entities often presents synchronization problems. Synchronous connectivity problems can be solved by considering a time delay between the data transfers enough for the IO data to be updated in the physical system. Asynchronous problems can be addressed by implementing formal programming methodologies such as state machines to copy the machine state to the virtual machine.

The user defines the desired positions and duration of the translation movements. The trajectory generator and PID controller use these data for each axis to obtain the trajectory the 3D animated model must follow. At the end of the trajectory, the router stops showing the positions in both Unity and the SoftPLC.

5.3 *Implementing Supervisory Control Functions*

Once the coordination of the virtual twin with the physical systems is achieved, the control, monitoring, diagnostics, and other state-of-the-art functions can be developed and implemented. The virtual system reproduces the behavior of the physical system through a virtual hybrid automaton and every time the CNC physical controller activates an actuator or reads a sensor, IO data arrays are read using Modbus TCP by the virtual controller to be compared with the status of the virtual IO (Park et al., 2019).

The virtual controller interacts with the models of the actuators, based on numerical simulations, to produce the virtual router XYZ Cartesian positions. A possible way to improve the control response of the physical router is to implement a multiple loop control strategy to provide feedback speed and position from the virtual router to the physical controller and from the physical router to the virtual controller. This scheme is depicted in Fig. 15.8. Digital twins of CPS can diagnose

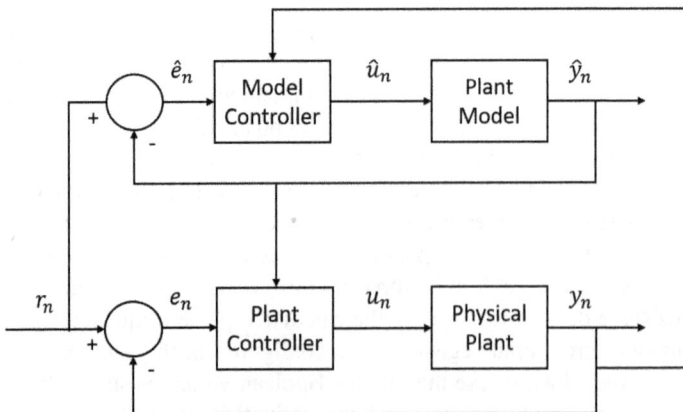

Fig. 15.8. Advanced control strategy with the virtual twin.

and locate faults in real time, providing expert process monitoring and control functions in manufacturing systems.

6. Conclusions

In Industry 4.0, the accessibility of information is decisive and involves higher degrees of automation of various functions. Implementing technological trends can improve the automation of the production processes Cyber-Physical Systems (CPSs) constitute one main enabling technology of Industry 4.0 and Smart Factories. A CPS is any device that integrates computing and communication capabilities to control and interact with a physical process. A CPS consists of a machine or process having its virtual twin that can model in 3D the shape of the physical system and simulate its dynamic response.

Digital technologies have allowed the virtualization of physical systems with high precision. Advanced manufacturing systems can be configured by including the digital twins and by integrating with other cyber physical machines. A basic approach to a CPS design process is described to implement basic and specialized control functions. The numerical simulation of the digital twin running in real time enhances supervisory control system functions. Optimization of the control of the physical system is possible to implement with virtual model reference control strategies. Fault detection and location in real time are possible with the inclusion of the physical twin and its virtual controller.

A virtual CNC router is presented as a case study to show the development of a digital twin using different software platforms such as Unity and CoDeSys, and the Modbus communications protocol for the interconnection of the virtual CNC router to the SoftPLC based virtual controller. The virtual twin with its controller is verified in a hardware in the loop configuration. It is then interconnected to the physical CNC router in a twin-in-the-loop scheme for potential operational benefits such as fault diagnostics and improved control.

Technological challenges regarding the design and development of digital twins and their incorporation into a CPS are research opportunities that require intensive collaboration between academia and industry.

References

Bradley, J.M. and Atkins, E.M. 2012. Toward continuous state-space regulation of coupled cyber-physical systems. *Proceedings of the IEEE* 100: 60–74.

Cimino, C., Ferretti, G. and Leva, A. 2020. The role of dynamics in digital twins and its problem-tailored representation. *IFAC–PapersOnLine* 53(2): 10556–10561.

Garcia-Martinez, M.A., Sanchez, I.Y. and Martell-Chavez, F. 2021. Development of Automated Virtual CNC Router for Application in a Remote Mechatronics Laboratory. 2021 International Conference on Electrical, Computer, Communications and Mechatronics Engineering (ICECCME).

Havard, V., Jeanne, B., Lacomblez, M. and Baudry, D. 2019. Digital twin and virtual reality: a co-simulation environment for design and assessment of industrial workstations. *Production & Manufacturing Research* 7(1): 472–489.

Hernandez-Vazquez, H., Sanchez, I.Y., Martell, F., Guzman, J.E. and Ortiz, R.A. 2022. Development of virtual router machine for modbus open connection. *In*: Flores Rodríguez K.L., Ramos Alvarado,

R., Barati, M., Segovia Tagle, V. and Velázquez González, R.S. (eds.). *Recent Trends in Sustainable Engineering. ICASAT 2021. Lecture Notes in Networks and Systems, 297.* Springer, Cham.

Iannino, V., Denker, J. and Colla, V. 2022. An application-oriented cyber-physical production optimisation system architecture for the steel industry. *IFAC–PapersOnLine* 5(2): 60–65.

Lee, E.A. and Seshia, S.A. 2011. *Introduction to Embedded Systems: A Cyber-Physical Systems Approach.* MIT Press.

Lee, J., Bagheri, B. and Kao, H. 2015. A cyber-physical systems architecture for industry 4.0-based manufacturing systems. *Manufacturing Letters* 3: 18–23.

Lu, Y., Liu, C., Wang, K., Huang, H. and Xu, X. 2020. Digital twin-driven smart manufacturing: connotation, reference model, applications and research issues. *Robotics and Computer-Integrated Manufacturing* 61: 101837.

Martell, F., López, J.M., Sánchez, I.Y., Paredes, C.A. and Pisano, E. 2023. Evaluation of the degree of automation and digitalization using a diagnostic and analysis tool for a methodological implementation of Industry 4.0. *Computers & Industrial Engineering* 177: 109097.

Park, H., Easwaran, A. and Andalam, S. 2019. TiLA: Twin-in-the-loop architecture for cyber-physical production systems. *IEEE 37th International Conference on Computer Design (ICCD)*, pp. 82–90.

Pivoto, D.G.S., de Almeida, L.F.F., de Rosa Righi, R., Rodrigues, J.J.P.C., Baratella Lugli, A. and Alberti, A.M. 2021. Cyber-physical systems architectures for industrial internet of things applications in industry 4.0: A literature review. *Journal of Manufacturing Systems* 58, Part A: 176–192.

Platzer, A. 2018. *Logical Foundations of Cyber-Physical Systems.* Springer International Publishing.

Qi, Q., Tao, F., Hu, T., Anwer, N., Liu, A., Wei, Y. and Wang, L. 2021. Enabling technologies and tools for digital twin. *Journal of Manufacturing Systems* 58, Part B: 3–21.

Samir, K., Maffei, A. and Onori, M.A. 2019. Real-time asset tracking: a starting point for digital twin implementation in manufacturing. *Procedia CIRP* 81: 719–723.

Index

For Product Safety Concerns and Information please contact our EU
representative GPSR@taylorandfrancis.com
Taylor & Francis Verlag GmbH, Kaufingerstraße 24, 80331 München, Germany